T0343287

Protein
Engineering
and
Design

Protein Engineering and Design

Edited by

Paul R. Carey

Department of Biochemistry
Case Western Reserve University
Cleveland, Ohio

ACADEMIC PRESS

San Diego   New York   Boston

London   Sydney   Tokyo   Toronto

Copyright © 1996 by ACADEMIC PRESS, INC.

Academic Press, Inc.
A Division of Harcourt Brace & Company
525 B Street, Suite 1900, San Diego, California 92101-4495

United Kingdom Edition published by
Academic Press Limited
24-28 Oval Road, London NW1 7DX

Library of Congress Cataloging-in-Publication Data

Protein engineering and design / edited by Paul Carey.
      p.   cm.
  Includes bibliographical references and index.
  ISBN 0-12-159640-0 (alk. paper)
  1. Protein engineering.   I. Carey. Paul (Paul R.)
TP248.65.P76P737 1996
660'.63--dc20   96-2000
                                    CIP

Printed and bound in the United Kingdom
Transferred to Digital Printing, 2011

Contents

Part I

PREDICTION

1 Strategies for the Design of Novel Proteins

MICHAEL H. HECHT

2   Computer Methods in Protein Modeling: An Overview

JIRI NOVOTNY

Part II

PRODUCTION

3   The Art of Expression: Sites and Strategies for Heterologous Expression

JASON R. ROSÉ AND CHARLES S. CRAIK

4 Methods for Expressing Recombinant Proteins in Yeast

VIVIAN L. MACKAY AND THOMAS KELLEHER

5 *In Vitro* Mutagenesis

THIERRY VERNET AND ROLAND BROUSSEAU

Part III

CHARACTERIZATION

6 Determination of Structures of Larger Proteins in Solution by Three- and Four-Dimensional Heteronuclear Magnetic Resonance Spectroscopy

G. MARIUS CLORE AND ANGELA M. GRONENBORN

7   A Consumer's Guide to Protein Crystallography

DAGMAR RINGE AND GREGORY A. PETSKO

8   Spectroscopic and Calorimetric Methods for Characterizing Proteins and Peptides

PAUL R. CAREY AND WITOLD K. SUREWICZ

Part IV

APPLICATIONS

9   The Design of Polymeric Biomaterials from Natural α-L-Amino Acids

ARUNA NATHAN AND JOACHIM KOHN

10   The Nicotinic Acetylcholine Receptor as a Model for a Superfamily of Ligand-Gated Ion Channel Proteins

K. E. MCLANE, S. J. M. DUNN, A. A. MANFREDI, B. M. CONTI-TRONCONI, AND M. A. RAFTERY

Contributors

Numbers in parentheses indicate the pages on which the authors' contributions begin.

Roland Brousseau (155), Biotechnology Research Institute, National Research Council, Montreal, Quebec, Canada H4P 2R2

Paul R. Carey (231), Department of Biochemistry, Case Western Reserve University, Cleveland, Ohio 44106

G. Marius Clore (181), Laboratory of Chemical Physics, National Institute of Diabetes and Digestive and Kidney Diseases, National Institutes of Health, Bethesda, Maryland 20892

B. M. Conti-Fine (289), Department of Biochemistry, University of Minnesota, St. Paul, Minnesota 55108; and Department of Pharmacology, University of Minnesota School of Medicine, Minneapolis, Minnesota 55455

Charles S. Craik (75), Department of Pharmaceutical Chemistry, University of California, San Francisco, San Francisco, California 94143

S. J. M. Dunn (289), Department of Pharmacology, Faculty of Medicine, University of Alberta, Edmonton, Alberta, Canada T6G 2HF

Angela M. Gronenborn (181), Laboratory of Chemical Physics, National Institute of Diabetes and Digestive and Kidney Diseases, National Institutes of Health, Bethesda, Maryland 20892

Michael H. Hecht (1), Department of Chemistry, Princeton University, Princeton, New Jersey 08544

Thomas Kelleher (105), Pilot Plant, ZymoGenetics, Inc., Seattle, Washington 98102

Joachim Kohn (265), Department of Chemistry, Rutgers University, New Brunswick, New Jersey 08903

Vivian L. MacKay (105), Protein Chemistry Department, ZymoGenetics, Inc., Seattle, Washington 98102

A. A. Manfredi[1] (289), Department of Biochemistry, University of Minnesota, St. Paul, Minnesota 55108; and Department of Pharmacology, University of Minnesota School of Medicine, Minneapolis, Minnesota 55455

K. E. McLane[2] (289), Department of Biochemistry, University of Minnesota, St. Paul, Minnesota 55108 and Department of Pharmacology, University of Minnesota School of Medicine, Minneapolis, Minnesota 55455

Aruna Nathan (265), Department of Chemistry, Rutgers University, New Brunswick, New Jersey 08903

Jiri Novotny (51), Department of Macromolecular Modeling, Bristol-Myers Squibb, Princeton, New Jersey 08543

Gregory A. Petsko (209), Departments of Biochemistry and Chemistry and Rosenstiel Basic Medical Sciences Research Center, Brandeis University, Waltham, Massachusetts 02254

M. A. Raftery (289), Department of Biochemistry, University of Minnesota, St. Paul, Minnesota 55108; and Department of Pharmacology, University of Minnesota School of Medicine, Minneapolis, Minnesota 55455

Dagmar Ringe (209), Departments of Biochemistry and Chemistry and Rosenstiel Basic Medical Sciences Research Center, Brandeis University, Waltham, Massachusetts 02254

Jason R. Rosé (75), Department of Pharmacology, University of California, San Francisco, San Francisco, California 94143

Witold K. Surewicz (231), Departments of Ophthamology and Biochemistry, University of Missouri, Columbia, Missouri 65212

Thierry Vernet (155), Institut de biologie structurale, Laboratoire d' Ingénierie des Macromoécules, CFA/CNRS, 38027 Grenoble, France

[1]Present Address: H.S. Rafaele Scientific Institute, Milan 20132, Italy.

[2]Present Address: Departments of Chemistry, Biochemistry, and Molecular Biology, College of Science and Engineering, Duluth, Minnesota 55812.

Preface

Protein engineering and design occupies a unique position in the development of late 20th-century science. It depends on not one but three scientific revolutions: the birth of genetic engineering and the burgeoning discoveries in molecular biology; the invention and development of "big machines," which have facilitated the preparation and analysis of proteins; and the ever increasing use of computers, which will make seminal contributions to our understanding of and ability to predict protein properties. Thus, just as a town at the confluence of three great rivers is virtually certain to occupy a prominent position in history and geography texts, so protein engineering and design, driven by three new technologies, will occupy a pivotal position in 21st-century science and technology.

The ability to reengineer existing proteins and to design protein elements *de novo* is a major feat in itself, but the most widespread impact will be in the new products and processes that result. Already, redesigned proteins and peptides are candidates for effective pharmaceuticals and biocompatible materials, and modified enzymes are used in large quantities in detergents and in industrial processes such as pulp bleaching in the making of paper. Proteins even have potential as the basic element of memory in a computer chip.

The multifaceted nature of protein engineering makes it a somewhat daunting topic for students or professional research specialists. Thus, the prime goal of this book is to introduce and define the techniques that constitute protein engineering and design. The reader will be able to obtain a rapid overview of the topics involved and how they contribute to the goal of designing and producing novel proteins and peptides with desired properties. By studying any chapter in depth, she or he can then develop an advanced understanding of the speciality involved.

The first two chapters come under the general heading Prediction and Design and deal with *de novo* protein design and computer modeling. These are followed by three contributions, under the heading Production, which set out the basic molecular biology of producing "mutant" proteins by expressing them from biological cells. The three chapters under Characterization summarize some of the major high technology approaches to protein analysis and characterization. These allow the protein crystallographer or spectroscopist to define, at the molecular level, the properties of the new material predicted by the theorist and produced by the molecular biologists and protein purification experts. In the laboratory, the sequence

PREDICTION–PRODUCTION–CHARACTERIZATION

can be repeated until the protein having the desired properties is produced. Finally, there are two chapters dealing with Applications. One of these outlines the uses of amino acid-based polymers in the materials sciences and thus constitutes an immediate application of peptide design. The final chapter deals with a different approach. Although this book emphasizes the progress that can be made combining all aspects of protein engineering, it is salutary to remember that many interesting biological systems are yet too complex to yield to this integrated approach. The last chapter, on the nicotinic acetylcholine receptor, deals with such a system and illustrates that a wealth of important detail can still be obtained by using molecular biology and protein chemistry.

Paul R. Carey

Part I

PREDICTION

1

Strategies for the Design of Novel Proteins

MICHAEL H. HECHT

Department of Chemistry
Princeton University
Princeton, New Jersey 08544-1009

I. INTRODUCTION

Why design novel proteins? Nature has provided an enormous number of natural proteins, which fold into a variety of different structures and carry out a bewildering diversity of functions. With so many natural proteins available to observe and manipulate, what can be gained by the design and characterization of novel proteins?

There are two main motivations for pursuing protein design. The first is based upon the assumption that a complete understanding of any natural system rests upon our ability to design a similar artificial system from first principles. Thus, our understanding of natural proteins—their folding pathways, thermodynamic stabilities, and catalytic properties—is enhanced (and ultimately tested) by our ability to design novel proteins with predetermined structures and properties. The second motivation for *de novo* protein design is a practical one: Current efforts to design simple structures represent the essential first steps toward designing novel macromolecules that will be "made to order" for solving important chemical or biochemical problems. Indeed, the ability to design proteins *de novo* has the potential to revolutionize fields of science and technology ranging from industrial catalysis to biomedical engineering.

Natural proteins are studied in terms of both their structures and their functions. Similarly, the emerging field of protein design aspires to control both the structures and the functions of novel macromolecules. The first section of this chapter will outline several different strategies for the design of novel protein structures. These strategies share the common goal of producing amino acid sequences that, although significantly different from naturally observed sequences, can nevertheless fold into three-dimensional structures that are both unique and stable. The second section of the chapter will discuss the design of protein function. Specifically, it will focus on the incorporation of novel activities into proteins that are themselves designed *de novo*.

II. STRATEGIES FOR THE DESIGN OF STRUCTURE

The central goal in designing novel protein structures is to devise an amino acid sequence that will adopt a unique and stable three-dimensional structure (Yue and Dill, 1992). To achieve this goal, the fundamental hurdle that must be overcome is the conformational entropy of the linear polymer chain. For example, if in a 100-residue protein each residue can exist in one of 10 different conformations, then the entropic cost of fixing the chain into a unique conformation is RT ln 10^{100} = 136 kcal/mol (Creighton, 1993). This represents a substantial amount of unfavorable free energy. Therefore, if a novel amino acid sequence is to be coerced into foregoing its conformational freedom, then the design must incorporate many favorable interactions. In essence, for a design to succeed, the favorable free energy associated with these designed interactions must outweigh the entropic cost of fixing the chain into a unique structure.

Several different strategies have been used to achieve this goal. Most of these strategies have expended considerable effort to maximize the strength and number of favorable interactions in the designed structure. Some have attempted simultaneously to reduce the entropic cost of folding by introducing covalent cross-links, which limit the number of conformational states accessible to the chain. In this section I will describe examples of several different strategies that have been employed to design novel protein structures.

A. Self-Assembly of Modular Units of Secondary Structure

Perhaps the simplest and most direct strategy for the design of novel protein structures is a modular approach in which a single unit of secondary structure (an α-helix or a β-strand) is synthesized as an individual segment of polypeptide chain. The amino acid sequence is designed to promote the association of several copies of the segment into a globular structure resembling that of a natural protein. Self-assembly is mediated by noncovalent interactions that are designed into the peptide sequence. Typically, the unit of secondary structure is designed to be amphiphilic so that the burial of hydrophobic surface area drives assembly into a compact globular structure. A schematic diagram of the self assembly of two peptides into an ordered structure appears in Fig. 1.

The modular approach has several features that render it an attractive first step toward protein design. The key advantage of this approach is its simplicity, both at the level of design and at the level of synthesis. In a self-assembling modular structure, the challenge in design is restricted to devising the sequence of a single unit of secondary structure. Since the segment is usually intended to assemble into homooligomers, the noncovalent interactions between segments can be repeated many times throughout the structure. Furthermore, since the individual segments are not linked by turns or loops, the modular strategy simplifies design by not requiring a detailed understanding of how specific turn sequences favor or disfavor chain reversals. Perhaps the most attractive feature of the modular approach is its accessibility at the synthetic level. The synthesis of peptides of an appropriate length to form individual α-helices or β-strands is usually quite straightforward, and modern solid-phase methods lend themselves to the production of pure samples in reasonably large quantities.

Unfortunately, the simplicity of the modular approach also limits its power as a general strategy for protein design. Self-assembling modules of secondary structure have several disadvantages compared to single-chain native-like proteins. The most significant of these disadvantages pertain to protein stability. As described above, the major barrier to successful protein design is the conformational entropy of the randomly fluctuating unfolded state. In the case of designed structures that assemble from several unlinked chains, the entropic cost of forming a desired unique structure is even greater than for the folding of a single polypeptide chain. Furthermore, the stability

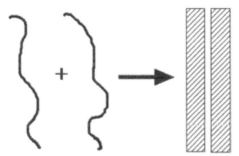

FIGURE 1 Schematic diagram showing the self-assembly of two peptides into an ordered dimeric structure.

of any structure that forms via intermolecular association will necessarily be concentration dependent. This is in marked contrast to single-chain proteins (natural or designed) where folding is *intra*molecular and therefore independent of concentration. In addition to issues of stability, the modular approach has the disadvantage of leading to simple repetitive structures. By optimizing the features of the individual modular element and designing it to self-assemble into multimers, one generally ends up with a structure that is more repetitive and more symmetric than most natural proteins.

Despite these shortcomings, modular designs mimic many of the structural and thermodynamic features common to native proteins and therefore represent an extremely important step toward the goal of protein design *de novo*.

The α-helix is an appealing modular building block because its structure is stabilized by hydrogen bonding within a given segment and because isolated helices can be stable in solution (Brown and Klee, 1971; Kim and Baldwin, 1984; Marqusee and Baldwin, 1987; Shoemaker *et al.*, 1987; Marqusee *et al.*, 1989). The simplest all-helical structures observed in nature are the coiled-coil and the 4-helix bundle (Weber and Salemme, 1980; Richardson, 1981), and these structures were the first to be designed successfully by the modular approach.

The coiled-coil is a relatively simple model system and has been used by several groups interested in structure and design (O'Shea *et al.*, 1989; Cohen and Parry, 1990; Hodges *et al.*, 1990; Hu *et al.*, 1990; O'Neil and DeGrado, 1990; O'Neil *et al.*, 1990; Oas *et al.*, 1990; Engel *et al.*, 1991; Lupas *et al.*, 1991; O'Shea *et al.*, 1991; Lovejoy *et al.*, 1992b, 1993; Harbury *et al.*, 1993). The structure of the coiled-coil is composed of two α-helices packed against one another in a parallel orientation(Crick, 1953; Pauling and Corey, 1953; O'Shea *et al.*, 1991). As pointed out initially by Crick (1953) as well as by Pauling and Corey (1953), if the two helices are tilted at an angle of 20° relative to one another and allowed to wrap around one another, then the resulting structure will have 3.5 residues/turn. Although this periodicity differs only slightly from the 3.6 residues/turn observed in the canonical α-helix, this difference is significant for design: When the periodicity is 3.5, the entire structure is repeated every seven residues. Therefore, the design of a coiled-coil structure can be simplified to the design of an appropriate heptad repeat.

Hodges and co-workers exploited this repetitive feature to design a number of coiled-coil structures composed of simple heptad repeats. A schematic diagram of their designed heptad, Lys-Leu-Glu-Ala-Leu-Glu-Gly, is shown in Fig. 2. The design of this sequence required simultaneous stabilization of both the α-helical secondary structure and the intermolecular interactions between the two helices. Thus in six of the seven positions residues were chosen from among those that are known to be good helix formers (Chou and Fasman, 1974, 1978). [Glycine is rarely found in α-helices but

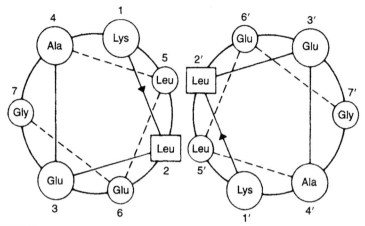

FIGURE 2   End-on view of two parallel α-helices forming a coiled-coil structure. A single heptad repeat is shown and the view is looking from the N-terminal end with the α-helices descending into the page. (Figure adopted from Hodges *et al.*, 1981.) [(Note: positions 2 and 5 are denoted 'a' and 'd,' respectively, in an alternative nomenclature that begins counting at the hydrophobic interface (Cohen and Parry, 1990).]

nonetheless was included in the initial design for synthetic reasons (Hodges *et al.*, 1981).] Leucines were used at positions 2 and 5 (see Fig. 2) to provide a hydrophobic interface along the entire length of the coiled coil. (Note: positions 2 and 5 are denoted "a" and "d" respectively in an alternative nomenclature that begins counting at the hydrophobic interface (Cohen and Parry, 1990.) Lys and Glu were used at positions 1 and 6, respectively, to facilitate favorable ionic interactions between the two helices. Although the features of this designed coiled-coil sequence are representative of heptad repeats found in the natural protein α-tropomyosin, only 25% of heptads found in tropomyosin have more than 2 residues in common with the designed heptad (Hodges *et al.*, 1990). Hodges and co-workers synthesized peptides containing several repeats of the heptad sequence shown in Fig. 2. The structures of the synthetic peptides were then characterized by circular dichroism spectroscopy; their oligomeric states were measured by gel filtration chromatography and sedimentation equilibrium. The results showed that, as specified by the design, the peptides dimerized to form coiled-coil structures. Furthermore, the designed structures were significantly more stable than the naturally occurring coiled-coil in tropomyosin (Hodges *et al.*, 1981; Lau *et al.*, 1984).

Having established that their simple repetitive design succeeded in producing the desired structure, Hodges *et al.* set out to identify those features that contribute to, or detract from, its stability. They first probed the effect of chain length. Polyheptapeptides were synthesized with the sequences Acetyl-(Lys-Leu-Glu-Ala-Leu-Glu-Gly)$_n$-Lys-amide where $n=1$–5. Thus, the properties of peptides having 8, 15, 22, 29, and 36 residues could be compared. These experiments showed that in this system, a minimum of 29 residues were required to promote self-assembly into coiled-coils (Lau et al., 1984).

Next, Hodges and co-workers probed the effect of varying the hydrophobic contacts at the dimer interface. They synthesized peptides consisting of 5 heptads in which the leucines at positions 2 and 5 in the central heptad (see Fig. 2) were replaced by Ile, Val, Ala, Phe, or Tyr. With the exceptions of Phe and Tyr, the stability of the coiled-coils correlated with the residue's hydrophobicity (Leu > Ile > Val > Ala) (Hodges et al., 1990; Zhou et al., 1992a,b). Phe and Tyr, although very hydrophobic, did not increase the stability as expected, presumably because they caused steric problems (Hodges et al., 1990). Further experiments compared the relative importance of the leucines at positions 2 and 5 of the heptad repeat (Zhou et al., 1992a,b; 1993, Zhu et al., 1993). By studying the effects of several Leu → Ala substitutions, Hodges and co-workers demonstrated that leucines at these two positions do not make equivalent contacts at the dimer interface. This result is consistent with both genetic and crystallographic studies of coiled coils in natural proteins (Hu et al., 1990; O'Shea et al., 1991).

Finally, Hodges and co-workers asked whether the stability of the designed structure could be enhanced by including a disulfide bond to link the two α-helices. Disulfide bonds have been engineered into the structures of several natural proteins (Villafranca et al., 1983; Perry and Wetzel, 1984; Sauer et al., 1986; Matsumura et al., 1989; Matsumura and Matthews, 1989), and this seemed like a reasonable strategy for enhancing the stability of designed proteins as well. They synthesized 35-mers having a cysteine at the second residue, and as expected, they found that formation of the interchain disulfide bond stabilized the coiled-coil.

However, they also observed an unexpected result: The magnitude of the stabilization afforded by the disulfide bond varied depending upon the inherent stability of a given designed sequence. Specifically, the greater the stability of the coiled-coil without the disulfide bond, the larger the contribution of the disulfide bond to protein stability. For example, a disulfide bond enhanced the stability of a coiled-coil containing leucines at all of its interface positions by 3.3 kcal/mol. However, when two of the leucines at the contact positions are replaced by phenylalanines, the peptide is less stable to start with, and formation of the disulfide bond enhances its stability by only 1.2 kcal/mol. Apparently, formation of the disulfide cross-link makes the coiled-coil structure more rigid and thereby interferes with its ability to accommodate the Leu → Phe substitution (Hodges et al., 1990). These results suggest that although the design of disulfide bonds can be a simple and powerful strategy for increasing the stability of novel proteins, the steric and geometric environment of the proposed disulfide bond must be considered carefully (Zhou et al., 1993). Indeed, proteins designed *without* disulfide bonds are likely to be more plastic and therefore more forgiving of the designer's mistakes.

Although the coiled-coils described above were designed to form homo-dimers, O'Shea *et al.* (1993) have demonstrated that heterodimeric coiled coils can also be designed *de novo*. They designed two peptides, called "ACID-p-1" and "BASE-p-1." These peptides were identical to one another at their nonpolar "a" and "d" positions and thus could have formed homo- or heterodimers with identical hydrophobic cores. However, at the "e" and "g" positions, which abut the hydrophobic core, "ACID-p-1" contained glutamic acids, whereas "BASE-p-1" contained lysines. In the desired het-erodimer intermolecular salt bridges would stabilize the complex, whereas in a homodimeric structure these favorable interactions would be replaced by repulsive interactions between similar charges. Characterization of the purified peptides showed that the heterodimeric structure was preferred over the homodimeric structures by at least 100,000-fold. Studies of the pH and ionic strength dependence of this "Peptide Velcro" (O'Shea *et al.*, 1993) showed that formation of the heterodimers occurs largely because of the electrostatic destabilization of the alternative homodimeric structures. These results illustrate that successful protein design depends not only on the "positive design" of desired structures but also on the "negative design" against alternative competing structures.

What determines the oligomeric state of a coiled-coil? The heptad re-peats of coiled-coils contain hydrophobic residues at positions a and d (posi-tions 2 and 5 in the nomenclature of Fig. 2), which mediate the interhelical contacts in the coiled-coil structure. However, hydrophobicity at positions a and d is not sufficient to ensure that a designed coiled-coil will actually form dimers. Specific interactions between side chains must also be considered in the design, as demonstrated by the crystal structure of the designed peptide "Coil-Ser" (Lovejoy *et al.*, 1993). This peptide was explicitly designed to form a dimeric coiled-coil with both α-helices running in the same direction (O'Neil and DeGrado, 1990). However, the crystal structure of Coil-Ser showed that the peptide formed trimers with one of the helices running in the opposite direction from the other two.

The role of interresidue contacts in determining the oligomeric state of coiled-coiled peptides was probed in detail using mutants of the GCN4 leucine zipper (Harbury *et al.*, 1993). In wild-type GCN4, the a positions of the four heptads are occupied by three valines and an asparagine, while the d positions are occupied by leucines in all four heptads (hence the name "leucine zipper"). The wild-type protein forms dimeric coiled-coils. How-ever, mutant forms of GCN4 can form dimers, trimers, or tetramers, de-pending upon the identity of the hydrophobic residue at the a and d posi-tions. For example, peptides having Ile at the a positions form dimers when the d position is Leu, but trimers when the d position is Ile. Furthermore, the a and d positions clearly play very different roles, since a peptide where a =

Ile and d = Leu forms dimers, while the converse peptide having a = Leu and d = Ile forms tetramers. Comparison of the crystal structure of the tetramer with that of the original GCN4 diimer shows that the local environments of the a and d residues differ in the two different oligomeric states: The packing geometry surrounding a positions in the dimer resembles the packing geometry surrounding d positions in the tetramer. Conversely the geometry surrounding d positions in the dimer resembles that surrounding a positions in the tetramer. Thus it is not surprising that mutations at the a positions exert different effects than mutations at the d positions.

These results demonstrate an important feature about design strategies that rely upon self-assembly of modular units of secondary structure. Since the individual elements are not linked into a single polypeptide chain, the oligomeric state of the assembly is specified solely by interactions between side chains. Therefore, to design novel proteins via the self-assembly of modular peptides, one must explicitly design interactions to favor the desired oligomeric state and disfavor competing alternatives.

Aside from coiled-coils, the simplest α-helical motif found among natural proteins is the four-helix bundle (Weber and Salemme 1980; Richardson, 1981). As with the coiled-coil, the four-helix bundle lends itself to a modular design strategy, wherein structures are assembled from individual units of secondary structure (α-helices, in this case). This modular strategy has been used with great success by DeGrado and co-workers, who employed it as the first step of their incremental approach toward the design of an extremely stable four-helix bundle (see Fig. 3) (Eisenberg et al., 1986; DeGrado et al., 1987, 1989; Ho and DeGrado, 1987; Regan and DeGrado, 1988; Osterhout et al., 1992). They initially designed α_1, a simple 16-residue amphiphilic helix containing leucine side chains on its hydrophobic face and glutamic acid and lysine side chains on its hydrophilic face. Circular dichroism (CD) and NMR spectroscopy demonstrated that α_1 is predominantly α-helical in aqueous solution (Ho and DeGrado, 1987; Osterhout et al., 1992). Furthermore, as expected for a molecule that assembles from modular components, the helicity is concentration-dependent. Size-exclusion chromatography and equilibrium sedimentation ultracentrifugation confirmed that α_1 assembles into tetramers in aqueous solution (Osterhout et al., 1992).

Attempts to crystallize α_1 have not been successful. However, a crystal structure has been determined for a truncated version of α_1 that is missing the first 4 residues (Hill et al., 1990). The crystal structure of this 12-residue peptide differs significantly from the solution structure observed for the intact 16-residue peptide (Osterhout et al., 1992). In the crystal, the 12-residue peptide forms both tetramers and hexamers. The hexameric association appears to be tighter than the tetrameric one. Furthermore, even in the tetrameric association, the crossing angles are more oblique than envisaged in the original design. The reasons why the 12-residue version of α_1

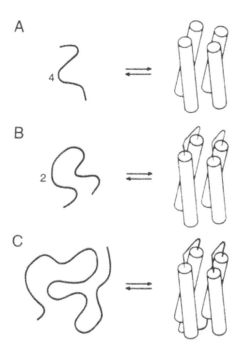

FIGURE 3 Schematic illustration of the incremental approach used by DeGrado and co-workers to design a four-helix bundle protein. (Figure adopted from Regan and DeGrado, 1988.) (A) α-1 is a 16-residue peptide that forms tetramers. Since individual helices are not covalently linked to one another, assembly is mediated entirely by noncovalent interactions. The sequence of α-1 is acetyl-Gly-Glu-Leu-Glu-Glu-Leu-Leu-Lys-Lys-Leu-Lys-Glu-Leu-Leu-Lys-Gly-amide. (B) α-2 is a 35-residue peptide that dimerizes to form a four-helix bundle. It contains two copies of the α-1 helix covalently linked by an interhelical loop, whose sequence is Pro-Arg-Arg. The N-terminus of α-2 is acetylated and the C-terminus is amidated. (C) α-4 is a 74-residue protein that forms a four-helix bundle by intramolecular folding. Its sequence contains four copies of α-1, punctuated by three Pro-Arg-Arg loops. An N-terminal Met is included to facilitate expression *in vivo*.

does not form a classic 4-helix bundle are not clear. However, it must be stressed that the crystal structure was determined for a truncated fragment, not for the intact peptide. Indeed, the intact 16-residue peptide originally designed to form a four-helix bundle could not possibly form the same structure as seen in the crystal; its longer hydrophobic stripe cannot be accommodated in the hydrophobic core of the hexamer (Hill *et al.*, 1990).

In order to favor the formation of four-helix bundles, the next step used by DeGrado and co-workers was to link together two adjacent 16-residue helices into an antiparallel arrangement (see Fig. 3). When Pro-Arg-Arg was used as the turn sequence, the resulting peptide, called α_2, dimerized to form the expected four-helix bundle. The final stage in their incremental approach involved constructing a synthetic gene so that all four helices could be linked together and expressed as a single polypeptide chain. The resulting protein, α_4, folds into a four-helix bundle that is significantly more stable than either the tetramer of α_1 or the dimer of α_2 (Regan and DeGrado, 1988).

The work of DeGrado and colleagues illustrates the strengths of the modular strategy. By starting with simple elements of secondary structure, they were able to isolate and optimize individual structural features. Thus the α-helical sequence was varied in the context of short peptides that are easily synthesized and manipulated. Likewise, the turn connecting two α-helices was varied in the α_2 peptide. Only after all these features had been optimized was the entire structure put together to make α_4 (see Fig. 3). This strategy allowed DeGrado's group to construct the first designed protein that folded into a stable globular conformation in aqueous solution. Moreover, by incrementally optimizing various features of the design, it was possible to generate a protein that is significantly more stable (ΔG = 22.5 kcal/mol) than most natural proteins (Regan and DeGrado, 1988). Further work on α_4 and its metal-binding derivatives is discussed below in Sections II,E and III,C, respectively.

Although dimeric coiled-coils and tetrameric helical bundles have received the most attention from protein designers, attempts have also been made to design higher order α-helical multimers. An octadecapeptide designed to form hexameric helical bundles was reported by Chin et al. (1992). The sequence of the octadecapeptide is EQLLKALEXLLKELLEKL, where X is either Trp or Phe. These sequences contain eight leucines in the hydrophobic face compared to six leucines in the α_4 sequence. Consequently, they would have a larger nonpolar surface that might favor hexamers rather than tetramers. The synthetic peptides are α-helical as shown by CD and they form oligomers as shown by the concentration dependence of this α-helicity. The oligomers bind a hydrophobic probe molecule consisting of a long aliphatic chain linked to an aromatic ring. This binding suggests that the oligomers have an interior hydrophobic cavity. The main evidence suggesting the oligomers are actually hexamers comes from size exclusion chromatography. However, since the oligomers have an interior cavity they do not resemble the compact globular proteins typically used to calibrate size exclusion columns. Consequently, the size exclusion behavior is difficult to interpret. If further characterization demonstrates that hexamers are the sole oligomeric form, then controlling the number of nonpolar residues in the hydrophobic face of an α-helix may prove to be a useful strategy for controlling the oligomeric state of self-assembled helical bundles.

Although most modular designs have focused on α-helical structures, some have attempted to construct β-sheet proteins that would assemble via the association of synthetic β-strands (Osterman and Kaiser, 1985; Altmann et al., 1986). β-structures are somewhat less modular than α-structures because individual β-strands are not stable in isolation; they must be linked together by inter-strand hydrogen bonds. (This is in contrast to α-helices, which are stabilized by hydrogen bonding within an individual helix.) However, sequences designed to form self-assembling β-strands and those designed to form self-assembling α-helices are similar in that both must be amphiphilic. Osterman and Kaiser incorporated β-strand amphiphilicity in their design of a series of peptides having the repeat Val-Glu-Val-Orn (Orn = ornithine). They varied the length of these peptides, and in some cases they protected the amino groups with the trifluoroacetyl group. Although some of their designed peptides formed large aggregates that were difficult to characterize, one of them, (Val-Glu-Val-Orn)$_3$-Val, appears to form an octamer under certain conditions.

The self-assembling modular structures described in this section share several features, but they also differ in many ways. Which features are essential for modular design? Are there any general rules for designing peptides that will form regular secondary structures and self-associate? Since self-association is typically mediated by hydrophobic interactions, it is clearly important for designed sequences to form secondary structures that are amphiphilic. Such amphiphilic secondary structures can form only if the amino acid sequence has a periodicity of polar and nonpolar residues that matches the fundamental repeat pattern of the particular type of secondary structure. Thus, for a peptide to form an amphiphilic β-strand, its sequence must be designed from alternating polar and nonpolar residues. Conversely, formation of an amphiphilic α-helix requires a sequence in which the nonpolar residues occur at every third or forth position. Experiments using model systems have demonstrated that the periodicity of polar and nonpolar amino acids clearly plays a dominant role in dictating structure and assembly (Brack and Spach, 1981; Kaiser and Kézdy, 1983; DeGrado and Lear, 1985; Altmann et al., 1986; Dado and Gellman, 1993). Is it necessary to also consider the intrinsic propensities of the amino acids? Must α-helical structures be composed of "good" helix formers and β-structures composed of "good" beta formers? In recent experiments in my laboratory, we have shown that the ability of these intrinsic propensities to dictate structure is clearly overwhelmed by the polar/nonpolar periodicity of the sequence. For example, if a sequence is composed of good helix formers but has a polar/nonpolar periodicity of 2, then it forms self-associating β-strands. Conversely, a sequence composed of good β-formers but having a periodicity of 3 or 4 readily forms a bundle of α-helices (Xiong et al., 1995). Thus it appears that as long as the periodicity of a sequence is rigorously controlled, self-assembling modular structures are likely to tolerate a wide variety of different sequences.

B. Ligand-Induced Assembly

A second strategy for the design of novel proteins employs a ligand (typically a metal ion) to induce the assembly of modular protein segments.

As diagrammed schematically in Fig. 4, a ligand binding site is designed into the proposed structure at the interface of several interacting segments. If the site has a high affinity for the ligand, then the favorable free energy associated with binding the ligand will be sufficient to overcome the entropic cost and drive the peptides to self-assemble.

In cases where the peptides are synthetic in origin, the binding site can be constructed from moieties other than those represented by the 20 naturally occurring amino acid side chains. For example, a nonnatural metal binding site was used by Lieberman and Sasaki (1991) to design a three-helix bundle, which was induced to assemble by the addition of iron(II). They synthesized a 15-residue amphiphilic α-helix with a bipyridine moiety attached to its N-terminus. Based on the known structures of tris-bipyridine metal complexes, Lieberman and Sasaki reasoned that in the presence of iron(II), the bipyridine groups at the ends of three α-helices would form a metal binding site. Thus, metal binding would induce the formation of a three-helix bundle as shown in Fig. 5. Characterization of the synthetic molecule by circular dichroism showed that although the isolated peptides were devoid of any regular secondary structure, the addition of iron(II) had induced the formation of a structure that was predominantly α-helical. Furthermore, gel filtration chromatography indicated that the folded molecule had the expected size in aqueous solution (Lieberman and Sasaki, 1991).

An experiment similar to that of Lieberman and Sasaki was carried out by Ghadiri *et al.* (1992a). They also attached a bipyridine group to the N-terminus of a 15-residue amphiphilic α-helix and showed that the addition of divalent metals such as Ni(II), Co(II), or Ru(II) induced the formation of a triple helical structure. Both groups used amino acid sequences com-

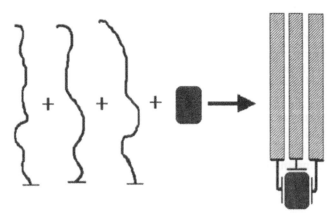

FIGURE 4   Schematic diagram showing ligand-induced assembly. In the absence of the ligand (typically a metal), the peptides are disordered. However, binding of the ligand induces formation of a well-ordered multimeric structure.

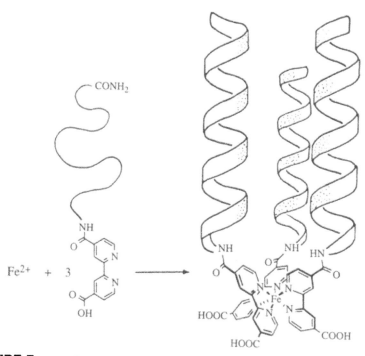

FIGURE 5   Ligand-induced assembly of a three-helix bundle. In the absence of metal, the peptides are monomeric and disordered. Addition of Fe(II) induces formation of a three-helix bundle. The sequence of the peptide is Ala-Glu-Gln-Leu-Leu-Gln-Glu-Ala-Glu-Gln-Leu-Leu-Gln-Glu-Leu. (Figure adopted from Lieberman and Sasaki, 1991.)

posed of good helix formers, but the design of Ghadiri *et al.* also included several possible intrahelical and interhelical electrostatic interactions.

Ghadiri's group then extended the metal-assisted self-assembly strategy from 3-helix bundles to 4-helix bundles (Ghadiri *et al.*, 1992b). They synthesized a 15-residue amphiphilic peptide with a pyridyl functionality at its N-terminus, reasoning that the affinity of ruthenium(II) for nitrogen-containing aromatic heterocyles would provide a strong driving force for the assembly of the four-helix bundle modeled in Fig. 6 (Ghadiri *et al.*, 1992b). The desired Ru(pyridyl peptide)$_4$ Cl$_2$ complex formed spontaneously upon the addition of Ru$_5$Cl$_{12}^{-2}$. Circular dichroism spectroscopy was used to measure α-helicity as a function of guanidine hydrochloride concentration for both the uncomplexed and the metal complexed peptide. These studies showed that in the absence of added metal, (i) the helical structure was only marginally stable and (ii) this stability was concentration-dependent. However, upon forming the desired metal complex, the α-helical structure be-

FIGURE 6   Ligand-induced assembly of a four-helix bundle. The top of the figure shows a computer-generated model of the four-helix bundle, which forms upon addition of Ruthenium(II). The metal binding site is formed by pyridyl groups attached to the N-terminus of the sequence as shown in the bottom of the figure. (Figure adopted from Ghadiri *et al.*, 1992b.)

came extremely stable (denaturation midpoint ~5.5 *M* GuHCl), and this stability was relatively insensitive to concentration (Ghadiri *et al.*, 1992b). These results demonstrate that the inherent instability and concentration-dependence of structures assembled from modular elements can be overcome, to some extent, by harnessing the favorable binding energy of an added ligand.

    Further work on the use of metals to stabilize novel proteins is discussed in Section III,C.

C. Assembly of Peptides via Covalent Cross-linking

As discussed above, the major obstacle in designing novel proteins is the conformational entropy of the polypeptide chain. When structures are assembled from several unlinked chains, this entropic barrier is all the more difficult to overcome. Thus, preorganizing the peptides by covalently linking them together is a powerful strategy to direct the formation of a desired structure (see Fig. 7).

Among the possible bonds that might be used to cross-link peptides, the only one used by nature is the disulfide bond. However, if a designed protein is made synthetically, then a variety of other cross-links are also possible. For example, Richardson, Erickson, and coworkers used a novel cross-linker called DAB in their initial design of Betabellin, an eight-stranded β-barrel intended to fold with a simple up-and-down topology (Unson *et al.,* 1984; Richardson and Richardson, 1987, 1989). Since DAB is divalent, its attachment to the solid-phase resin would permit the simultaneous synthesis of two identical peptides from a single point of origin. In the model originally proposed for Betabellin (shown in Fig. 8), the cross-link constrains the locations of the 2 C-termini and thereby preorganizes the two 31-residue polypeptides in very close proximity to one another. Unfortunately, the large nonpolar DAB moiety decreased Betabellin's solubility in aqueous solution. Therefore, later versions were designed without this cross-link. Instead, the two chains were linked by a disulfide bond. Betabellin 14D, HSLTASIkaLTIHVQakTATCQVkaYTVHISE, includes D-amino acids (small letters) to facilitate type I′ β-turns (Yan and Erickson, 1994). Circular dichroism studies of the disulfide dimer indicate it is predominately β-sheet. The thermal melt is cooperative and the NMR spectrum shows reasonable dispersion indicating that the iterative design and redesign process has produced a molecule that captures many of the features of native proteins. (See also Betadoublet in section E.)

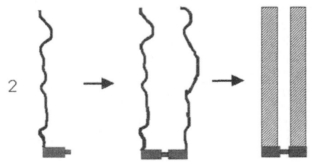

FIGURE 7 Schematic diagram showing the assembly of a structure that is stabilized by covalent cross-links.

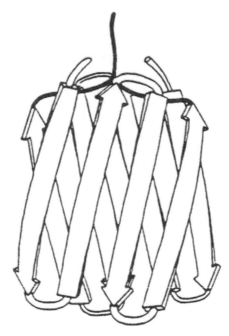

FIGURE 8 Ribbon diagram of the tertiary structure proposed for Betabellin. The cross-linker is shown at the top. Two synthetic peptides are joined to the cross-linker; each of these peptides is intended to from a four-stranded, antiparallel β-sheet. (Figure, courtesy of Jane Richardson.)

Synthetic peptides can also be linked together without the use of either disulfide bonds or artificial cross-linkers. The side-chain amino group of lysine (or ornithine) and the side-chain carboxylic acid groups of Glu or Asp can form peptide bonds with each other or with the N- and C-termini of the main chain. Such cross-linking leads to branched structures that are topologically quite different from ribosomally produced natural proteins. However, an advantage of these structures is that the interchain cross-links constrain the positions of the chains relative to one another, and thereby reduce the entropic cost of forming a unique structure.

An example of this strategy is the Chymohelizyme design of Stewart and co-workers (Hahn *et al.*, 1990). In this parallel 4-helix bundle, four synthetic polypeptide chains were joined via peptide bonds linking amino groups on lysine or ornithine side chains with the C-terminal carboxylate of the neighboring main chain. This is shown schematically in Fig. 9. Note that this strategy is well suited for designing structures composed of either parallel or antiparallel elements of secondary structure. This is in marked contrast to natural proteins, which cannot form parallel structures without an intervening antiparallel segment. The parallel 4-helix bundle of Chymohelizyme was designed to possess the enzymatic activity of

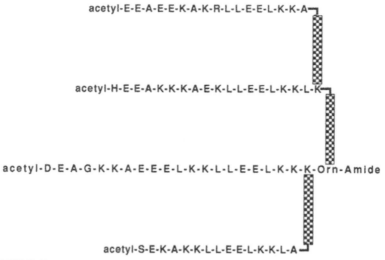

acetyl-E-E-A-E-E-K-A-K-R-L-L-E-E-L-K-K-A

acetyl-H-E-E-A-K-K-K-A-E-K-L-L-E-E-L-K-K-L-K

acetyl-D-E-A-G-K-K-A-E-E-E-L-K-K-L-L-E-E-L-K-K-K-Orn-Amide

acetyl-S-E-K-A-K-K-L-L-E-E-L-K-K-L-A

FIGURE 9 The amino acid sequence and cross-linking pattern of Chymohelizyme (Hahn *et al.*, 1990). In this branched structure, four synthetic chains are joined by peptide bonds linking amino groups on lysine or ornithine side chains to the C-terminal carboxylate of the neighboring main chain. The lysine and ornithine side chains used in the cross-linking are depicted as checkered boxes.

chymotrypsin in a structural context that is entirely different from that of the natural enzyme. The functional aspects of this design will be discussed in Section III,D, which describes the *de novo* design of protein function.

D. Assembly of Peptides on a Synthetic Template

An extreme example of peptides that are covalently constrained to adopt a particular structure are the template-assembled synthetic proteins (TASPs) developed by Mutter and co-workers (Mutter, 1985, 1988; Mutter *et al.*, 1986, 1988a,b, 1989, 1992; Mutter and Vuilleumier, 1989; Altmann and Mutter, 1990; Tuchscherer *et al.*, 1992). Rather than using turns and loops to link elements of secondary structure as natural proteins do, assembly is achieved by linking elements of secondary structure to an artificial template as shown in Fig. 10. In contrast to natural linear proteins, where α-helices and β-strands are positioned into correct relative orientations by a specific folding process, the helices and strands of the TASP molecule are directed into the desired conformation by the structure of the template.

The templates that Mutter's group employs are themselves oligopeptides. A typical example is the cyclic decamer:

Acetyl-Cys-Lys-Ala-Lys-Pro-Gly-Lys-Ala-Lys-Cys-amide.

FIGURE 10   Schematic diagram of a template-assembled synthetic protein (TASP). Covalent linkage of the peptides to the template preorganizes the structure and thereby facilitates folding. (Figure adapted from Altmann *et al.*, 1986.)

This template was designed to form two antiparallel β-strands connected by a Pro-Gly turn at one end and a disulfide bond at the other. Because it has a cyclic structure, this molecule is relatively constrained and so one can be fairly certain which way the individual side chains will point. This feature is extremely important because the ε-amino groups of the four lysine side chains serve as the attachment sites for the several elements of secondary structure. For example, in the design of a parallel four-helix bundle, an amphiphilic α-helix having the sequence Glu-Ala-Leu-Glu-Lys-Ala-Leu-Lys-Glu-Ala-Leu-Ala-Lys-Leu-Gly was attached to the ε-amino group on each of the four lysine side chains in the cyclic peptide shown above. A general feature of TASPs is that the individual peptides show marginal secondary structure that is concentration-dependent, but once they are joined to the template, their structures tend to be quite stable and independent of concentration.

One of the possible limitations of the TASP strategy is that it might be difficult to attach several different peptides to a given template: Any synthetic step leading to the attachment of a peptide to one lysine on the template would presumably lead to coupling at all four lysines. Mutter and co-workers have overcome this limitation by using different protecting groups on lysines destined to receive different peptides (Mutter *et al.*, 1989). This and other modifications of the basic TASP approach have allowed them to construct a variety of different topologies including βαβ and 4α/4β structures.

The structure-inducing properties of a template can be used to generate TASPs that mimic properties of natural proteins in ways not possible with individual peptides. For example, short isolated peptides typically are not useful as antigens because their structures are not stable in solution and are rapidly degraded *in vivo*. Therefore, prior to their use as antigens, peptides are usually linked to large carrier proteins. Mutter and colleagues have constructed a TASP molecule that mimics the antigenic properties of a large natural protein. Specifically, they joined four copies of an α-helical sequence from the HLA-A2 protein to the cyclic template shown above. The resulting molecule was then used without carrier protein to raise antibodies capable of recognizing the natural HLA-A2 antigen (Tuchscherer *et al.*, 1992). Thus the TASP strategy was used to constrain the structure of synthetic peptides so that they could effectively mimic key properties of a much larger natural protein.

The type of template that can be used to induce assembly is not limited to oligopeptides. Sasaki and Kaiser used a porphyrin as a template for the attachment of four copies of a synthetic amphiphilic α-helix as diagrammed in Fig. 11 (Sasaki and Kaiser, 1989, 1990). The resulting heme-protein, called Helichrome, is quite stable: The midpoint for its denaturation by guanidine hydrochloride is 5.2 M and the free energy stabilizing the folded structure in the absence of denaturant was extrapolated to be −4.4 kcal/ mol. These values are similar to those observed for typical natural proteins. However, in the absence of the porphyrin group, the isolated peptides do not form a stable structure (Sasaki and Kaiser, 1989, 1990). Thus, as with the TASPs described above, the template contributes significantly to the structure and stability of the designed molecule. Helichrome was designed not only to form a stable globular structure, but also to possess enzymatic activity. The aniline hydroxylase activity of this protein is discussed in Section III,D on the design of catalytically active proteins.

E. Linear Polypeptides That Fold into Globular Structures

The ultimate goal in designing novel structures is to construct single-chain proteins that fold into defined three-dimensional structures without

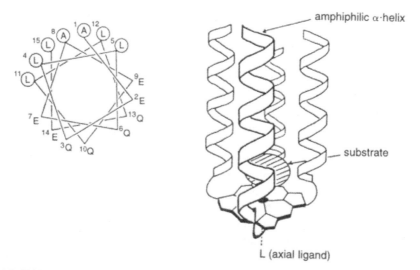

FIGURE 11 Proposed structure of Helichrome. The sequence of the amphiphilic α-helix is shown as a helix wheel diagram on the left. The modeled structure of the entire four-helix bundle is shown on the right. (Figure adopted from Sasaki and Kaiser, 1989.)

the assistance of templates or covalent cross-links. However, as described above, folding a linear polymer into a unique globular structure occurs only at a significant cost in conformational entropy, and so design strategies that rely only upon chain folding are likely to be somewhat more difficult than those that incorporate covalent linkages. Furthermore, although structures that use covalent cross-links (or tightly bound ligands) can be stabilized via the design of a few very strong interactions, the stability of a native-like protein may require the explicit design of a myriad of more subtle interactions. Nonetheless, significant progress has been made toward the design of native-like protein structures.

The first example of a single-chain polypeptide successfully designed to fold into a stable globular structure was the α_4 structure of DeGrado and co-workers (Regan and DeGrado, 1988). As described above in Section II,A, α_4 represents the final stage in a stepwise approach that began with the synthesis of a 16-residue peptide, α_1, which self-associates into a 4-helix bundle (see Fig. 3). In α_4 these helices are joined by Pro-Arg-Arg turns and the complete 74-amino acid sequence is expressed *in vivo* from a synthetic gene. The resulting molecule is stabilized relative to the unfolded form by a free energy of 22.5 kcal/mol. It is significantly more stable than most natural proteins, which are stabilized by only 5 to 10 kcal/mol (Regan and De-Grado, 1988).

The extreme stability of α_4 suggests that its structure has been "idealized." Indeed, in designing α_4, DeGrado and co-workers employed a "minimalist" approach (DeGrado *et al.*, 1989). In other words, the complexity of the design challenge was simplified by focusing on a few features known to be critically important for the stability of natural proteins. These features were optimized and then repeated several times throughout the structure.

What were these minimalist features in the design of α_4 that accounted for its success? Probably the most important feature stabilizing α_4 is the extreme amphiphilicity of its α-helices. The solvent exposed and interior faces of the helices were unambiguously determined by using charged residues (Glu or Lys) at every surface position and an extremely hydrophobic side chain (Leu) at every buried position. When the four α-helices pack against one another, a total of 24 leucines can be removed from contact with water. Since the burial of hydrophobic surface area is generally accepted as the major driving force in protein folding (Kauzmann, 1959; Dill, 1990), it is likely that the key feature responsible for the remarkable stability of α_4 is the formation of an extremely hydrophobic core.

The design of α_4 incorporated several additional features that presumably also contribute to the stability of the protein. For example, the charged side chains on the exposed faces of the helices were spaced appropriately for forming a large number of Glu^-.....Lys^+ salt bridges and the helix dipole (Wada, 1976; Hol *et al.*, 1978) was neutralized by incorporating negatively charged glutamates near the N-terminus of each helix and positively charged lysines near the C-termini. In addition, all of the N-cap and C-cap residues (Richardson and Richardson, 1988) at the ends of the helices are glycines, which allow for increased flexibility at the helix/turn junctions. The turns themselves start with the strong helix breaker, proline, thereby fixing where each helix must end. Finally, the length of the turns, 3 residues, favors the formation of a chain reversal by disrupting the register of nonpolar residues (i.e., the hydrophobic stripe) and thereby preventing the continuation of a long single amphiphilic helix (Ho and DeGrado, 1987; Richardson and Richardson, 1989).

Although the minimalist approach has proven enormously successful, it nevertheless has some drawbacks. By minimizing the complexity of a design, one invariably ends up with sequences that are more repetitive and structures that are simpler than their naturally occurring counterparts. These repetitive sequences may favor structures that are inherently more plastic and less precisely determined than native protein structures. Indeed, De-Grado et $al.$ have reported that despite its exceptional stability, "α_4 has a fairly dynamic structure in which the main chain is stably folded the great majority of the time but the side chains are less well ordered than in a native protein" (DeGrado et $al.$, 1991). This plasticity presumably results from the ability of the leucine side chains in the interior to pack in a number of different low energy conformations. Thus, because of its repetitive sequence, α_4 "lacks the complementary packing of small and large side chains that is characteristic of native proteins," and which may be necessary to fix the side chains into a rigid structure (DeGrado et $al.$, 1991).

Raleigh and DeGrado subsequently modified the sequence of the four-helix bundle to make it less repetitive and thus more likely to form a well-ordered structure (Raleigh and DeGrado, 1992). They replaced several of the buried leucines with alternative nonpolar side chains such as Val, Phe, Trp, and Ile, which are more conformationally constrained than Leu and also can introduce some shape complementarity into the helix–helix packing (Raleigh and DeGrado, 1992). The amino acid substitutions were made in the context of the α_2 peptide, which readily dimerizes into four helix bundles, as described above in Section II,A. Characterization of the new peptide by CD and NMR suggests that at low temperatures its interior side chains are more ordered than in the original version of α_2. When the temperature is raised, a transition occurs to a less ordered structure that resembles that of the original α_2 sequence. Thus, by departing from their original minimalist sequence, DeGrado and co-workers constructed a molecule whose structural and thermodynamic properties more closely resemble those of natural proteins. An alternative strategy to reduce the flexibility of interior side chains involves the incorporation of metal binding sites into novel proteins (Handel et $al.$, 1993). This approach is described below in Section III,C.

An alternative strategy to the minimalist approach was used by Hecht et $al.$ to design Felix, a four-helix bundle with a nonrepeating sequence (Hecht *et al.*, 1990). In the design of Felix (for Four hELIX), our major goal was to produce a native-like protein capable of forming a unique three-dimensional structure, stabilized by specific interactions between different side chains. For this reason we felt it imperative to avoid simple repeating sequences. Although Felix is not homologous to any known proteins, its sequence is native-like in that (i) each of the four helices is different, (ii) the helices themselves are nonrepetitive, and (iii) 19 of the 20 naturally occurring amino acids are included in its 79 amino acid sequence (Hecht *et al.*, 1990).

All four α-helices in Felix were designed to be amphiphilic, although less so than the helices in α_4 and more like those typically observed in known protein structures. In designing the sequences of the four helices, the residues at each specific location along a helix were chosen from among those most frequently found at that location in the helices of natural proteins (Richardson and Richardson, 1988). Once the individual helices had been chosen, the interhelical packing of the entire four-helix bundle was modeled and the amino acid sequence was modified to remove lumps or fill holes. Although the internal packing of a protein is presumably one of the most critical factors for a successful design, it is also one of the most difficult to model. Finally, a single intramolelcular disulfide bond intended to connect Cys_{11} in the first helix with Cys_{71} in the fourth helix was included for two reasons: (i) to reduce the conformational entropy of the chain and thereby stabilize the folded structure and (ii) to facilitate experiments that would probe the orientation of the first and fourth helices relative to one another. Figure 12 shows the final sequence of Felix and a ribbon diagram of its proposed structure.

Protein design involves a delicate balance between stabilizing a designed structure and destabilizing competing structures. The process of designing *against* alternative structures is sometimes called "negative design" (Hecht *et al.*, 1990). Several features of negative design were incorporated into the Felix sequence. For example, in designing the α-helices, it was important to avoid sequences containing alternating hydrophobic and hydrophilic residues, since such sequences would favor formation of β-sheets (Brack and Spach, 1981). We also incorporated negative design when planning the connections between successive helices. If these connections did not produce chain reversals, then the helices would continue through their designed breaks, leading to a hairpin of two double-length helices rather than a four-helix bundle. To design against this "wrong" structure, we chose turn lengths such that if the helices tried to continue through the designed breaks, the hydrophobic stripe would necessarily be shifted to a different face, thus eliminating helix amphiphilicity (see Fig. 3 in Hecht *et al.*, 1990.) A final example of negative design concerns the two possible "mirror-symmetric" structures that four-helix bundles can assume: If the N-terminal helix goes up the bundle, then the second helix can either be to its right or its left (Efimov, 1982). Both "right-turning" and "left-turning" structures exist

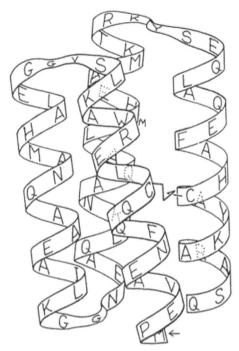

FIGURE 12   Ribbon drawing of the sequence and proposed structure of Felix. The Cys_{11}–Cys_{71} disulfide bond is indicated. (Figure adopted from Hecht *et al.*, 1990.)

among natural proteins and it was important that Felix be designed for one of these topologies and against the other. Consequently, the hydrophobic faces of the four helices were arranged so that they would be complementary to one another and pack tightly in a left-turning structure, but would be sterically incompatible in the competing mirror-symmetric structure. Furthermore, the left-turning topology was favored by including the Cys_{11}-Cys_{71} disulfide bond. In the designed left-turning structure, Cys_{11} and Cys_{71} would point toward one another allowing an intramolecular disulfide to form, whereas in the competing right-turning structure these two side chains would point away from one another and the bond could not form.

We constructed a synthetic gene to encode Felix and expressed the protein at high levels in *Escherichia coli*. The protein was purified and subjected to a variety of physical characterizations with the following results: (i) Size-exclusion chromatography indicated that Felix folds into a compact globular structure that is monomeric in aqueous solution; (ii) Circular dichroism spectroscopy demonstrated that the protein was predominantly α-helical; (iii) Formation of the Cys_{11}-Cys_{71} disulfide bond showed that regions of the chain distant in sequence were in fact close to one another in the three-dimensional structure, and thus α-helix No. 1 was packed against α-helix

No. 4 (see Fig. 12); (iv) This disulfide bond was intramolecular rather than intermolecular, demonstrating that the two cysteines pointed toward one another, which is indicative of the designed left-turning bundle and not the alternative right-turning bundle; (v) Fluorescence spectroscopy indicated that the single tryptophan at position 15 was buried in a nonpolar environment as expected; and (vi) The quenching of this fluorescence suggested that the tryptophan is very close to the Cys_{11}-Cys_{71} disulfide bond, as would be the case when this region of the protein is in an α-helical conformation. Taken together, these results rule out several alternative structures and indicate that Felix adopts a folded structure similar to the designed model shown in Fig. 12.

Although it appears that Felix folds into an approximately correct structure, it is not very stable. Denaturation experiments show that the folded conformation is stabilized by <1 kcal/mol relative to the unfolded state. This marginal stability demonstrates that although we may understand some of the determinants of protein folding, much remains to be learned about the design of native-like proteins. In all likelihood, the central feature that has not been designed correctly in Felix is the center of the protein itself. The specific interactions between hydrophobic side chains in the interior of a protein may be essential for the formation of a unique structure, but they are also extremely difficult to design. Indeed, the low stability of Felix probably indicates that it does not maintain a unique, well-packed hydrophobic core. Instead, its interior side chains may fluctuate, as has been observed for the molten globule state of some natural proteins (Ohgushi and Wada, 1983; Dolgikh et al., 1984; Kuwajima, 1989).

Although four-helix bundles represent an appealing model system, attempts to design novel proteins have not been limited to these structures. Another structure commonly observed among natural proteins is the eight-stranded α/β barrel typified by the structure of triose phosphate isomerase (Banner et al., 1975). Goraj et al. (1990) designed an amino acid sequence for a novel α/β barrel called Octarellin. Like α_4, Octarellin was designed to form an idealized structure composed of several repeats of a single structural unit. In this case the structural unit is a turn–βstrand–turn–αhelix. A 32-residue sequence (DARS-GLVVYL-GKRPDSG-TARELLRHLVAEG) for this structural unit was designed based upon an analysis of naturally occurring α/β barrels. A gene encoding this sequence was synthesized and expressed in E. coli as a monomer and as direct repeats of 2–12 units. Preliminary structural characterization was carried out on proteins with 7, 8, or 9 repeats. All three proteins yielded CD spectra consistent with the presence of some ordered structure. Further analysis by urea–gradient gel electrophoresis (Creighton, 1979), suggested that under certain conditions (in Tris–acetate pH 4, but not in Na-acetate at the same pH) the eightfold repeated protein underwent a cooperative 2-state transition between a compact globular structure and a urea-induced unfolded form. Interestingly, the sevenfold and ninefold repeating sequences, which have no natural counter-parts, did not undergo this transition. More recently, Tanaka *et al.* (1994) also designed novel TIM barrels. These proteins underwent cooperative transitions, but NMR, broad thermal melts, and ANS-binding suggest they are not fully native-like.

Recently, Quinn *et al.*, (1994) designed a novel β-sheet protein called Betadoublet. Earlier work in the Richardson lab on Betabellin (see Section II,C above) suggested that β-sheet proteins might be considerably more difficult to design than α-helical proteins. Much of this difficulty stems from the poor solubility of model β systems in aqueous solvents. To circumvent this problem, the amino acid sequence of Betadoublet contains a charged residue in each of the turns and in most of the β-strands. Betadoublet was designed to contain two identical β-sheets that could be linked by a single disulfide bond. Each of the two sheets consists of four strands joined togeth-er by type I' tight turns. The hydrophobic core was modeled by initially setting all side-chain χ angles to those found in natural proteins (Ponder and Richards, 1987), and then using small probe dot surfaces (Richardson and Richardson, 1987) to visualize interresidue packing. Quinn *et al.* have ex-pressed the designed protein as a fusion protein in *E. coli*, and the purified Betadoublet has been partially characterized. CD, Raman, and NMR spec-troscopies all indicate that the secondary structure of Betadoublet is pre-dominantly β. Furthermore, temperature-dependent CD experiments show (i) that Betadoublet can be thermally unfolded and (ii) that this unfolding is cooperative. Cooperative thermal denaturation is a hallmark of natural pro-teins and suggests that the modeling process may have succeeded in generat-ing a protein that possesses some native-like features.

The designed proteins described above are all based on structural motifs frequently observed among natural proteins. In a departure from this ap-proach, Ptitsyn and co-workers (Fedorov *et al.*, 1992) attempted to design a protein that would fold into a three-dimensional structure not previously seen among natural proteins. This structure, the "open sandwich," consists of a four-stranded β-sheet packed against two antiparallel α-helices, as shown in Fig. 13. In designing the sequence for their novel protein (which they call Albebetin), Ptitsyn and co-workers included valine residues at the i, i+2, and i+4 positions of the intended β-strands, and leucine residues at the i, i+3, i+4, and i+7 positions of the intended α-helices. This sequence pattern favors formation of amphiphilic elements of secondary structure capable of burying their hydrophobic faces against one another. To ensure that β-strands 1 and 3 would lie at the edges of the sheet, Fedorov *et al.* (1992) made the sequences of these strands somewhat more polar than those of strands 2 and 4. The interior packing of Albebetin was designed on the basis of CPK models and the surface was designed to be sufficiently polar to prevent aggregation.

A somewhat unusual approach was taken in the production and charac-

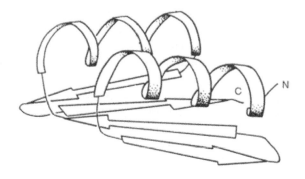

<pre>
1   35
M-D-P-G-D-P-E-C-L-E-Q-L-L-R-R-L-G-G-S-V-E-V-E-V-T-G-G-T-V-H-V-E-V-S-P-

    |-------α-helix-------| |-β-strand-| |β-strand-|

36   73
E-D-P-G-D-P-E-C-L-E-Q-L-L-R-R-L-G-G-S-V-E-V-E-V-T-G-G-T-V-H-V-E-V-S-P-E-D-R

    |-------α-helix-------| |-β-strand-| |β-strand-|
</pre>

FIGURE 13   Ribbon diagram of the proposed structure of Albebetin. The proposed α-helices and β-strands are underlined in the amino acid sequence. (Figure adopted from Fedorov *et al.*, 1992.)

terization of Albebetin. In order to avoid the possibility that Albebetin would either aggregate into insoluble inclusion bodies or be degraded *in vivo*, Fedorov *et al.* (1992) chose not to express the protein *in vivo*. Instead, a synthetic gene encoding Albebetin was transcribed and translated *in vitro* using a cell-free wheat germ system. Although this system produces only nanogram quantities of protein, the incorporation of radio-labeled amino acids facilitated partial characterization of this small quantity. Size exclusion chromatography demonstrated that Albebetin folds into a monomeric structure of approximately the correct size and urea-gradient electrophoresis showed that in the absence of denaturant, Albebetin forms a compact structure. Moreover, this structure undergoes cooperative unfolding with a midpoint of 5.5 *M* urea. Finally, proteolysis experiments demonstrated that under native conditions, Albebetin is relatively insensitive to digestion by trypsin, but in 4 *M* urea it is readily degraded. These results suggest that it folds into a compact protease-resistant structure. Although more physical characterizations are required to verify that the designed sequence actually folds into an αββ open sandwich, the initial results are quite encouraging. Furthermore, the strategy of characterizing small quantities of radiolabeled protein synthesized *in vitro* is generally applicable and should allow for the rapid screening of large numbers of designed sequences.

The encouraging results from the experiments described in the previous sections have motivated several protein design workshops sponsored by

EMBO and held at the EMBL in Heidelberg. At one of these workshops, 23 participants used computer modeling tools to design five novel proteins: Shpilka, a sandwich of two four-stranded β-sheets; Grendel, a four-helical membrane anchor; Fingerclasp, a dimer of inter-digitating ββα units; Aida, an antibody binding surface; and Leather, a minimal NAD-binding domain (Sander *et al.*, 1992). The amino acid sequences and proposed structures of these five model proteins have been "placed in the public domain" (Sander *et al.*, 1992), but no experimental verifications of the designs have been published to date.

F. Protein Design by Binary Patterning of Polar and Nonpolar Amino Acids

The folded structures of natural proteins can be compared to three-dimensional jigsaw puzzles in which the individual pieces are the amino acid side chains (Ponder and Richards, 1987; Taylor, 1992). The complementary packing of these pieces gives rise to the van der Waals interactions that contribute to the stability of native protein structures. These structures are further stabilized by numerous other interactions including hydrogen bonds, electrostatic effects, hydrophobic interactions, and the intrinsic propensities of different amino acids to be found in particular types of secondary structure (Dill, 1990). Which of these interactions must be designed explicitly? Must every contact in the jigsaw puzzle be specified *a priori*?

Current work in my own laboratory focuses on identifying those features of a protein structure that are essential for successful design. These essential features are suggested by the familiar properties of natural proteins structures. In these structures, two unifying themes can be observed: (i) Soluble proteins invariably fold into structures that maximize the burial of hydrophobic surface area while simultaneously exposing polar side chains; and (ii) these structures typically contain an abundance of hydrogen bonded secondary structure (α-helices and β-strands). Taken together, the dual constraints of forming regular secondary structure while at the same time burying hydrophobic side-chains (and exposing hydrophilic ones) have significant implications for protein design: They require not only that a designed sequence be arranged to form α-helices and/or β-strands, but moreover, that these helices and/or strands be *amphiphilic*. Thus a sequence must be designed with a periodicity of polar and nonpolar residues that matches the helical repeat for that type of secondary structure. For example, to design a β-sheet protein, the sequence must be composed of alternating hydrophobic and hydrophilic residues; for an α-helical protein, the periodicity of hydrophobic and hydrophilic residues must approximate the 3.6-residue repeat that is characteristic of α-helices.

We have developed a general strategy for protein design based upon the assumption that the ability of a sequence to form highly amphiphilic secondary structures is not only *necessary* for the formation of structure but moreover is actually *sufficient* to drive a designed polypeptide chain to fold into a compact native-like structure. Our strategy assumes that the formation of stably folded structures does not require the explicit design of specific inter-residue contacts. In other words, the precise packing of the three-dimensional jigsaw puzzle need not be specified *a priori*. Only the sequence *locations* of the hydrophobic and hydrophilic residues must be specified explicitly; the precise *identities* of the hydrophobic (or hydrophilic) residues are not constrained and can be varied extensively.

We have applied this strategy to the design of four-helix bundle proteins (Kamtekar *et al.*, 1993). We have constructed a vast collection of amino acid sequences that satisfy the above constraint. Each sequence is different, yet they all share the precise locations of polar and nonpolar residues. Amino acid sequence degeneracy is introduced into this collection by rigorously controlling the "flavor" (i.e., hydrophilic versus hydrophobic) of each residue while simultaneously allowing their identities to vary. This was accomplished by exploiting the organization of codons in the genetic code. Genes were synthesized in which the degenerate codon NAN is used wherever a polar amino acid is required, and the degenerate codon NTN is used wherever a nonpolar amino acid is required (N represents a mixture of A, G, T, and C.) By using these degenerate codons, sequence positions requiring polar amino acids can be filled by Glu, Asp, Lys, Asn, Gln, or His; positions requiring nonpolar residues can be filled by Phe, Leu, Ile, Met, or Val.

We have used this approach to construct a large collection of novel proteins, each of which is designed to fold into a four-helix bundle. The sequences differ from one another, yet they all have their polar and nonpolar amino acids in an appropriate periodicity for generating extremely amphiphilic α-helices punctuated by interhelical turns. A schematic of this pattern is shown in Fig. 14 as a head-on representation of a four-helix bundle. Generic polar residues are shown as white circles and generic nonpolar residues are shown as black circles. The interhelical turns do not play a dominant role in dictating structure (Brunet *et al.*, 1993) and are represented by arrows.

We have characterized ~50 members of this collection and have demonstrated that 60% (29 of 49) of our novel sequences actually fold into structures that are compact and α-helical. These results imply that a simple binary code of polar and nonpolar amino acids arranged in the appropriate pattern may suffice as an initial step toward the design of vast collections of *de novo* proteins (Kamtekar *et al.*, 1993).

III. STRATEGIES FOR THE DESIGN OF FUNCTION

Ultimately, the goal of protein design is to produce novel macromolecules that not only fold into predetermined structures, but moreover,

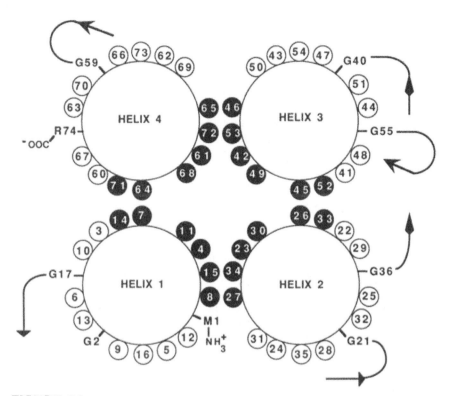

FIGURE 14 Head-on representation of a four-helix bundle. Individual helices are shown in the helix wheel representation using a repeat of 3.6 residues per turn. The binary code shows generic polar residues as white circles and generic nonpolar residues as black circles. The identities of N-cap and C-cap residues at the ends of each helix are shown explicitly. The interhelical turns are represented by arrows. (Figure adopted from Kamtekar *et al.*, 1993.)

possess interesting and useful functions. Most attempts at the design of function have focused on binding and/or catalysis. To achieve these functions, two very different approaches have been pursued: (i) "retrofitting" natural proteins to engineer substantial changes in function; and (ii) designing functional proteins entirely *de novo*. Since the purpose of this chapter is to review strategies for the design of *novel* proteins, most of what follows will focus on the second of these two approaches. However, I will first present a (very) brief discussion of retrofitting natural proteins.

A. Novel Functions by Retrofitting Natural Proteins

Natural proteins, which have evolved to perform particular biological functions, can now be retrofitted with new activities. The possibility of taking a macromolecule that evolved over billions of years and modifying it to perform a function of our own choosing can be an extremely appealing and profitable enterprise. Nature has provided us with many structural scaffolds that are quite robust and able to tolerate a wide variety of sequence modifications (Hampsey et al., 1988; Lim and Sauer, 1989; Poteete et al., 1992). As a result of this tolerance, it is possible to tinker with natural structures and thereby engineer proteins with altered specificities and/or activities. A full summary of successful attempts to modify natural proteins is beyond the scope of this chapter. However, the following examples illustrate the range of possibilities for engineering proteins with altered specificities:

(i) Alteration of DNA-binding specificity: Phage 434 repressor was changed to recognize the DNA operator site of phage P22 in preference to its natural binding site (Wharton and Ptashne, 1985).

(ii) Alteration of cofactor specificity: Glutathione reductase was changed from an enzyme requiring NADP to one that prefers NAD (Scrutton et al., 1990).

(iii) Alteration of substrate specificity: Trypsin was changed from a protease that cleaves after lysine and arginine residues to a chymotrypsin-like enzyme that cleaves after large hydrophobic residues (Hedstrom et al., 1992).

The ability to retrofit natural proteins can extend well beyond modifications that merely alter specificity. Indeed, entirely novel activities can be appended onto the structures of preexisting protein scaffolds. Novel functions engineered into natural proteins include these examples:

(i) Metal binding: A novel binding site for transition metals was grafted into E. coli thioredoxin (Hellinga et al., 1991) and a binding site for calcium was introduced into human lysozyme (Kuroki et al., 1989; Inaka et al., 1991). The calcium site was constructed by replacing Gln_{86} and Ala_{92} with aspartic acid side chains. The resulting metallo-lysozyme is both more active and more thermostable than the wild-type enzyme.

(ii) Site-specific DNA-cleavage: A 52-residue DNA-binding fragment of the Hin recombinase protein was converted into a sequence specific DNA-cleaving protein via the covalent attachment of the iron chelating agent, EDTA (Sluka et al., 1987).

To date, the most powerful and widely applicable strategy for retrofitting natural proteins is the design of catalytic antibodies (Lerner et al., 1991). Antibodies are ideal proteins upon which to graft novel functions because, in a sense, they have evolved to be retrofitted: The antibody's Greek key β-barrel structure forms a stable framework that tolerates an enormous variety of hypervariable loops. By varying these loops, the immune system is constantly generating novel binding specificities. Over the past several years, immunologists and chemists have managed to coax the mammalian immune system into generating antibodies that not only bind specific antigens, but moreover, catalyze a variety of chemical reactions. Central to the design of catalytic antibodies has been the use of antigens that resemble the transition state of a reaction. Antibodies raised against transition state analogues can significantly hasten the rate of a chemical reaction by stabilizing the rate-determining transition state on a reaction pathway. The insight of chemists in choosing appropriate antigens combined with the capacity of the immune system to respond to virtually any challenge has led to the development of hundreds of catalytic antibodies, including some that catalyze reactions for which no natural enzyme is known. The design and properties of catalytic antibodies are reviewed by Lerner *et al.* (1991).

B. *De Novo* Design of Binding and Catalysis—Early History

The design of novel proteins capable of binding specific ligands was pioneered by the work of Gutte and co-workers. In 1979 they reported the design, synthesis, and characterization of a 34-residue peptide that interacts with nucleic acids (Gutte *et al.*, 1979). The sequence of the peptide was de-signed to form a $\beta\beta\alpha$ structure that would bind the trinucleotide 2′-methyl -guanine-adenine-adenine, as modeled in Fig. 15. In this model, the side chains of Thr_{12}, Gln_{14}, and Gln_{16} hydrogen bond to the base moieties; Phe_1, Phe_3, and Tyr_5 stack or intercalate with the bases; and Lys_{28} and His_{32} form stabilizing salt bridges with the phosphates of the trinucleotide ligand. A disulfide bond connecting Cys_{10} with Cys_{33} was included to enhance the stability of the proposed structure (Gutte *et al.*, 1979; Jaenicke *et al.*, 1980).

The 34-residue peptide was synthesized by the solid-phase method, and both the 34-residue monomer and a 68-residue disulfide-linked dimer were isolated. However, for the dimer the authors did not determine which of several possible interchain disulfide bonds were present (Gutte *et al.*, 1979; Jaenicke *et al.*, 1980). Characterization of the synthetic material demon-strated that the 68-residue dimer binds 2′-citidine monophosphate with a K_d of 5×10^{-6} M, similar to the K_d observed for the natural protein, ribonuclease A. To test whether the peptide actually interacts with the in-tended ligand, Gutte and co-workers incubated their synthetic material with tRNA-Phe, which contains the trinucleotide 2′-methyl-GAA as its anti-codon sequence. They found that the 68-residue dimeric peptide digested the tRNA with a specific activity that was 2.5% that of the natural protein ribonuclease A. However, since the designed 34-residue monomer was 25-fold less active than the disulfide-linked dimer, the modeled structure shown in Fig. 15 is apparently *not* responsible for this activity. Furthermore, circular dichroism spectroscopy showed that in contrast to the structure modeled in Fig. 15, neither the monomeric nor the dimeric peptides con-tained any α-helical secondary structure (Gutte *et al.*, 1979; Jaenicke *et al.*,

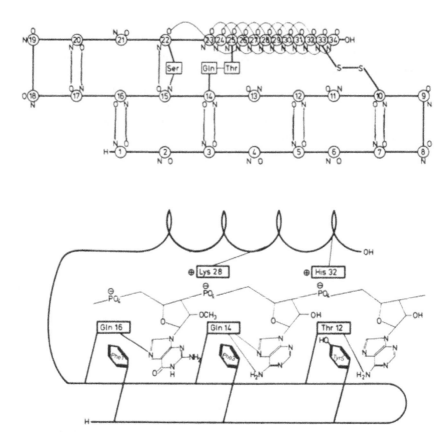

FIGURE 15   Proposed ββα secondary structure of a synthetic 34-residue polypeptide (top); and modeled interactions with the trinucleotide 2′-methyl-GAA (bottom). The amino acid sequence of the peptide is F-T-F-T-Y-T-D-P-N-C-Q-T-G-Q-G-Q-N-P-N-G-I-S-E-P-T-A-A-K-V-A-Q-A-H-C-A. (Figure adopted from Gutte *et al.*, 1979.)

1980). Despite these uncertainties about the exact three-dimensional structure of these peptides, Gutte and co-workers clearly demonstrated that it is possible to construct entirely novel proteins that demonstrate desired binding and catalytic activities.

In addition to their work on proteins that interact with nucleic acids, Gutte and colleagues also designed and synthesized a 24-residue peptide intended to bind the insecticide DDT (Moser *et al.*, 1983, 1987; Klauser *et al.*, 1991). The peptide was designed to form a 4-stranded anti-parallel β-pleated sheet based upon the prediction methods of Chou and Fasman, and of Levitt. Application of these methods to the design of β structures tends to bias toward extremely hydrophobic sequences. Furthermore, since the DDT ligand is itself a lipophilic molecule, the peptide was designed with a binding site that was also hydrophobic. For these reasons, the final sequence of the peptide (MTFIRPNVGAMSNFYHYPNIIITF) was extremely hydrophobic and consequently not soluble in water. However, the peptide was soluble in 50% ethanol and in this solvent it binds DDT with an apparent dissociation constant of 9×10^{-7} M (Klauser et al., 1991). A related 24-residue peptide having the same composition but a scrambled sequence binds DDT with an affinity ~600-fold lower (Klauser et al., 1991). CD spectroscopy indicates that the secondary structure of the DDT binding peptide is predominantly β-sheet. Although the peptide has been crystallized (Moser et al., 1983), the crystals are not suitable for structural analysis and it is not known whether the designed molecule actually folds into the proposed structure.

C. Incorporation of Binding Sites into *de Novo* Proteins

The successful design of proteins that fold into predetermined structures sets the stage for the eventual design of novel binding sites and catalytic activities (Regan, 1993). Following the design of α_4, two groups of researchers modified the sequence of this very stable four-helix bundle in order to incorporate metal binding sites (Handel and DeGrado, 1990; Regan and Clarke, 1990). Both groups chose to design a tetrahedral binding site for Zn(II). In the design of Regan and Clarke (1990), the site is formed by a combination of histidine and cysteine side chains, as is found in several natural proteins such as alcohol dehydrogenase (Vallee and Auld, 1990) and the zinc-finger domains (Parraga et al., 1988; Lee et al., 1989). The particular locations in α_4 most appropriate for the placement of the new His and Cys side chains were chosen by a computer program explicitly written for this purpose. A model of the original α_4 protein was used as the starting structure, and each residue in this structure was replaced by both His and Cys to find combinations of side chains that could form a tetrahedral Zn(II) binding site. The site that was finally chosen is formed by Cys_{21} and His_{25} from the second α-helix and Cys_{47} and His_{51} from the third α-helix. A gene encoding the metal-binding variant of α_4 was constructed and the protein was expressed in *E. coli*. Characterization of the purified protein demonstrated that: (i) Zn(II) is bound with a dissociation constant of 2.5×10^{-8} M; (ii) The binding site is wholly contained within a single protein monomer; (iii) The secondary structure of the protein is unchanged upon metal binding; (iv) Binding of Zn(II) increases protein stability; and (v) The cysteine sulfhydryls are essential for binding. Furthermore, the absorption spectrum of the Co(II)-substituted protein is consistent with binding to a tetrahedral site formed by two histidine and two cysteine side chains (Regan and Clarke, 1990).

Handel and DeGrado (1990) also designed a Zn(II) binding site by modifying the sequence of α_4. However, rather than forming a tetrahedral binding site by explicitly placing His or Cys side chains at all four corners, they chose to approximate the binding site of carbonic anhydrase, which has His residues at three corners of a tetrahedron. The remaining ligand was intended to be a solvent molecule. Thus, Handel and DeGrado designed their metallo-protein with an eye toward the future incorporation of catalytic activity. The metal binding variant was synthesized by solid phase methods in both the α_4 context and the α_2 context. (As described above, α_4 is a 74 amino acid sequence that folds into a four-helix bundle, whereas α_2 is a half-length peptide that must dimerize to form a four-helix bundle.) In each case, two of the three histidine side chains were incorporated within a single α-helix, while the third His side chain came from the neighboring helix. The binding of Zn(II) to the novel protein occurs with no significant change in secondary structure. However, when metal is bound a number of resonances in the NMR spectrum become sharper and less degenerate. This suggests that Zn(II) binding induces the protein to assume a more ordered structure that is less like a molten globule and more like the native state of natural proteins (Handel and DeGrado, 1990; DeGrado *et al.*, 1991).

Handel *et al.* (1993) have extended this work by designing a variant of α_4 that has two metal binding sites. Each of the two sites contains three histidine side chains and so the protein is called H6-α_4. This protein was compared to the original α_4 and to the version having only a single metal binding site (H3-α_4). NMR showed a doubling of resonance for H6-α_4 compared to H3-α_4 demonstrating that the two metal sites in H6-α_4 are nonequivalent. Furthermore, the binding of zinc to H6-α_4 caused a dramatic increase in the dispersion of chemical shifts for the leucine methyl protons and for the α proton resonances. This indicates that upon binding metal, the environments of these protons become well defined and less flexible—more native-like and less molten globule-like. However, even the H6-α_4 protein with two bound zinc ions is not truly native-like. It has a relatively high affinity ($K_d = 80$ μM) for the hydrophobic dye 1-anilino-8-naphthalenesulfonate (ANS). This binding indicates that a nonpolar pocket is accessible to the dye and is diagnostic for molten-globule behavior (Kuwajima, 1989).

As discussed in Section II,B, metal binding has been used as a strategy to induce structure by directing the assembly of modular protein segments (see Fig. 4). The three- and four-helix bundles designed by Sasaki's and Ghadiri's groups are excellent examples of this strategy and were described in Section II,B (see Figs. 5 and 6). More recently, Ghadiri and Case (1993) built upon their original design and constructed a three-helical bundle that has a site for ruthenium(II) at one end and a site for copper(II) at the other. The ruthenium(II) is bound by three bipyridyl groups at the N-termini of the helices (Ghadiri *et al.*, 1992a), while the copper(II) is bound by three histidine side chains at the C-termini of the helices. The synthetic protein was shown to bind both metals with each metal binding specifically to its designed site. The correct assembly of three peptides and two different metals into a unique heterodinuclear metalloprotein is an impressive result and demonstrates considerable control over the self-assembly process.

Construction of novel metallo-proteins has not been limited to α-helical structures. Pessi *et al.* (1993) designed a metal binding site into a β-sheet framework based on the variable heavy domain of immunoglobulin McPC603. Their designed protein, which they call the "minibody," is 61 residues long. Its amino acid sequence and proposed structure are shown in Fig. 16. The sequence of the designed protein differs from McPC603 in three respects: (i) Side chains that are buried in the intact McPC603 antibody but would be exposed in the minibody were mutated from nonpolar to polar residues; (ii) the 7-residue loop connecting β-strands two and three was replaced by two glycines because this loop makes significant contacts with parts of the antibody not present in the minibody; (iii) a metal binding site was introduced by incorporating three histidines into hypervariable loops H1 and H2. The designed metallo-protein was synthesized by chemical methods (Bianchi *et al.*, 1993). Size exclusion chromatography of the purified product showed it to be monomeric and circular dichroism spectroscopy showed it to be predominantly β-structure. The minibody binds zinc and the

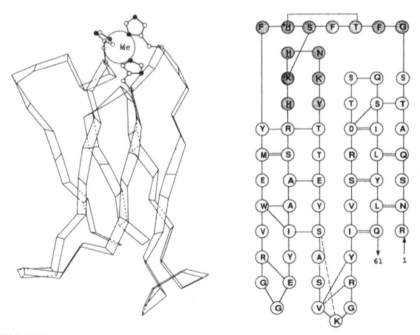

FIGURE 16   (Left) Schematic representation of the proposed structure of the minibody. The metal-binding site and three ligating histidines are shown at the top. (Right) Amino acid sequence of the minibody. The metal-binding loop (corresponding to hypervariable loops H1 and H2 in the intact antibody) are shaded and the three ligating histidines are underlined. A mutation of His-13 (marked with asterisk) to Ala reduces metal binding by a factor of four. Putative main chain H-bonds are indicated by dotted lines. (Figure adopted from Bianchi *et al.*, 1993.)

pH dependence of this binding indicates that histidine side chains are involved. A lower limit for the binding of zinc was estimated at 10^{-6} M and the order of affinities for different metals, $Cu > Zn \gg Cd > Co$, mirrors that of carbonic anhydrase.

The design of binding sites into *de novo* proteins need not be limited to metal sites. For example, DeGrado and coworkers built a heme binding site into their α-2 system (Choma *et al.* 1994). A model of a 4-helix bundle composed of an α-2 dimer was used as a starting point for the design of a novel heme binding site. Based on extensive computer modeling, the original α-2 sequence was modified. Leu25 was changed to His, Leu22 and Leu29 to Val, and Leu10 to Ala. Additionally, the two α-2 peptides were linked by a disulfide bond. The resulting design was called VAVH25(S-S). Initially, the peptide was synthesized backwards, and so Choma *et al.* were able to study both the designed VAVH25(S-S) peptide and the "retro" peptide. Both peptides bound heme. Suprisingly, the retro peptide bound heme more tightly. Furthermore, the retro protein turned bright pink, similar to natural heme proteins, while the designed VAVH25(S-S) was yellow-pink. These experiments demonstrate (i) that heme proteins can readily be constructed *de novo* and (ii) that the detailed structure of the binding site may not need to be designed *a priori*. Further support for this second conclusion comes from our own work with *de novo* proteins designed by binary pattering of polar and nonpolar amino acids (see Section II,F). Although the "binary code" proteins were not explicitly designed to bind heme, ~25% of them indeed bind, turn bright pink, and have spectroscopic properties reminiscent of natural heme-proteins (S. Kamtekar and M. H. Hecht, unpublished observations).

De novo heme binding sites need not be limited to one heme per protein. In an attempt to build a "molecular maquette" of the multi-heme proteins involved in respiration and photosynthesis, Robertson *et al.* (1994) constructed a novel four helix bundle composed of two 62-residue peptides. The assembled macromolecule accommodates four parallel hemes with spectroscopic and electrochemical properties that resemble those of natural multiheme proteins. Particularly striking is the observation of heme–heme electrochemical cooperativity, indicating that *de novo* proteins cannot only be made to bind cofactors, but moreover to mimic some of the complex properties found among highly evolved natural proteins.

D. Design of Catalytically Active Proteins

Two of the proteins described in Section II of this chapter were explicitly designed to be catalytically active. The Chymohelizyme of Hahn *et al.* (1990) was designed as an esterase with a specificity similar to chymotrypsin, and the Helichrome of Sasaki and Kaiser (1989) was designed as an aniline hydroxylase. The strategies used in designing these two proteins had several features in common. In both cases the designed catalytic sites were based on sites found in natural enzymes: The catalytic triad designed into Chymohelizyme mimics that of chymotrypsin, and the porphyrin ring of Helichrome occurs in the active sites of natural heme-proteins. In both designs a native-like catalytic site was used as the cornerstone for constructing a hydrophobic binding pocket at the interface of four parallel α-helices. Both designs also used artificial cross-links to preorganize the structure and position a binding pocket close to the catalytically active moieties. In Chymohelizyme, the helices were linked by peptide bonds between side-chain amino groups and main chain carboxylates (see Fig. 9). In Helichrome, the helices were attached to the porphyrin template (see Fig. 11). The Fe(III) complex of Helichrome hydroxylates aniline to form p-aminophenol with an affinity (K_m = 5 mM) and rate (k_{cat} = 0.02 min^{-1}) similar to those observed for natural hemeproteins, such as hemoglobin. The template alone was inactive, demonstrating that the designed binding pocket at the interface of the four helices is essential for activity.

In an attempt to devise an active site from "first principles" rather than by mimicking the site of a known enzyme, Benner and co-workers designed a novel enzyme that catalyzes the decarboxylation of oxaloacetate (Benner *et al.*, 1992; Johnsson *et al.*, 1990, 1993). Whereas all of the known oxaloacetate decarboxylases require either magnesium or manganese as a cofactor, the novel enzyme was designed to catalyze decarboxylation via a metal-independent mechanism in which an active site lysine forms a Schiff's base with the substrate. For this mechanism to be effective, the lysine must act as a nucleophile and must therefore be unprotonated. Consequently, enzyme activity at neutral pH requires that the pK of the catalytic lysine be lowered substantially from the pK of 10.5 typical of normal lysine side chains. In their design of an artificial oxaloacetate decarboxylase (ART-OAD), Benner and co-workers lowered the pK of a lysine by placing it in close proximity to several other lysine side chains on the polar face of an amphiphilic α-helix, as shown in Fig. 17. Their intention was for the central lysine to act as the nucleophile while the bracketing lysine would be positively charged and thus facilitate binding to the negatively charged substrate.

ART-OAD was synthesized by solid phase methods. (The peptide is called Oxaldie in later publications.) CD and NMR showed the peptide to be predominantly α-helical. This secondary structure was concentration-dependent, demonstrating that the amphiphilic α-helical modules assembled, as expected, into a multimeric (probably tetrameric) structure. The synthetic peptide displays Michaelis-Menton kinetics and a substrate specificity that prefers oxaloacetate over acetoacetate. Its K_m (14 mM) is similar to that of the natural enzyme (K_m = 4 mM). The k_{cat} (0.4 min^{-1}) of the synthetic peptide, although ~5 orders of magnitude slower than that of the

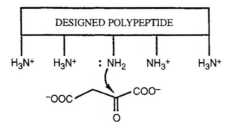

NH$_2$-Leu-Ala-Lys-Leu-Leu-Lys-Ala-Leu-
Ala-Lys-Leu-Leu-Lys-Lys-CONH$_2$

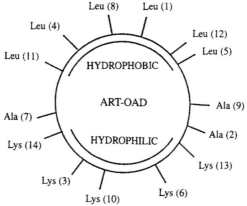

FIGURE 17   Schematic drawing of a ART-OAD (also called Oxaldie, Johnsson *et al.,* 1990), a peptide designed to catalyze the decarboxylation of oxaloacetate. The helical wheel representation demonstrates that the peptide can form an amphiphilic α-helix, with the lysine side chains all facing in the same direction. (Figure adopted from Johnsson *et al.,* 1990.)

natural metallo-enzyme, is still ∼900-fold faster than spontaneous decarboxylation (Benner *et al.,* 1992; Johnsson *et al.,* 1990, 1993).

What functional group(s) is responsible for the activity of this novel enzyme? Titration experiments showed that one amino group had a pK of 7.2. Although it was tempting to assign this lowered pK to the central lysine, later experiments showed that it belonged to the N-terminal amino group. Thus although ART-OAD catalyzes the decarboxylation of oxaloacetate, it does not do so exactly according to the original plan. Nonetheless, some features of the original design clearly contribute to catalytic activity. The lysine side chains almost certainly facilitate substrate binding and the central lysine has enhanced ability to serve as an active site nucleophile. These features are substantiated by the finding that an N-terminal acetylated version of ART-OAD (also called Oxaldie 2), while not quite as active as the original molecule, is nonetheless catalytically active (K_m = 48 mM; k_{cat} = 0.5 min^{-1}). Furthermore, mutations that disrupt the α-helicity of the peptide diminish catalytic activity, indicating that the designed structure is an important prerequisite for the observed activity (Johnsson *et al.*, 1993).

E. Membrane Proteins and Ion Channels

Membrane proteins play essential roles in many biological processes, and the ability to construct novel membrane proteins that possess predetermined structures and "tailor-made" activities is certain to have enormous impact. However, the design of a membrane protein challenges would-be designers with obstacles beyond those already encountered in the design of water-soluble proteins. The interactions that drive the formation of protein structure in membranes differ from those that dominate the folding of water soluble proteins. For example, the hydrophobic effect, which requires proteins in aqueous environments to have nonpolar insides and polar outsides, makes very different demands on proteins immersed in environments that are themselves nonpolar. Furthermore, once an appropriate sequence for a membrane protein is actually designed, constructing the novel protein can be fraught with technical difficulties not encountered in synthesizing and/or expressing water-soluble proteins: Since membrane proteins are not soluble in standard aqueous buffers, they typically form inclusion bodies when expressed *in vivo* or insoluble precipitates when synthesized *in vitro*. Moreover, the transfer of a novel membrane protein from a water-insoluble aggregate into a membrane-like environment will almost certainly require a conformational change that may or may not be kinetically facile.

Despite these challenges, several membrane proteins have been designed and characterized. Montal and co-workers designed a series of novel proteins that mimic the essential features of natural ion channels (Montal, 1990; Montal *et al.*, 1990; Oiki *et al.*, 1990; Grove *et al.*, 1991). These synthetic pore proteins, called Synporins, were constructed by using the template-assisted strategy described by Mutter (Oiki *et al.*, 1990), and summarized in Section II,D of this chapter. In one example of this approach, four copies of a 23-residue α-helix were tethered to a carrier template (Montal *et al.*, 1990). The sequence of the 23-mer was taken from the δ subunit of the acetylcholine receptor of *Torpedo californica*. The template was a 9-residue peptide originally designed by Mutter and co-workers (Mutter, 1988) to form a β-strand–turn–β-strand with four lysine side chains (two before the turn and two after it), all pointing in the same direction. Thus, when the four 23-mers are attached to this template they are constrained to be parallel to one another. Since the amino acid sequence of the 23-mer can form an amphiphilic α-helix, the Synporin design assumes that linkage to the template would direct the formation of a parallel four helix bundle, as modeled in Fig. 18. This model is supported by the results of circular dichroism experi-

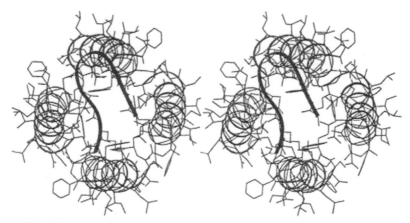

FIGURE 18   Stereo representation of the proposed structure of a Synporin (Montal *et al.*, 1990). The peptide template is in the foreground and the four parallel α-helices project into the page. The central pore is clearly visible. (Figure courtesy of J. Schiffer and M. Montal.)

ments, which show that the structure of the intact 101-residue synporin is predominantly α-helical. According to the design, the nonpolar side chains of the helices should form an external lipophilic surface, which interacts with the lipid bilayer, while the polar side chains of the helices should form a central pore. Characterization of the synthetic protein shows that it indeed forms ion channels. These channels (i) show single-channel conductance; (ii) can discriminate between different cations; (iii) can open and close in the millisecond time range; and (iv) are sensitive to local anesthetic channel blockers. Thus, the Synporin reproduces several features that are characteristic of the authentic acetylcholine receptor (Montal *et al.*, 1990).

In another example of the Synporin strategy, Montal's group constructed an analogue of a calcium channel. This synthetic protein was built on the same template used in the previous study, but the 22-residue α-helical sequence used in this study was taken from the dihydropyridine-sensitive calcium channel. The resulting 97-residue synthetic protein mimics many features of the authentic channel, including the ability to distinguish between enantiomers that act as activators (agonists) or blockers (antagonists) (Grove *et al.*, 1991). Since the α-helical sequence used in this Synporin is found in natural channel proteins from skeletal muscle, cardiac muscle, brain, and aorta, Montal *et al.* suggest that their synthetic four-helix bundle may serve as a "molecular blueprint" for the pore-forming structures of all of these voltage-gated calcium channels (Grove *et al.*, 1991).

The occurrence of helical bundle structures in many natural ion channels prompted Lear *et al.* to design synthetic ion channels comprising simple amphiphilic α-helices (Lear *et al.*, 1988; DeGrado and Lear, 1990). These designed channels differ from those of Montal *et al.* in two important respects: (i) The sequences were *not* based upon the sequences of natural ion channels, but were designed entirely *de novo;* and (ii) instead of relying upon an artificial template, the synthetic peptides were designed to self-assemble into multimeric structures surrounding a central hydrophilic pore. The "minimalist" approach (DeGrado *et al.,* 1989) led to the design a 21-residue sequence that is extremely simple, but nevertheless captures the essential features of natural ion channels. Repeating sequences composed of only two amino acids (Leu and Ser) were designed to form amphiphilic helices that would assemble into multimeric structures. Two different versions of the design were constructed. In each case, the 21-residue sequence contained 3 heptad repeats. However, one version had two serines per heptad $\{(LSLLLSL)_3\}$, while the other had three $\{(LSSLLSL)_3\}$. Computer modeling suggests that the synthetic peptide $(LSLLLSL)_3$ forms a tetramer of α-helices, whereas $(LSSLLSL)_3$ forms hexamers. In both model structures, the helices span the membrane and enclose a central pore that is lined by serine side chains. However, as shown in Fig. 19, the central pore in the modeled

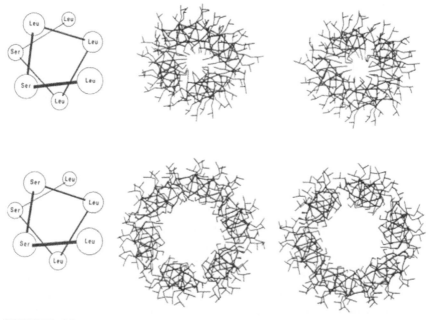

FIGURE 19   Proposed models of the ion channels designed by Lear, Wasserman, and DeGrado (1988). The top panel shows the $(LSLLLSL)_3$ sequence, and the bottom panel shows the $(LSSLLSL)_3$ sequence. In each case a single heptad repeat is depicted in the helical wheel projection on the left, and a stereo representation of the ion pore is shown on the right. For the $(LSLLLSL)_3$ sequence, the pore is modeled as a bundle of four α-helices (top); for the $(LSSLLSL)_3$ sequence, the pore is modeled as a bundle of six α-helices (bottom). Note that the $(LSLLLSL)_3$ tetramer generates a smaller pore than the $(LSSLLSL)_3$ hexamer. (Figure adopted from Lear *et al.,* 1988).

tetramer is significantly smaller than in the modeled hexamer. Consistent with these modeled structures, the ion channel formed by the $(LSLLLSL)_3$ peptide conducts only protons, whereas the channel formed by the $(LSSLLSL)_3$ conducts larger ions as well.

Additional support for helical models shown in Fig. 19 was provided by constructing a variant of the $(LSLLLSL)_3$ sequence in which the central leucine in each of the three heptads is replaced by the conformationally constrained amino acid α-aminoisobutyric acid (Aib). Since Aib possesses two α-methyl groups, it is conformationally constrained and predisposes peptides to form α-helices. CD spectroscopy of the Aib-modified peptide showed, as expected, that it was predominantly α-helical in phospholipid vesicles. Since the modified peptide demonstrates nearly identical conductance properties to the original sequence, these results suggest that the $(LSLLLSL)_3$ peptide is also α-helical in its active state (DeGrado and Lear, 1990). Crystallization has been reported for both the $(LSLLLSL)_3$ sequence and the Aib-modified sequence (Lovejoy et al., 1992a). Although the high-resolution structure of these crystals has not yet been reported, the space group and unit cell dimensions of the crystals are consistent with the tetrameric structure shown in Fig. 19.

Following this work, the synthetic ion channel was preorganized into a four-helix bundle via the template-assisted strategy. In collaborative work by the groups of DeGrado, Lear, and Groves four copies of the Aib-containing peptide were covalently linked to the rigid C4 symmetric molecule tetraphenylporphyrin (Akerfeldt et al., 1992). A model of the resulting protein, called Tetraphilin is shown in Fig. 20. The cross-linked material resembles the original self-assembling material in that both form proton-selective ion channels. However, the cross-linked channels differ from the original channels in three respects: Cross-linking (i) increases the lifetime of the major conductive state by >25-fold; (ii) causes channels to form with a linear (rather than higher order) concentration dependence; and (iii) causes channel formation to be nearly independent of voltage (Akerfeldt et al., 1992). The authors have postulated a model to explain these differences. According to this model the parent peptide (without the cross-link) can exist in an equilibrium between the trans-membrane four-helix bundle and an alternative orientation in which the monomeric amphiphilic helices are oriented with their helical axes parallel to the membrane surface. Increasing the concentration of helices or applying a trans-membrane voltage favors the assembly of trans-membrane tetramers and thereby increases pore activity. In the case of Tetraphilin, where the four helices are cross-linked to the porphyrin template, the nonproductive structure cannot compete with the pore structure. Therefore the pores have longer lifetimes and are not sensitive to either voltage or peptide concentration. Because the template-assisted strategy constrains the locations of the peptides, it can direct the molecule toward the desired structure and function.

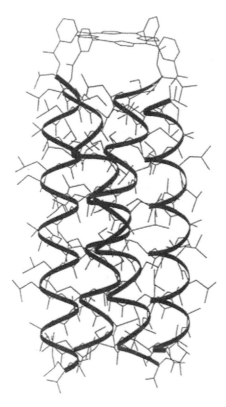

FIGURE 20 Proposed model of the Tetraphilin structure. The tetraphenylporphyrin template is at the top of the figure, and the four parallel α-helices are shown as ribbons projecting down from it. The ion channel is enclosed by the four α-helices. (Figure adopted from Akerfeldt et al., 1992.)

Membrane proteins are notoriously difficult to work with because of their insolubility in aqueous buffers. Typically, they can only be solubilized in the presence of small-molecule detergents. These protein/detergent complexes are suitable for many purposes, but with few notable exceptions (Deisenhofer et al., 1985) they rarely yield crystals suitable for structure determination by X-ray diffraction. This shortcoming presumably results from a level of disorder imposed by the small detergent molecules surrounding the protein. To circumvent this problem, Schafmeister et al. (1993) designed a large-molecule detergent in the form of an amphiphilic α-helix. The new peptide, called a peptitergent, has the sequence acetyl-EELLKQALQQAQQLLQQAQELAKK-amide. The hydrophilic face of the α-helix contains mostly Gln residues, with Lys and Glu side chains near the N- and C-termini, respectively, to neutralize the helix dipole. The hydrophobic face was designed with alanines projecting from the center, and leucines at the edges, thereby generating a wide flat nonpolar surface capable of interacting with, and solubilizing, a range of different membrane proteins. The 30 Å length of this face is sufficient to cover the membrane-spanning region of integral membrane proteins (Schafmeister *et al.* (1993). To test the ability of the peptitergent to solubilize membrane proteins, experiments were performed using bacteriorhodopsin and bovine rhodopsin. In both cases, small-molecule detergents were diluted away and the peptitergent enabled the protein to remain in solution under conditions where omission of the peptitergent led to 100% precipitation. The crystal structure of the peptitergent (without added membrane protein) was solved and it forms a four-helix bundle. As predicted by the design, the α-helices pack flat surface to flat surface, suggesting that the observed activity of the peptide reflects the structural features incorporated into its design.

F. Design of New Materials

The roles of proteins in biological systems include both structure and function. In those cases where the protein is an enzyme and catalyzes a chemical transformation, the structure serves as the scaffold upon which the function is built. However, in other cases the structural properties themselves dictate function. For example, the different structural properties of the fibrous proteins α-keratin, β-keratin, collagen, and elastin dictate the different functional properties of hair, silk, skin, and elastic ligaments, respectively (Voet and Voet, 1990). Similarly, for the design of novel polypeptide materials, the structural properties of a polypeptide may suffice to endow it with a new function.

Protein-like polymeric materials have been designed and constructed both by recombinant methods *in vivo* (Tirrell *et al.*, 1991; Creel *et al.*, 1991; Zhang *et al.*, 1992; McGrath *et al.*, 1992; Krejchi *et al.*, 1994) and via synthetic methods *in vitro*. Some of these novel polymers are designed to expand and contract in response to easily manipulated conditions such as temperature, pH, or ionic strength. Changes in length can then be used to achieve mechano-chemical coupling. In one example, Urry and co-workers constructed a polymer that can lift weights in response to a drop in pH. This polymer has the sequence $(Val-Pro-Gly-Xxx-Gly)_n$, where Xxx is Val or Glu at a ratio of 4:1. (Thus it contains 4 Glu's per 100 residues.) At neutral pH the glutamates are charged, but when the pH is lowered to 3, the Glu side chains are converted to neutral COOH groups, thus permitting the hydrophobic collapse or contraction of the polymer. This material is able to pick up and set down weights that are 1000 times greater than the weight of the contracting polymer (Urry, 1988a,b; 1990; Luan *et al.*, 1991.)

Self-assembly is a theme common to both materials science and protein folding. Ghadiri *et al.* (1993) designed a new material using protein-like modules that self-assemble into hollow nanotubes. The designed peptide has the sequence $cyclo[-(D-Ala-Glu-D-Ala-Gln)_2]$, which can adopt a flat ring-shaped conformation (see Fig. 21a). At neutral pH the glutamate side chains

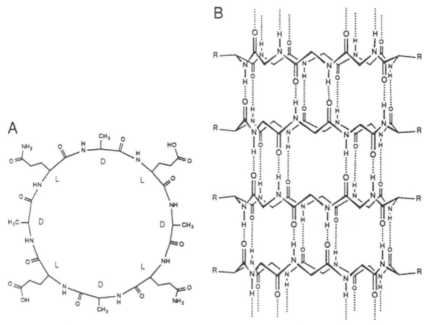

FIGURE 21   (a) Chemical structure of the designed circular peptide. (b) Four rings stacked on one another to generate hollow tubes. The peptide chains in each layer run anti-parallel to the neighboring layers and the hydrogen bonding is between layers. (Figure adopted from Ghadiri *et al.*, 1993.)

are negatively charged and the cyclic peptides repel one another. However, at low pH, the carboxylate groups become protonated and the repulsion is replaced by intermolecular hydrogen bonding between the flat rings (see Fig. 21b). Thus, acidification of the medium triggers the spontaneous self-assembly of peptide rings into long hollow tubes hundreds of nanometers long and having internal diameters dictated by the size of the cyclic peptide (Ghadiri *et al.*, 1993, 1994).

IV. CONCLUSION

I began this chapter by asserting two main motivations for pursuing protein design: (i) Design is the ultimate test of our understanding of natural proteins; and (ii) current design efforts represent the first step toward a new generation of novel macromolecules that will have practical applications in industry and biomedicine. Are these motivations justified by the designed proteins described in this chapter? How well has our understanding of natural proteins stood up to the test of design and how well have current designs laid the groundwork for the eventual construction of practical macromolecules?

As a test of our understanding of natural proteins, the design endeavor has been quite successful. This is not to say that current design demonstrates that we understand all there is to know about proteins. Quite the contrary; attempts to design proteins *de novo* have revealed some features about proteins that we understand better than we heretofore realized, as well as other features that we only now realize how little we understand. For example, we apparently understand a great deal about the making simple α-helical proteins, as demonstrated by the many different helical structures described in this chapter. On the other hand, we have also found that there is a significant difference between the design of a compact stable structure and the successful design of a protein in which all the interior atoms are fixed into unique locations (Betz *et al.*, 1993). Molten globules seem to be much easier to design than native-like structures. In a sense, the first step—obtaining compact and stable structures—has been easier than expected. However, the difficulty of the second step—the design of uniquely determined interiors—has shown us that there are central features in natural proteins whose importance we did not fully appreciate prior to our attempts to design them.

In terms of the practical goal of designing functionally useful proteins, again the first steps have been remarkably successful while the later steps may be more elusive. It now seems relatively easy to generate macromolecules that possess some low level of activity, as demonstrated by several examples described in the previous sections. However, none of the current generation of novel proteins approach the levels of either specificity or activity produced routinely by nature. Perhaps our shortcomings here mirror our shortcomings in the design of structure. Perhaps it is easy to produce some activity because it is easy to direct atoms into "more or less" the right place. However, obtaining a level of activity comparable to that produced by nature may require us to pin down the positions of atoms with far greater precision than we are yet capable of doing. Nature's success at making "practical" macromolecules is the product of innumerable trials and errors conducted over 4 billion years. In our attempts to design novel macromolecules, we can learn a great deal from natural systems; it is hoped that these lessons will enable us to succeed with fewer errors and in far less time.

ACKNOWLEDGMENTS

Thanks go to Alexandra van Geel, whose undergraduate Junior Project Research helped to lay the ground work for this chapter. Thanks also go to Judith Swan for advice on the manuscript.

REFERENCES

Akerfeldt, K. S., Kim, R. M., Camac, D., Groves, J. T., Lear, J. D., DeGrado, W. F. (1992) *J. Amer. Chem. Soc.* **114**, 9656–9657.

Altmann, K.-H., Florsheimer, A., Mutter, M. (1986) *Int. J. Peptide Protein Res.* 27, 314–319.

Altmann, K.-H., and Mutter, M. (1990) *Int. J. Biochem.* 22, 947–956.

Banner, O. W., Bloomer, A. C., Petsko, G. A., Phillips, D. C., Pogsa, C. I., Wilson, I. A., Corran, P. H., Forth, A. J., Milman, J. D., Offord, R. E., Priddle, J. D., and Waley, S. G. (1975) *Nature (London)* 255, 609–614.

Benner, S. A., Johnson, K., and Allemann, R. K. (1992) Poster presented at the 21st annual meeting of the Federation of European Biochemical Societies (FEBS), August 9–14, 1992, Dublin, Ireland.

Betz, S. F., Raleigh, D. P., and DeGrado, W. F. (1993) *Curr. Opinion Struct. Biol.* 3, 601–610.

Bianchi, E., Sollazzo, M., Tramontano, A., and Pessi, A. (1993) *Int. J. Peptide Protein Res.* 41, 385–393.

Brack, A., and Spach, G. (1981) *J. Amer. Chem. Soc.* 103, 6319–6323.

Brown, J. E., and Klee, W. A. (1971) *Biochemistry* 10, 470–476.

Brunet, A. P., Huang, E. S., Huffine, M. E., Loeb, J. E., Weltman, R. J., and Hecht, M. H. (1993) *Nature (London)* 364, 355–358.

Chin, T. M., Berndt, K. D., and Yang, N. C. (1992) *J. Amer. Chem. Soc.* 114, 2279–2280.

Choma, C. T., Lear, J. D., Nelson, M. J., Dutton, P. L., Robertson, D. E., and DeGrado, W. F. (1994) *J. Amer. Chem. Soc.* 116, 856–865.

Chou, P. Y., and Fasman, G. D. (1974) *Biochemistry* 13, 211–222.

Chou, P. Y., and Fasman, G. D. (1978) *Annu. Rev. Biochem.* 47, 251–276.

Cohen, C., and Parry, D. A. D. (1990) *Proteins Struct. Function Genet.* 7, 1–15.

Creel, H. S., Fornier, M. J., Mason, T. L., and Tirrell, D. A. (1991) *Macromolecules* 24, 1213–1214.

Creighton, T. E. (1979) *J. Mol. Biol.* 129, 235–264.

Creighton, T. E. (1993) *Proteins: Structures and Molecular Principles* (2nd ed.). Freeman, New York.

Crick, F. H. C. (1953) *Acta Crystallogr.* 6, 689–697.

Dado, G. P., and Gellman, S. H. (1993) *J. Amer. Chem. Soc.* 115, 12609–12610.

DeGrado, W. F., and Lear, J. D. (1985) *J. Amer. Chem. Soc.* 107, 7684–7689.

DeGrado, W. F., and Lear, J. D. (1990) *Biopolymers* 29, 205–213.

DeGrado, W. F., Raleigh, D. P., and Handel, T. (1991) *Curr. Opinion Struct. Biol.* 1, 984–993.

DeGrado, W. F., Regan, L., and Ho, S. P. (1987) *Cold Spring Harbor Symp. Quant. Biol.* 52, 521–526.

DeGrado, W. F., Wasserman, Z. R., and Lear, J. D. (1989) *Science* 243, 622–628.

Deisenhofer, J., Epp, O., Miki, K., Huber, R., and Michel, H. (1985) *Nature (London)* 318, 618.

Dill, K. A. (1990) *Biochemistry* 29, 7133–7155.

Dolgikh, D. A., Kolomiets, A. P., Bolotina, I. A., and Ptitsyn, O. B. (1984) *FEBS Lett.* 165, 88–92.

Efimov, A. V. (1982) *Mol. Biol.* 16, 271–281.

Eisenberg, D., Wilcox, W., Eshita, S. M., Pryciak, P. M., Ho, S. P., and DeGrado, W. F. (1986) *Proteins Struct. Function Genet.* 1, 16–22.

Engel, M., Williams, R. W., and Erickson, B. W. (1991) *Biochemistry* 30, 3161–3169.

Fedorov, A. N., Dolgikh, D. A., Chemeris, V. V., Chernov, B. K., Finkelstein, A. V., Schulga, A. A., Alakhov, Y. B., Kirpichnikov, M. P., and Ptitsyn, O. B. (1992) *J. Mol. Biol.* 225, 927–931.

Ghadiri, M. R., and Case, M. A. (1993) *Angewandte Chemie (English Edition)* 32, 1594–1597.

Ghadiri, M. R., Granja, J. R., and Buehler, L. K. (1994) *Nature (London)* 369, 301–304.

Ghadiri, M. R., Granja, J. R., Milligan, R. A., McRee, D. E., and Khazanovich, N. (1993) *Nature (London)* 366, 324–327.

Ghadiri, M. R., Soares, C., and Choi, C. (1992a) *J. Amer. Chem. Soc.* 114, 825–831.

Ghadiri, M. R., Soares, C., and Choi, C. (1992b) *J. Amer. Chem. Soc.* 114, 4000–4002.

Goraj, K., Renard, A., and Martial, J. A. (1990) *Protein Engineering* 3, 259–266.

Grove, A., Tomich, J. M., and Montal, M. (1991) *Proc. Natl. Acad. Sci. U.S.A.* 88, 6418–6422.

Gutte, B., Daumigen, M., and Wittschieber, E. (1979) *Nature (London)* 281, 650–655.

Hahn, K. W., Klis, W. A., and Stewart, J. M. (1990) *Science* **248**, 1544–1547.

Hampsey, D. M., Das, G., and Sherman, F. (1988) *FEBS Lett.* **231**, 275–283.

Handel, T. M., and DeGrado, W. F. (1990) *J. Amer. Chem. Soc.* **112**, 6710–6711.

Handel, T. M., Williams, S. A., And DeGrado, W. F. (1993) *Science* **261**, 879–885.

Harbury, P. B., Zhang, T., Kim, P. S., and Alber, T. (1993) *Science* **262**, 1401–1407.

Hecht, M. H., Richardson, J. S., Richardson, D. C., and Ogden, R. C. (1990) *Science* **249**, 884–891.

Hedstrom, L., Szilagyi, L., and Rutter, W. J. (1992) *Science* **255**, 1249–1253.

Hellinga, H. W., Caradonna, J. P., and Richards, F. M. (1991) *J. Mol. Biol.* **222**, 787–803.

Hill, C. P., Anderson, D. H., Wesson, L., DeGrado, W. F., and Eisenberg, D. (1990) *Science* **249**, 543–546.

Ho, S. P., and DeGrado, W. F. (1987) *J. Amer. Chem. Soc.* **109**, 6751–6758.

Hodges, R. S., Saund, A. K., Chong, P. C. S., St.-Pierre, S. A., and Reid, R. E. (1981) *J. Biol. Chem.* **256**, 1214–1224.

Hodges, R. S., Zhou, N. E., Kay, C. M., and Semchuk, P. D. (1990) *Peptide Res.* **3**, 123–137.

Hol, W. G. J., Duijnen, P. T., and Benerdsen, H. J. C. (1978) *Nature (London)* **273**, 443–446.

Hu, J. C., O'Shea, E. K., Kim, P. S., and Sauer, R. T. (1990) *Science* **250**, 1400–1403.

Inaka, K., Kuroki, R., Kikuchi, M., and Matsushima, M. (1991) *J. Biol. Chem.* **266**, 20666–20671.

Jaenicke, R., Gutte, B., Glatter, U., Strassburger, W., and Wollmer, A. (1980) *FEBS Lett.* **114**, 161–164.

Johnsson, K., Allemann, R. K., and Benner, S. A. (1990) In *Molecular Mechanisms in Bioorganic Processes* (C. Bleasdale and B. T. Golding, Eds.), pp. 166–187. Cambridge Royal Society of Chemistry.

Johnsson, K., Allemann, R. K., Widmer, H., and Benner, S. A. (1993) *Nature (London)* **365**, 530–532.

Kaiser, E. T., and Kézdy, F. J. (1983) *Proc. Natl. Acad. Sci. U.S.A.* **80**, 1137–1143.

Kamtekar, S., Schiffer, J. M., Xiong, H., Babik, J. M., and Hecht, M. H. (1993) *Science* **262**, 1680–1685.

Kauzmann, W. (1959) *Adv. Protein Chem.* **14**, 1–63.

Kim, P. S., and Baldwin, R. L. (1984) *Nature (London)* **307**, 329–334.

Klauser, S., Gantner, D., Salgam, P., and Gutte, B. (1991) *Biochem. Biophys. Res. Commun.* **179**, 1212–1219.

Krejchi, M. T., Atkins, E. D. T., Waddon, A. J., Fournier, M. J., Mason, T. L., and Tirrell, D. A. (1994) *Science* **265**, 1427–1432.

Kuroki, R., Taniyama, Y., Seko, C., Nakamura, H., Kikuchi, M., and Ikehara, M. (1989) *Proc. Natl. Acad. Sci. U.S.A.* **86**, 6903–6907.

Kuwajima, K. (1989) *Proteins Struct. Function Genet.* **6**, 87–103.

Lau, S. Y. M., Taneja, A. K., and Hodges, R. S. (1984) *J. Biol. Chem.* **259**, 13253–13261.

Lear, J. D., Wasserman, Z. R., and DeGrado, W. F. (1988) *Science* **240**, 1177–1181.

Lee, M. S., Gippert, G. P., Soman, K. V., Case, D. A., and Wright, P. E. (1989) *Science* **245**, 635–637.

Lerner, R. A., Benkovic, S. J., and Schultz, P. G. (1991) *Science* **252**, 659–667.

Lieberman, M., and Sasaki, T. (1991) *J. Amer. Chem. Soc.* **113**, 1470–1471.

Lim, W. A., and Sauer, R. T. (1989) *Nature (London)* **339**, 31–36.

Lovejoy, B., Choe, S., Cascio, D., McRorie, D. K., DeGrado, W. F., and Eisenberg, D. (1993) *Science* **259**, 1288–1293.

Lovejoy, B., Akerfeldt, K. S., DeGrado, W. F., and Eisenberg, D. (1992a) *Protein Sci.* **1**, 1073–1077.

Lovejoy, B., Le, T. C., Lüthy, R., Cascio, D., O'Neil, K. T., DeGrado, W. F., and Eisenberg, D. (1992b) *Protein Sci.* **1**, 956–957.

Luan, C. H., Parker, T. M., Prasad, K. U., and Urry, D. W. (1991) *Biopolymers* **31**, 465–475.

Lupas, A., Dyke, M. V., and Stock, J. (1991) *Science* **251**, 1–3.

Marqusee, S., and Baldwin, R. L. (1987) *Proc. Natl. Acad. Sci. U.S.A.* **84**, 8898–8902.

Marqusee, S., Robbins, V. H., and Baldwin, R. L. (1989) *Proc. Natl. Acad. Sci. U.S.A.* **86**, 5286–5290.

Matsumura, M., Becktel, W. J., Levitt, M., and Matthews, B. W. (1989) *Proc. Natl. Acad. Sci. U.S.A.* **86**, 6562–6566.

Matsumura, M., and Matthews, B. W. (1989) *Science* **243**, 792–794.

McGrath, K. P., and Fournier, M. J., Mason, T. L., and Tirrell, D. A. (1992) *J. Amer. Chem. Soc.* **114**, 727–733.

Montal, M. (1990) *FASEB J.* **4**, 2623–2635.

Montal, M., Montal, M. S., and Tomich, J. M. (1990) *Proc. Natl. Acad. Sci. U.S.A.* **87**, 6929–6933.

Moser, R., Frey, S., Münger, K., Hehlgans, T., Klauser, S., Langen, H., Winnacker, E.-L., Mertz, R., and Gutte, B. (1987) *Protein Engineering* **1**, 339–343.

Moser, R., Thomas, R. M., and Gutte, B. (1983) *FEBS Lett.* **157**, 247–251.

Mutter, M. (1985) *Angew. Chem. Int.* **24**, 639–653.

Mutter, M. (1988) *TIBS* **13**, 260–265.

Mutter, M., Altmann, E., Altmann, K.-H., Hersperger, R., Koziej, P., Nebel, K., Tuchscherer, G., Vuilleumier, S., Gremllich, H.-U., and Müller, K. (1988a) *Helvetica Chim. Acta* **71**, 835–847.

Mutter, M., Altmann, K.-H., Tuchscherer, G., and Vuilleumier, S. (1988b) *Tetrahedron* **44**, 771–785.

Mutter, M., Altmann, K.-H., and Vorherr, T. (1986) *Z. Naturforsch.* **41(B)**, 1315–1322.

Mutter, M., Hersperger, R., Gubernator, K., and Müller, K. (1989) *Proteins Struct. Funct. Genet.* **5**, 13–21.

Mutter, M., Tuchscherer, G. G., Miller, C., Altmann, K.-H., Carey, R. I., Wyss, D. F., Labhardt, A. M., and Rivier, J. E. (1992) *J. Amer. Chem. Soc.* **114**, 1463–1470.

Mutter, M., and Vuilleumier, S. (1989) *Angew. Chem. Int.* **28**, 535–554.

O'Neil, K. T., and DeGrado, W. F. (1990) *Science* **250**, 646–651.

O'Neil, K. T., Hoess, R. H., and DeGrado, W. F. (1990) *Science* **249**, 774–778.

O'Shea, E. K., Klemm, J. D., Kim, P. S., and Alber, T. (1991) *Science* **254**, 539–544.

O'Shea, E. K., Lumb, K. J., and Kim, P. S. (1993) *Current Biol.* **3**, 658–667.

O'Shea, E. K., Rutkowski, R., and Kim, P. S. (1989) *Science* **243**, 538–542.

Oas, T. G., McIntosh, L. P., O'Shea, E. K., Dahlquist, F. W., and Kim, P. S. (1990) *Biochemistry* **29**, 2891–2894.

Ohgushi, M., and Wada, A. (1983) *FEBS Lett.* **164**, 21–24.

Oiki, S., Madison, V., and Montal, M. (1990) *Proteins Struct. Funct. Genet.* **8**, 226–236.

Osterhout, J. J., Handel, T., Na, G., Toumadje, A., Long, R. C., Connolly, P. J., Hoch, J. C., Johnson, W. C., Live, D., and DeGrado, W. F. (1992) *J. Amer. Chem. Soc.* **114**, 331–337.

Osterman, D. G., and Kaiser, E. T. (1985) *J. Cell Biochem.* **29**, 57–72.

Parraga, G., Horvath, S. J., Eisen, A., Taylor, W. E., Houd, L., Young, E. T., and Klevitt, R. E. (1988) *Science* **241**, 1489–1492.

Pauling, L., and Corey, R. B. (1953) *Nature (London)* **171**, 59–61.

Perry, L. J., and Wetzel, R. (1984) *Science* **226**, 555–557.

Pessi, A., Bianchi, E., Crameri, A., Venturini, S., Tramontano, A., and Sollazzo, M. (1993) *Nature (London)* **362**, 367–369.

Ponder, J. W., and Richards, F. M. (1987) *J. Mol. Biol.* **193**, 775–791.

Poteete, A. R., Rennell, D., and Bouvier, S. E. (1992) *Proteins Struct. Funct. Genet.* **13**, 38–40.

Quinn, T. P., Tweedy, N. B., Williams, R. W., Richardson, J. S., and Richardson, D. C. (1994) *Proc. Natl. Acad. Sci. U.S.A.* **9**, 8747–8751.

Raleigh, D. P., and DeGrado, W. F. (1992) *J. Amer. Chem. Soc.* **114**, 10079–10081.

Regan, L. (1993) *Annu. Rev. Biophys. Biomol. Struct.* **22**, 257–281.

Regan, L., and Clarke, N. D. (1990) *Biochemistry* **29**, 10878–10883.

Regan, L., and DeGrado, W. F. (1988) *Science* **241**, 976–978.

Richardson, J. S. (1981) *Adv. Protein Chem.* **34**, 167–339.

Richardson, J. S., and Richardson, D. C. (1987) In *Protein Engineering* (D. L. Oxander and C. F. Fox, Eds.), pp. 149–163. Liss, New York.

Richardson, J. S., and Richardson, D. C. (1988) *Science* **240**, 1648–1651.

Richardson, J. S., and Richardson, D. C. (1989) *TIBS* **14**, 304–309.

Robertson, D. E., Farid, R. S., Moser, C. C., Urbauer, J. L., Mulholland, S. E., Pidikiti, R., Lear, J. D., Wand, A. J., DeGrado, W. F., and Dutton, P. L. (1994) *Nature (London)* **368**, 425–432.

Sander, C., Vriend, G., Bazan, F., Horovitz, A., Nakamura, H., Ribas, L., Finkelstein, A. V., Lockhart, A., Merkl, R., Perry, L. J., Emery, S. C., Gaboriaud, C., Marks, C., Moult, J., Verlinde, C., Eberhard, M., Elofsson, A., Hubbard, T. J. P., Regan, L., Banks, J., Jappelli, R., Lesk, A. M., and Tramontano, A. (1992) *Proteins Struct. Funct. Genet.* **12**, 105–110.

Sasaki, T., and Kaiser, E. T. (1989) *J. Amer. Chem. Soc.* **111**, 380–381.

Sasaki, T., and Kaiser, E. T. (1990) *Biopolymers* **29**, 79–88.

Sauer, R. T., Hehir, K., Stearman, R. S., Weiss, M. A., Jeitler-Nilsson, A., Suchanek, E. G., and Pabo, C. O. (1986) *Biochemistry* **25**, 5992–5998.

Schafmeister, C. E., Miercke, L. J. W., and Stroud, R. M. (1993) *Science* **262**, 734–738.

Scrutton, N. S., Berry, A., and Perham, R. N. (1990) *Nature (London)* **343**, 38–43.

Shoemaker, K. R., Kim, P. S., York, E. J., Stewart, J. M., and Baldwin, R. L. (1987) *Nature (London)* **326**, 563–567.

Sluka, J. P., Horvath, S. J., Bruist, M. F., Simon, M. I., and Dervan, P. B. (1987) *Science* **238**, 1129–1132.

Tanaka, T., Kimura, H., Hayashi, M., Fujiyoshi, Y., Fukuhara, K-I., and Nakamura, H. (1994) *Protein Science* **3**, 419–427.

Taylor, W. (1992) *Nature (London)* **356**, 478–480.

Tirrell, D. A., Fournier, M. J., and Mason, T. L. (1991) *Curr. Opinion Struct. Biol.* **1**, 638–641.

Tuchscherer, G., Servis, C., Corradin, G., Blum, U., Rivier, J., and Mutter, M. (1992) *Protein Sci.* **1**, 1377–1386.

Unson, C. G., Erickson, B. W., Richardson, D. C., and Richardson, J. S. (1984) *Fed. Proc.* **4**, 1837.

Urry, D. W. (1988a) *J. Protein Chem.* **7**, 1–114.

Urry, D. W. (1988b) *Alabama J. Med. Sci.* **25**, 475–485.

Urry, D. W. (1990) In *Protein Folding: Deciphering the Second Half of The Genetic Code* (L. Gierasch and J. King, Eds.), pp. 63–67. American Association for the Advancement of Science.

Vallee, B. L., and Auld, D. S. (1990) *Proc. Natl. Acad. Sci. U.S.A.* **87**, 220–224.

Villafranca, J. E., Howell, E. E., Voet, D. H., Strobel, M. S., Ogden, R. C., Abelson, J. N., and Kraut, J. (1983) *Science* **222**, 782–788.

Voet, D., and Voet, J. G. (1990) *Biochemistry.* Wiley, New York.

Wada, A. (1976) *Adv. Biophys.* **9**, 1–63.

Weber, P. C., and Salemme, F. R. (1980) *Nature (London)* **287**, 82–84.

Wharton, R. P., and Ptashne, M. (1985) *Nature (London)* **316**, 601–605.

Xiong, H., Buckwalter, B. L., Shieh, H. M., and Hecht, M. H. (1995) *Proc. Natl. Acad. Sci. U.S.A.* **92**, 6349–6353.

Yan, Y., and Erickson, B. W. (1994) *Protein Science* **3**, 1069–1073.

Yue, K., and Dill, K. A. (1992) *Proc. Natl. Acad. Sci. U.S.A.* **89**, 4163–4167.

Zhang, G., Fournier, M. J., Mason, T. L., and Tirrell, D. A. (1992) *Macromolecules* **25**, 3601–3603.

Zhou, N. E., Kay, C. M., and Hodges, R. S. (1992a) *J. Biol. Chem.* **267**, 2664–2670.

Zhou, N. E., Kay, C. M., and Hodges, R. S. (1992b) *Biochemistry* **31**, 5739–5746.

Zhou, N. E., Kay, C. M., and Hodges, R. S. (1993) *Biochemistry* **32**, 3178–3187.

Zhu, B. Y., Zhou, N. E., Kay, C. M., and Hodges, R. S. (1993) *Protein Sci.* **2**, 383–394.

2

Computer Methods in Protein Modeling: An Overview

JIRI NOVOTNY

Department of Macromolecular Modeling
Bristol-Myers Squibb
Princeton, New Jersey 08543-4000

I. INTRODUCTION

Computers have now become an indispensable tool and, indeed, an epitome of molecular biology. There are two essentially different ways at which computer modeling is done: a subjective, intuitive, interactive computer graphics and an objective, computational analysis based on energetic criteria and other mathematically formulated biophysical concepts. Casually, one can speak of the following levels of computer modeling approaches.

On the most basic level, computers provide a convenient viewing interface to scientists contemplating atomic details of a protein molecule. Even this "basic" level requires a sophisticated technology such as powerful color graphics engines and a sufficient processing power to generate two-dimensional or stereo images the viewer can operate in real time by dials or mice.

The next level consists of software programs that allow "modeling by hand," i.e., definition of atomic groups that form parts of the polypeptide chain (e.g., α-helices, loops); independent manipulation of these parts (e.g., splicing of helices or loops from one structure to another); definition of torsional angles (i.e., four successive atoms connected by rotatable bonds) and their values (ideally by direct dialing); automatic building of polypeptides from selected short segments found in a protein data banks; and simi lar operations based on vector and solid geometry formulae.

Protein Engineering and Design
51

Yet a higher level provides for an objective evaluation of protein energetics and stereochemistry via a potential function, a forcefield, i.e., a set of equations that define an optimal state of a protein polymer. The equations form the heart of molecular mechanics programs that also contain energy minimization and dynamical simulation routines.

Technically speaking, protein structures are sets of three-dimensional Cartesian coordinates (x,y,z) of thousands of N, C, S, O, and H atoms that constitute them. Any convenient and meaningful manipulation of this large set of data can only be achieved on a computer. This is true both for simple translations, rotations, and zooming in the real space, and for the more abstract and involved operations such as generation and examination of molecular surfaces, simulation of atomic movements, least-squares superpositions of molecules, and other molecular modeling procedures per se.

Computer modeling sessions start with a set of Cartesian coordinates specifying positions for all the atoms of the molecule (at the very extreme, when a design of a new protein is attempted *ab initio,* the starting coordinates default to zeros). At the end of the session, we have obtained a modified set of atomic coordinates. The modeling process itself may be quite involved and its practical aspects include hardware, software, algorithms, biophysical concepts, database management, prediction schemes, sampling procedures, simulation protocols, protein structure analysis, and much more. It follows that a person practicing molecular modeling should have some familiarity with all of these.

The structure of this chapter loosely follows the above-mentioned threefold scheme of computer modeling. From a historical perspective, the computer-aided modeling-by-hand emerged first, and is summarized in the following section; it is supplemented by a brief outline of scientific hardware evolution, with some discussion of computing workstyles. Next, we focus on molecular mechanics, biophysics of proteins, and basic concepts of structure modeling. Secondary structure prediction, dynamical simulation techniques, and conformational searchers come next. The chapter is closed by a discussion of methods used to gauge the correctness of modeled structures.

An overview chapter such as this cannot give an exhaustive coverage of the field. Literary citations mainly focus on reviews, landmark papers introducing new concepts and algorithms, important practical examples, and samplings of alternative techniques. For a more thorough introduction to molecular design and modeling you may consult, e.g., the two *Methods in Enzymology* monographs edited by John J. Langone (1991).

II. EARLY METHODS OF COMPUTER MODELING "BY HAND"

Proteins are flexible polymers and once their amino acid sequence (primary structure) is determined, the three-dimensional path of the polypeptide chain is obtained by specifying pairs of backbone torsional angles (the so-called Ramachandran angles) C-N-Cα-C (ϕ) and N-Cα-C-N (ψ) for each of the amino acid residues. The complete tertiary structure requires that all the side-chain torsional angles χ_n be specified as well. Protein modeling is then synonymous with selection of backbone and side-chain torsional angles. The very first computer modeling algorithms focused on procedures that allowed one to display polypeptides on the screen of a computer terminal and allowed for interactive manipulation of torsions and larger structural modules.

The early pioneers of protein modeling algorithms were Diamond (1966, program BUILDER), Feldmann (1976), Jones (1985, program FRODO first described in 1978) Langridge (Langride and MacEwan, 1965), and Levinthal (program PAGGRAPH). In 1966, Levinthal described "molecular model building by computer" in the now classic article in *Scientific American*. An even earlier milestone of (noninteractive) computer graphics was an article, by Johnson (1965), describing generation of "crystal structure illustrations" using the "Oak Ridge Terminal Ellipsoid Package," ORTEP. The early modeling software was oriented towards tasks such as chain closure, molecule build-up from building blocks (such as Langridge's contruction of the DNA sugar-phosphate backbone), and exploration of conformational space by manual changes of torsional angle values. In 1972, Katz and Levinthal reviewed the hardware and software aspects of molecular structure presentation, manipulation, and structure fitting into electron density contours. As with so many other aspects of protein structural science, the early development of computer graphics was often driven by technical requirements of protein X-ray crystallography techniques.

One of the first examples of hardware dedicated to interactive graphics was the ADAGE terminal (Katz and Levinthal, 1972). Attached to a mainframe computer such as IBM 360, the terminal had its own digital processor, an analog matrix multiplier, and a vector generator capable of producing a display of over 5000 vectors at flicker-free display rates. The most popular graphics terminal of the late 1970s and early 1980s was the Evans and Sutherland Picture System. With the advent and rapid progression of microprocessors (in 1971, by INTEL) and the never-ending power increase and size miniaturization of the microprocessor chips, the dedicated graphics terminal attached to a computer host has been superseded (by about 1985) by a workstation combining both the central processing unit and the graphics integrated circuit board(s) into a single box. At the time of writing, the most popular graphics workstations are the Sun, Silicon Graphics IRIS series, and the IBM RS/6000. The most widespread modeling software of today are the commercial and academic program packages such as INSIGHT (Biosym Technologies, Inc., San Diego), QUANTA (Molecular Simulations, Inc., Pasadena), GRAMPS (O'Donnell and Olson, 1985), and MIDAS (University of California, San Francisco).

III. HARDWARE AND COMPUTATIONAL WORKSTYLES

Computer hardware design and evolution have always impacted on structural science in a very direct way. The milestones of scientific computing can be summarized as follows. By about 1943, the first tube-based scientific computer, the "electronic numerator, integrator and calculator" (ENIAC) was developed at the University of Pennsylvania (J. Presper Eckert, John W. Mauchly, and John von Neumann). The machine lacked any software and was programmed for individual computing tasks by manual rewiring of circuitry. The first software language, "formula translation" (FORTRAN) was developed by 1956. The invention of transistors, by J. Bardeen, W. H. Brattain, and W. Shockley at Bell Laboratories in 1948, sparked a major hardware revolution that unfolded in the 1950s. The first integrated circuit was developed in 1959 in the Fairchild Corp., and the chip-based trend of computing culminated in microprocessor development in 1971 (INTEL Co.).

The scientific computing machines of the 1960s and 1970s were the IBM 360 and IBM 3090. In the late 1970s and the 1980s, Digital Corp. VAX systems prevailed but by 1985, the graphics workstation revolution replaced the VAXes with Suns and Silicon Graphics IRISes. During these decades, supercomputers were being designed and built primarily by Seymour Cray first at the Control Data Corp. in the 1970s (CDC 7600, CYBER 205), and then at the Cray Research Co. (Cray-2, Cray X-MP, Cray Y-MP). The Japanese-built supercomputers NEC, Fujitsu, and HITAC never made much of an impact in the Western scientific world. The perspective of scientific supercomputing at a moderate cost has led to development of minisupercomputers, machines with processing powers between the workstations and the megaFLOPS-delivering supercomputers: ARDENT, CONVEX, STELLAR, ALLIANT, etc. (FLOPS stands for *floating point operation,* such as decimal point arithmetics, *per second;* it is a standard measure of computer processing power).

The much publicized Connection Machine, CM-1, of Thinking Machine Co., founded by Daniel Hillis, introduced the extreme concept of massively parallel (MPP) architecture, a computer built of thousands of miniature (one-bit) processors intercommunicating via a hypercube network. The most recent MPP machines, INTEL, CM-2, NCUBE, etc., combine a smaller number of larger processors with more conventional networking schemes. It is still a matter of debate whether the massively parallel architectures alone will deliver the largest computing power of the future necessary for solving, e.g., the protein folding problem, by brute force. So far, the human (i.e., software development) cost of massively parallel computing has been rather high.

Currently, a protein designer has at his/her disposal workstations priced at $\approx$ \$20,000–\$50,000 with central processing units (CPUs) operating at the speed of 50 MHz (20 ns/operation) and delivering some 16 MFLOPS, with graphics boards capable of handling as many as 250,000–1,000,000 vectors/s and disk storage capacity of 1–2 gigabytes (GB), and a possibility to access a supercomputer (via the public computer network such as BIT-NET or INTERNET), e.g., the Cray Research C-90 with 16 processors, each operating at a speed of 4 ns and delivering $\approx$ 650 MFLOPS (10 GFLOPS total). The cost of supercomputing remains high ($\approx$ \$500–\$1000/hour or $\approx$ \$3,000,000 for a two-processor Cray Y-MP2E with about \$100,000/year maintenance costs) and an average protein designer might be forced to do without supercomputing, although some of the important design methods (free energy perturbation, Monte Carlo searches) do require large processing power. It is important to realize, however, that a workstation of today, with four processors delivering 30 MFLOPS of computing power, would be a top-of-the line supercomputer of the late 1970s. So far, no one has yet seen the end of the trend of increasing computing power (by $\approx$ 100% every 10–12 months) and decreasing hardware costs. Thus, the future of computer modeling is wide open, and the outlook is rosy.

IV. MOLECULAR MECHANICS, A CONTROVERSIAL PARADIGM OF PROTEIN FOLDING

In order to build and manipulate protein structures in the computer, the chemical and geometrical aspects of the structure such as bonds, angles, torsions, and atomic radii have to be mathematically expressed and encoded in a software program. The field dealing with development and usage of such programs has become known as molecular mechanics.

Molecular mechanics has its origin in X-ray diffraction of proteins. There, the Fourier-tranformed diffraction data (i.e., the electron density map) is initially fitted with an amino acid sequence. The crude three-dimensional protein model is then refined, stereochemistry is regularized, bad atomic overlaps are corrected, etc. For the purpose of model refinement, Levitt and Lifson (1969) developed a set of formulae that explicitly described potential energy of an atom in a protein. The total energy of the system could be obtained by summing over all the atoms. First and second derivatives of this empirical energy potential could be calculated from the formulae, and the total energy of the molecule could be minimized by moving all the atoms along the potential energy gradient. This computational tool has proven to be an efficient and elegant way of regularizing protein structures.

It is important to realize that the potential energy function of Levitt and Lifson—and the other early force fields such as ECEPP (Warme and Scheraga, 1974), Hagler's formulations (Hagler *et al.*, 1974) embodied in Biosym's DISCOVER, Hermans and McQueen (1974), Allinger's (1977) MM2, the CHARMM (Brooks *et al.*, 1983), AMBER (Weiner *et al.*, 1986), and

GROMOS (van Gunsteren and Berendsen, 1987) potentials—were nothing but a means of building an abstract wire model of a molecule. The potential specifies how long (d_0) are the wires that represent atomic bonds, how strong they are (K_{bond}), and how the stretch (harmonically, for mathematical convenience):

$$E_{bond} = K_{bond}(d - d_0)^2; \qquad (1)$$

what are the correct values of angles connecting three atoms (θ_0), and how soft the angles are (K_{angle}):

$$E_{angle} = K_{angle}(\theta - \theta_0)^2; \qquad (2)$$

how easy is it to turn the torsions (K_ϕ) and where the torsional minima lie:

$$E_{torsion} = K_\phi(1 - \cos n\phi). \qquad (3)$$

The molecular mechanics forcefield also makes sure that double bonds are planar and that proper values for the other, "improper" torsions, ω, (e.g, chiral atom stereochemistry) are enforced:

$$E_{improper} = K_\omega(\omega - \omega_0)^2; \qquad (4)$$

how big are the "balls" (defined by radii, r) that represent atoms, and how hard they are:

$$E_{vdW} = \frac{A_{ij}}{r_{ij}^{12}} - \frac{B_{ij}}{r_{ij}^6}. \qquad (5)$$

Equation (5) implies that nonbonded atoms interact by pairwise Lennard–Jones potentials. In the equation, r_{ij} stands for distance separating the ith and the jth atoms. Their interaction consists of a weak attraction (London-dispersion forces falling off with the 6th power of distance) and a steep repulsive van der Waals barrier (the $\frac{1}{r^{12}}$ term). Pairwise interactions are also used to describe electrostatic interactions between electrically charged atoms and formally neutral but dipolar groups with measurable partial charges, Q_i, Q_j (ϵ is the dielectric constant),

$$E_{elec} = \frac{Q_i Q_j}{4\pi\epsilon r_{ij}}. \qquad (6)$$

It was quickly realized that the utility of the potential energy function was much broader than a mere real space refinement. Via energy minimization and, better still, Monte Carlo conformational searches (Tanaka and Scheraga, 1975) people hoped to arrive at the "natural" structure of protein models; after all, the natural structure was the lowest energy structure. Taken at the face value, an exact potential energy function and powerful computers could calculate correct structures and, implicitly, the biological properties of proteins.

One of the early papers on computer modeling of proteins that fostered such hopes was that of Levitt and Warshel (1975). It described computer simulation of protein folding based on "a new and very simple representation of protein conformation . . . together with energy minimization and thermalization." "Under certain conditions," the paper claimed, "the method succeeds in 'renaturing' bovine pancreatic trypsin inhibitor from an open-chain conformation into a folded conformation close to that of the native molecule." Papers with similar themes were published by the Scheraga (Burgess and Scheraga, 1975; Tanaka and Scheraga, 1975) and the Kuntz (Kuntz *et al.*, 1976) groups.

After more than 20 years, the challenge and excitement of this proposition are still with us, essentially unsolved. What is it that has been missing from the molecular mechanics forcefield approach?

First, a potential energy of a protein *in vacuo* is not a very good approximation for the free energy of the biological system: "the protein in physiological solution." The molecular mechanics force field was really meant to describe a mechanical ball-and-stick model built by the crystallographers: impenetrable spheres of atoms on rigid bonds and somewhat more flexible angles, with rotatable torsional degrees of freedom. Second, even if the potential function was correct, the problem of finding the global energy minimum is an enormous one. Third, the original work of Levitt and Warshel and Kuntz *et al.* as well as later works had to adopt significant simplifications of the protein molecule. In Levitt and Warshel (1975), for example, the protein had only α and β carbons (placed on a virtual bond of varying lengths such that the "β carbon" was centered on the side-chain centroid) and a single torsion between the two pseudoresidues.

An insightful comment of the Levitt and Warshel paper, entitled "On the Formation of Protein Tertiary Structure on a Computer," was published by Hagler and Honig in 1978. They pointed out that some of the compact conformations achieved with the simplified protein representations were not sequence specific, that is, could be readily obtained from poly-Gly-poly-Ala sequences. Moreover, the compact structures obtained from computer simulations never had the proper chain threading (topology) expected for the native trypsin inhibitor.

Thus, as early as the middle of 1970s, the protein modeling field became defined by the sharp conceptual polarity between grand concepts and oversimplifications, the requirement for rigor, and the need to develop empirical schemes that work. We will now follow these themes at some length.

V. WHAT DO WE KNOW ABOUT FOLDING?

Computer modeling of any value, instead of merely cranking torsions and building imaginary wire models, ultimately has to face the biophysics and thermodynamics of protein folding in aqueous solution. As a starting premise, we know that it is the amino acid sequence that determines, via the poorly understood "stereochemical code," the three-dimensional structure of a protein (Anfinsen, 1973). From this viewpoint, the design problems have two extreme formulations. The first is: given an amino acid sequence of a protein, find its correct fold. The second is: find an amino acid sequence that folds into a given three-dimensional shape. The primary-to-tertiary folding problem often has less extreme formulations. For example, only certain parts of a protein need to be modeled such as the six hypervariable loops of an antibody that form the antigen-combining site (Chothia *et al.*, 1986, 1989; Bruccoleri *et al.*, 1988); or the given amino acid sequence is known to be similar to a sequence of another protein of known three-dimensional structure, so that one only needs to do "homology modeling": assuming conservation of the fold (Epstein, 1964, 1966), one fits the amino acid sequence in question into the known fold, adjusting some details but retaining the gross features of the fold (Browne *et al.*, 1969; Shotton and Hartley, 1970; McLachlan and Shotton, 1971; McLachlan, 1972; Warme *et al.*, 1974; Padlan *et al.*, 1976; Blundell *et al.*, 1978; Feldmann *et al.*, 1978; Greer, 1980, 1981).

An in-depth biophysical discussion of the nature of the stereochemical code does not quite belong in this chapter, but cannot be completely avoided either. Suffice to say that (1) the stereochemical code is degenerate in the sense that many different sequences adopt identical folds (Epstein, 1966); (2) folding motifs (homologous protein families) are but a few (Levitt and Chothia, 1976) and we are perhaps beginning to know them all; (3) native folds are essentially determined by solvation characteristics of the 20 different amino acid side chains, most notably the electrically charged and the nonpolar, hydrophobic residues; hydrophobicity, or solvation characteristics of nonpolar atoms, is likely to be the main driving force of protein folding (Dill, 1990), and the concept of hydrophobicity (Kauzmann, 1959; Tanford, 1979; Sharp, 1991; Murphy *et al.*, 1990) figures prominently in any conceivable protein modeling procedure or algorithm; (4) solvation/desolvation effects of polar atoms impact mostly on electrostatic phenomena that are as critical to the specificity of protein folds as hydrophobicity is to protein stability. The complicated nature of protein electrostatics is briefly reviewed in the following section.

VI. ELECTROSTATICS OF PROTEIN STRUCTURES

Proteins are polyelectrolytes containing formally charged atomic groups, as well as polarized neutral groups (such as the peptide bond) with significant dipole moments and appreciable partial atomic charges. In the classical framework of molecular mechanics, electrostatics has been dealt with as a collection of pairwise Coulombic interactions [Eq. (6)]. This treatment, although computationally convenient, is inadequate mainly because of the neglect of the difference in dielectric constants of the protein ($\epsilon \approx 2\text{--}4$) and the solvent ($\epsilon = 80$). The irregular dielectric boundary created by the protein surface provides for an involved biophysical problem. On protein folding, two important electrostatic effects take place: (1) creation of new clusters of (partially) charged atoms in the low dielectric constant environment of the protein interior; these charges tend to stabilize themselves by establishing favorable Coulombic interactions, most notably hydrogen bonds; (2) desolvation of charged groups; this is proportional to the square of the atomic charge, and so the desolvation process is always unfavorable, and often stronger than the net Coulombic attraction (Novotny and Sharp, 1992).

The precise matching of charged groups with charged groups of opposite sign is an important example of chemical complementarity characteristic of protein structures. Specificity imposed by this charge–charge matching can, however, only be achieved by paying a price in free energy of desolvation that decreases structural stability. The challenge presented to protein modeling is to reproduce the quality and the magnitudes of all the above electrostatic effects. The molecular mechanics Coulombic treatment being unsatisfactory, and the increase in computing power making it possible to implement more complete (and more complex) treatments, new methods are being intensely sought (Shire *et al.*, 1974; Warshel *et al.*, 1984; Zauhar and Morgan, 1985; Rashin and Namboodiri, 1987; Harvey, 1989). One of the most accurate and the most widely used alternative treatments is calculation of the electrostatic potential generated by the protein atoms in aqueous solution with use of the finite difference approximation to the Poisson–Boltzmann equation (Warwicker and Watson, 1982) as implemented, e.g., in the program DelPhi (Gilson *et al.*, 1988; Jean-Charles *et al.*, 1990; Sharp and Honig, 1990b).

Electrostatic fields are calculated with the Poisson–Boltzmann (PB) equation as

$$\nabla^2 \phi(r) - \kappa^2 \sin h[\phi(r)] = -4\epsilon(r)\pi\rho(r), \tag{7}$$

where $\epsilon(r)$, $\phi(r)$, and $\rho(r)$ are the dielectric constant, electrostatic potential, and molecular charge distributions as functions of the position r, and κ is the Debye–Hückel parameter, proportional to the square root of the ionic strength. To solve the Poisson–Boltzmann equation, the molecule and a region of surrounding solvent are mapped onto a cubic lattice. The grid scale can be chosen to leave a 10-Å border between the protein surface and the edge of the rectangular grid, and giving a lattice spacing ≈ 1 points/Å.

Solutions to the potential at all the interior lattice points are obtained using the finite difference algorithm combined with overrelaxation (Davis and McCammon, 1989; Nicholls *et al.*, 1991). Usually, the linear form of the

Poisson–Boltzmann equation is used, which is sufficiently accurate for low or moderately charged molecules at physiological ionic strength (0.15 M) (Sharp and Honig, 1990a). A single potential calculation typically requires about 2 min of a Silicon Graphics IRIS processor time, for a protein consisting of ≈ 5300 atoms, including hydrogens.

VII. *DE NOVO* DESIGN OF PROTEINS: FOLDING IN THE COMPUTER

The most impressive protein design results have been obtained in the reverse folding problem, i.e., design of an amino acid sequence to fit a fold. Long and stable α-helical sequences were designed that, by all accounts, folded into antiparallel four-helical bundles (DeGrado *et al.*, 1989; Hecht *et al.*, 1990). Design principles of this protein class have proven to be relatively straightforward, requiring (i) sequences with α-helical propensity and (ii) amphipathic distribution of residues. Properly chosen sequences have organized themselves into α helices of predetermined length and surface polarity: one half polar, the other nonpolar. The nonpolar surfaces collapsed spontaneously onto themselves to "precipitate" out of the aqueous solution. In the course of this collapse, short connecting loops among helices dictated the up-and-down topology of such a helical bundle protein.

An entirely *de novo* prediction of a protein fold from an amino acid sequence usually proceeds in several steps. (i) Secondary structure elements (α helices and β strands) have to be identified, as accurately as possible, in the amino acid sequence. (ii) Having the helices and sheets identified, the folding problem is reduced to that of assembling the secondary structure building blocks into a plausible fold, following empirical folding rules (Richardson, 1981; Richardson and Richardson, 1989; Chothia, 1984; Levitt and Chothia, 1976, e.g., connectivity of ααα, αβα, βββ, etc.; pieces should be right-handed; core should be well packed; protein surface should be minimized, etc.). (iii) The final model (or a set of models) will be adjusted in details, regularized, and run through various algorithmic filters to gauge its correctness.

For each of the above steps, there are computer-modeling aids available, and they will be briefly reviewed here. Secondary structure prediction protocols (see Fasman, 1989, for a review) are either statistical or neural networks (pattern recognition; Qian and Sejnowski, 1988; Risler *et al.*, 1988; Holley and Karplus, 1989; Kneller *et al.*, 1990). For structural assembly, still more an art than a codified science, there are various aids such as fragment buildup approach based on the Protein Data Bank (Jones and Thirup, 1986); libraries of amino acid rotamers preferred in natural proteins (Ponder and Richards, 1987; Summers and Karplus, 1991); and computer-based mechanical tools ranging from simple devices (dial- and/or mouse-driven routines) that manipulate torsions and protein pieces, to the very sophisticated ones such as the various docking machines (under development at NIH and Duke University). As to the filters, these are topological algorithms based on connectivity and pattern recognition principles (Cohen *et al.*, 1979, 1982, 1983), surface properties (nonpolar/polar atom ratio, solvation parameters, Novotny *et al.*, 1984, 1988; Eisenberg and McLachlan, 1986; Ooi *et al.*, 1987), side-chain environmental "imprints" (Bowie *et al.*, 1991; Lüthy *et al.*, 1992), and database-derived functionals (Sippl, 1990; Hendlich *et al.*, 1990) also related to the side-chain environment.

VIII. SECONDARY STRUCTURE PREDICTION

Because of their stereochemistry (e.g., presence or absence of β carbons, side chain branching) and solvation characteristics, types of amino acid residues prefer certain types of secondary structures. Chou and Fasman (1974a,b) developed comprehensive secondary structure propensities for all the 20 naturally occurring amino acids. The propensities were calculated from frequencies of occurrence of amino acids in the secondary structures, α-helix, β-sheet, and β-turn. The frequency of an amino acid i in, say, α helices is obtained by dividing its occurrence in the helices by its total occurrence in proteins: $f_{\alpha i} = n_{\alpha i}/n_i$. This frequence, divided by the average fraction of residues found in α helices, $\langle f_\alpha \rangle$, gives the propensitites (probabilities of occurrence):

$$P_\alpha = f_{\alpha i}/\langle f_\alpha \rangle. \tag{8}$$

In the original papers of Chou and Fasman, the basic statistics were complemented by additional rules determining beginnings and ends of α helices and propagation of helical signals. These additional rules reflected the fact that the α-helical signal has often been underpredicted in actual helices, and that helical termini have well-developed side-chain preferences different from those of the middle of the helices. All these features can be eventually understood from the structural specifics of the helices (Richardson and Richardson, 1988; Presta and Rose, 1988.).

In an early blind test of various secondary structure prediction algorithms (Schulz *et al.*, 1974), the Chou and Fasman algorithm scored as one of the best, with about 70% correct secondary structure assignment. This probably represents the very limit one can achieve with no other information but the amino acid sequence. The mere 70% probability of a correct prediction is likely to reflect the fact that stability of helices, sheets, and turns is dependent on nonlocal interactions involving residues that are not contiguous in amino acid sequence. One telling example of this is the finding, by Kabsch and Sander (1984), of the same pentapeptide sequence (Val-Asn-Thr-Phe-Val) in either the α or the β conformation in two different proteins.

Various attempts have been made to use pattern recognition (neural networks) methods in the hope of improving on secondary structure predictions. Pattern recognition algorithms that take advantage of regular structural repeats found in the α/β types of folds have probably been most successful (Cohen *et al.*, 1983). A very simple way of aiding the prediction is to display smoothed sequence patterns, including secondary structure propensities but adding properties such as hydrophobicity and electric charge distribution. Hydrophobicity maxima have been shown to be good predictors of secondary structure location (Rose and Roy, 1980), whereas hydrophilic segments are often associated with surface loops (Kuntz, 1972). The smoothed patterns can be compared with those calculated on sequences of known secondary structures, thus sharpening the prediction and better resolving ambiguities (Novotny and Auffray, 1984; see Schulz, 1988, for a review). A number of secondary structure prediction programs are available, based on algorithms developed by Chou and Fasman (1974a), Garnier *et al.* (1978), Robson (1974), and others.

IX. SURFACES AND VOLUMES

"Biology starts at protein surfaces" has been an adage of Prof. Fred M. Richards, and computer-based interpretation of protein structures in terms of surfaces and volumes (Richards, 1977) is an important component of protein design. It is commonplace to think of chemical structures in terms of surfaces and volumes. A mathematical representation of these molecular attributes starting from a set of Cartesian coordinates specifying atomic centers is, however, a challenging problem. In 1971, Lee and Richards constructed the solvent accessible surface of a protein and, for the first time, investigated its properties. A later work of Chothia (1975) and others established the link between surface properties and biophysical determinants of protein folds. The molecular dot-surface diagrams, so abundant in present literature, are due to an algorithm invented by Connolly (1983). Richmond (1984) developed an analytical surface representation based on differential geometry that allowed for surface minimization. Surface topology plays an important role in specific protein–protein interactions. For example, antigenic epitopes on proteins coincide with surface regions accessible to a large (radius = 10 Å) spherical probe (Novotny *et al.*, 1986; Jin *et al.*, 1992).

Richards (1974) developed the first computer program that partitioned the space occupied by a protein into atomic (Voronoi) polyhedra, and determined local packing density of proteins. Volume (Connolly, 1985; Stouch and Jurs, 1986; Lewis, 1989) and surface partitioning is useful in various computerized docking procedures (Kuntz and Crippen, 1979; Connolly, 1986; Gregoret and Cohen, 1990, 1991; Jiang and Kim, 1991; Cherfils *et al.*, 1991; Meng *et al.*, 1992) that determine the best shape complementarity between protein/ligand pairs.

X. MOLECULAR DYNAMICS SIMULATIONS

The 1977 article by McCammon, Gelin, and Karplus marked the first application of molecular dynamics simulation techniques to protein structures. Since then, the method has become widespread, and has found its applications in X-ray structure refinement (Brünger *et al.*, 1987), NMR structure solution (Kaptein *et al.*, 1985), binding energy calculations, and conformational searches. Several excellent reviews are available (see, e.g., McCammon and Harvey, 1987; Karplus and Petsko, 1990).

The molecular mechanics forcefield [Eqs. (1)–(6)] describes the potential energy of a rigid protein, as it were, at $0°$ kelvin. At room (and any other non-zero) temperature, however, the total energy of the molecule is divided between its potential energy and the kinetic energy of thermal motions. On the computer, the motion can be obtained directly from the molecular mechanics potential by (1) calculating the derivatives (i.e., velocities, forces and mass-normalized forces, i.e., accelerations) of the potential energy equation for each atom in the system at time t_0 and (2) by obtaining new positions for all the atoms at a time $t_0 + \Delta t$ by integration. In order to obtain smooth trajectories, the integration step must be very small, on the order of a femtosecond (s^{-15}). A molecular dynamics trajectory consists of a computer file containing a succession of "frames (snapshots)," i.e., immediate atomic positions and their velocities.

The most straightforward use of dynamical simulations is in reproducing high frequency atomic vibrations and similar physical phenomena (crystallographic B factors, aromatic ring flips, methyl group rotamer intercoversions, etc.), as well as low frequency "hinge bending" modes (Brooks and Karplus, 1989). The empirical potential forcefield can also be coupled with a random drag force (potential of mean force, Langevin equation) to implement Brownian dynamics and simulate molecular diffusion processes. A nice example of Brownian dynamics is the simulation of superoxide anion diffusion to the enzyme superoxide dismutase (Sharp *et al.*, 1987). The electrostatic field generated by the enzyme impacts on the superoxide diffusion in such a way that the approach toward the substrate binding site is enhanced, compared to a random diffusion.

Dynamical simulations have often been used to sample conformational space (Bruccoleri and Karplus, 1990). This application of the method is more problematic, due to the fact that the conformational space contains many relatively high energy barriers that the simulated structure can cross only at very high temperatures. Dynamics tends to sample the immediate vicinity of the local minimum the structure occupied at the beginning of the simulation. The rest of the conformational space remains poorly sampled. Simulated annealing (Kirkpatrick *et al.*, 1983) and the quenched dynamics techniques intend to remedy this problem by implementing a series of "heating" steps, during which the structure is excited into new and unexplored regions of space, followed by successive cooling (i.e., energy minimization) steps. Although the quenched dynamics is known to be the most efficient energy minimization protocol, it still remains doubtful whether the sampling of the space is thorough and adequate.

Casual practitioners of molecular dynamics should be aware of the importance of technical parameters essential to successful dynamical runs. One of them is the nonbonded cutoff distance. The long-range molecular mechanics forces, van der Waals and electrostatic interactions [Eqs. (5) and (6)], should, in theory, be evaluated to infinity. In practice, this is impossible and current simulations employ a finite distance limit (e.g., 8 Å) beyond which a long-range interaction of any pair of atoms is neglected. Needless to say, such practice is arbitrary and must be applied judiciously. Another practical requisite of dynamical simulations is the periodic boundary condition. Current computer power limits the size of simulated systems to tens of thousands of atoms (i.e., 0.1 attomol) at most; yet, the purpose of the simulations often is to obtain statistical averages applicable to an infinite number of molecules. In order to achieve this objective, and also to conserve the total mass and energy of the system, the simulated molecules are surrounded by boundaries that act as symmetry operations. In a typical periodic boundary, a molecule that exits the system by a translational movement across the boundary to the left immediately enters the system from the right at a symmetrical position, and with an identical velocity vector.

XI. FREE ENERGY PERTURBATION

Free energy perturbation (Mezei *et al.*, 1978; Chandrasekhar *et al.*, 1985; Lybrand *et al.*, 1985, 1986; Rao *et al.*, 1987) uses molecular dynamics techniques to calculate free energy values (e.g., binding constants of ligand protein complexes) that may not be easily accessible experimentally. A typical perturbation calculation generates statistical averages, samplings, along a fictitious thermodynamical trajectory leading from one state to a different state. The original state may be a ligand bound to a protein; the end state may be a chemically related ligand bound to a protein. Two different conformations of a side chain in a protein may be another example of the two limiting states. To describe the course of a simulation, let us consider a series of simulations aimed at the calculation of the solvation free energy of ethanol, ΔG_{Et}, based on that of methanol, ΔG_{Mt}. The simulation starts with a snapshot sample of methanol molecule in water. Gradually, a methyl group is introduced in a series of "infinitesimal" steps, and at each point of this fictitious chemical transmutation trajectory, the system is thoroughly equilibrated by molecular dynamics. The final free energy of the transition, $\Delta G_{Mt} \rightarrow \Delta G_{Et}$ is obtained as the sum of the free energy increments obtained in each of the intermediate steps. The success of the free energy perturbation method depends on the infinitesimality of the process and, implicitly, on the completeness of sampling and equilibration. In practice, it is impossible to achieve perfect perturbation trajectories and simulation errors may be larger than the calculated free energy differences. A good discussion of applications and technical details of the method can be found in Straatsma and McCammon (1992).

XII. KNOWLEDGE-BASED APPROACHES TO MODEL BUILDING

Perhaps the most important lesson learned from the several hundred protein three-dimensional structures deposited with the Brookhaven Protein Data Bank (Bernstein et al., 1977) is their modularity. Only two types of secondary structure (α helices and β sheets) are common. These organize themselves into four folding types (all-α, all-β, α intertwined with β, and α and β separate) (Levitt and Chothia, 1976). Within the four folding types, supersecondary structures recur, such as the Greek key topology of antiparallel sandwiches or the typical βαβ right-handed motif (Richardson, 1981). On a larger scale, high-molecular-weight proteins are made of domains. On a smaller scale, short protein segments such as loops contain a limited number of key residues that determine the conformation of the entire loop (Leszczynski and Rose, 1986; Chothia et al., 1986; Chothia and Lesk, 1987). Knowledge-based modeling algorithms that combine patterns from all of the above structural levels into a single program package have been described, e.g., by Blundell et al. (1987) and Rooman and Wodak (1988). In loop modeling, polypeptide backbones of modeled loops can be obtained by comparing the amino acid sequence of the model loop to the sequences of "canonical" templates (see below), and by splicing the selected canonical backbone onto the underlying structural framework.

Implicit in the canonical motif concept (Chothia and Lesk, 1987) are two assumptions, namely (1) that canonical conformations are largely independent of the framework structure and (2) that the Brookhaven Database is sufficiently complete to permit unambiguous identification of the correct canonical motifs. Although the assumptions are well-founded first order approximations, neither of them is generally valid. For example, Tramontano et al. (1990) described a framework residue in immunoglobulins that strongly impacts on the conformation of the neighboring loop.

An average root-mean-square (rms) difference among canonical backbones of the same type (i.e., backbones assumed to be identical) is $\approx$ 1–1.5 Å, the maximum accuracy limit one can possibly achieve with canonical motif modeling. Models of various antibody combining sites were assembled from canonical loop motifs that well captured the main conformational features of the native structures (Chothia *et al.*, 1986, 1989).

Jones and Thirup (1986) noticed that, in order to find a collection of loops with conformations identical to a predefined template, it was often sufficient to search for a close match of the N-terminal and C-terminal dipeptides on which the loop is grafted. Such loop transfers can now be done automatically using the interactive graphic program INSIGHT/ HOMOLOGY (Biosym Technologies, Inc.) or QUANTA (Molecular Simulations, Inc.). The INSIGHT/HOMOLOGY backbone splicing protocol refers to the terminal peptides as the "preflex" and "postflex" residues, respectively, or "tails"; the loop spanning in between the tails is being referred to as the "flex" region of the template. Thus, the HOMOLOGY procedure requires the structures of the two tail-matching framework endpoints (C-and N-termini) at input, and it positions the loop tails on these endpoints using an rms best fit of the corresponding backbone atoms in the two structures.

XIII. CONFORMATIONAL SEARCHES

As discussed earlier, protein conformation is determined by values of backbone and side-chain torsional angles ϕ (C-N-Cα-C), ω (N-Cα-C-N), and χ (N-Cα-CβCγ, Cα-Cβ-Cγ-Cδ, and so on). Three-dimensional modeling involves finding specific values for all these torsional degrees of freedom. A particularly promising approach is therefore to use automatic computer algorithms that either generate a representative sampling of possible loop structures (Fine *et al.*, 1986) or uniformly sample the complete conformational space of a polypeptide chain segment or its side chains (Moult and James, 1986; Snow and Amzel, 1986; Bruccoleri and Karplus, 1987; Lee and Levitt, 1991). Theoretically, all the backbone and side-chain conformations compatible with the rest of the protein structure can be generated. The lowest free-energy conformation should correspond to the naturally occurring one. In practice, technical problems associated with an exhaustive sampling of conformational space restrict searches to short polypeptide segments only.

The CONGEN search algorithm of Bruccoleri and Karplus (1987) uses a selectable angular grid to sample conformational degrees of freedom in a given polypeptide segment using the amino- and carboxy-terminals as fixed endpoints. In the searches, both *cis* and *trans* proline peptide bonds are considered. For a loop *n* amino acid residues long, sampling runs on *n*-3 peptide units are combined with the modified Go and Scheraga (1970) chain closure protocol (Bruccoleri and Karplus, 1985) which, taking advantage of the fact that loop endpoints are fixed, generates an analytical solution for the structure of the remaining tripeptide. Conformations that severely violate van der Waals barriers are rejected in the course of sampling. Thus, the program samples uniformly the complete conformational space of the loop, and retains all the conformations that are compatible with the rest of the structure and have acceptable potential energies. These energies are used to distinguish between "good" and "bad" conformations and depend critically on short-range interactions between loops and, consequently, on the order of loop construction. From the set of conformations generated for each loop, "the best one" is selected to be incorporated into the final model based on potential energy and other criteria, e.g., the size of solvent exposed surface (Bruccoleri *et al.*, 1988).

Bruccoleri and Karplus (1987) demonstrated that the collection of conformations found by CONGEN sampling on a 30° (or finer) angular grid always contains a conformation(s) corresponding to the natural one (rms < 1 Å), and Bruccoleri *et al.* (1988) developed a protocol that selects, with a high probability, this natural conformation from the complete set of discrete conformational space samplings. Nevertheless, errors are known to occur in this step and the lack of an absolutely reliably loop selection protocol is currently the major limitation associated with the use of this method.

XIV. THE FREE ENERGY PROBLEM

Replacing the molecular mechanics forcefield with a more realistic free energy forcefield is one of the most important current research topics, essential to the computer modeling theory. Whereas Eqs. (1)–(6) describe energy of a rigid protein model *in vacuo,* a free energy potential aims at describing energetics of a complete system protein–water (solute–solvent). Thus, solvent effects (hydrophobicity, electrostatic desolvation, and other effects related to the energetic difference of solute–solute, as opposed to solute–solvent, interactions) are taken into account.

A molecular dynamics simulation in the presence of water molecules (Levitt and Sharon, 1988) is, in essence, an explicit free energy treatment. Such simulations, however, are often computationally prohibitive. Even long dynamical trajectories (hundreds of ps) may not be long enough to afford sufficient statistical sampling of the system and yield reliable averages to physical values that are directly related to experimental observables. Empirical approaches that describe solvent effects implicitly (Wesson and Eisenberg, 1992; Schiffer *et al.*, 1992) may be easier to implement computationally but are less accurate in principle.

The necessity of free energy potentials was dramatically demonstrated by a computer experiment in which fictitious, incorrectly folded proteins were generated (Novotny *et al.*, 1984, 1988). Two protein domains of the same length (113 amino acid residues) but very different folds (α-helical hemerythrin and β-sheeted immunoglobulin domain) formed the basis of the experiment. Keeping the backbones of the two proteins intact, their side chains were exchanged so that the α-helical backbone received the side chains from the β-sheeted protein and vice versa. The misfolded proteins were easily energy minimized and, based on the *in vacuo* energy alone, could not be distinguished from their correctly folded counterparts. The most conspicuous irregularity of the misfolded proteins was their higher surface exposure of hydrophobic residues. This observation led directly to development of algorithms (Bryant and Amzel, 1987; Novotny *et al.*, 1988; Hendlich *et al.*, 1990; Chiche *et al.*, 1990; Holm and Sander, 1992; Lüthy *et al.*, 1992; Maiorov and Crippen, 1992), solvation parameters (Eisenberg and McLachlan, 1986; Ooi *et al.*, 1987; Villa *et al.*, 1991), and empirical free energy potentials (Novotny *et al.*, 1989; Williams *et al.*, 1991; Nicholls *et al.*, 1991; Wilson *et al.*, 1991; Horton and Lewis, 1992) that have been used with success in gauging the correctness of protein models. These efforts, however, have only begun, and their future is wide open.

XV. FUTURE TRENDS

1. For the forseeable future, we will continue to live in a hardware wonderland. Every 10 months or so, a new generation of graphics workstations will enter the market, with faster CPUs and more powerful graphics. It will soon be possible to buy workstations of today's power for the price of a MacIntosh ($\approx$ $5000), and in a box of MacIntosh dimensions. We will also continue to see more sophisticated and more user-friendly commercial modeling software packages. These, however, are likely to remain expensive (comparable in price of that of the hardware on which they run).

2. With proliferation of easy push-button modeling protocols and software, we can expect an inflation of computer-built protein models of various accuracy. This trend will make it even more more important to push the horizon of our theoretical knowledge, improve accuracy of our modeling tools, and develop sensitive model-gauging protocols. At the same time, technological (hardware) advances will make it increasingly more possible to approach some of our modeling problems quantum-chemically (Warshel and Levitt, 1976; Zheng *et al.*, 1988; Bash *et al.*, 1991; Bajorath *et al.*, 1991a,b; Rao and Singh, 1991).

3. The busiest areas of modeling theory will most likely be electrostatics, free energy, and database searches. We will probably soon have enough X-ray crystallographic protein structures in the Brookhaven Protein Data Bank (Bernstein *et al.*, 1977) to build a virtually complete library of supersecondary structure motifs, and an elementary dictionary of loop shapes.

4. One of the most important aspects of future modeling will remain a frequent visit to a "wet" protein chemistry lab where our colleagues, experts in gene cloning and protein expression, are preparing chimaeric proteins, mutants, and other animals of the growing menagerie of computer-designed monsters. It is essential that our computer-based ideas be tested and computer modeling procedures improved based on experimental experience. After all, computer modeling is but an extension of the same line of chemical thought and reasoning that leads experimenters to draw formulas on paper, and biophysicists to relate properties of compounds to physical observables.

Note added in proof. New developments have occurred in computer modeling since this chapter was submitted, perhaps the most conspicuous of them being the speed-up of computer workstation clocks to about 90 MHz (11 ns/operation; e.g., the Silicon Graphics R8000 chip), with 200 MHz (5 ns/operation) promised for chips appearing on the market later this year (the Silicon Graphics R10000).

REFERENCES

Allinger, N. L. (1977) *J. Amer. Chem. Soc.* **99**, 8127–8134.

Anfinsen, C. (1973). *Science* **185**, 862–864.

Bajorath, J., Kitson, D. H., Kraut, J., and Hagler, A. T. (1991a) *Proteins* **11**, 1–12.

Bajorath, J., Kraut, J., Li, Z., Kitson, D. H., and Hagler, A. T. (1991b). *Proc. Natl. Acad. Sci. U.S.A.* **88**, 6423–6426.

Bash, P. A., Field, M. J., Davenport, R. C., Petsko, G. A., Ringe, D., and Karplus, M. (1991) *Biochemistry* **30**, 5826–5832.

Bernstein, F. C., Koetzle, T. F., Williams, G. J. B., Meyer, E. F., Brice, M. D., Rodgers, J. R., Kennard, O., Shimanouchi, T., and Tasumi, M. (1977) *J. Mol. Biol.* **112**, 535–542.

Blundell, T. L., Bedarkar, S., Rinderknecht, E., and Humbel, R. E. (1978) *Proc. Natl. Acad. Sci. U.S.A.* **75**, 180–184.

Blundell, T. L., Sibanda, B. L., Sternberg, M. J. E., and Thornton, J. M. (1987) *Nature (London)* **326**, 347–352.

Bowie, J. U., Lüthy, R., and Eisenberg, D. (1991) *Science* **253**, 164–170.

Brooks, B. R., Bruccoleri, R. E., Olafson, B. D., States, D. J., Swaminathan, S., and Karplus, M. (1983) *J. Comput. Chem.* **4**, 187–217.

Brooks, C. L., and Karplus, M. (1989). *J. Mol. Biol.* **208**, 159–181.

Browne, W. J., North, A. C. T., Phillips, D. C., Brew, K., Vanaman, T. C., and Hill, R. L. (1969) *J. Mol. Biol.* **42**, 65–86.

Bruccoleri, R. E., Haber, E., and Novotny, J. (1988) *Nature (London)* **335**, 564–568.

Bruccoleri, R. E., and Karplus, M. (1987) *Macromolecules* **18**, 2767–2773.

Bruccoleri, R. E., and Karplus, M. (1987) *Biopolymers* **26**, 137–168.

Bruccoleri, R. E., and Karplus, M. (1990) *Biopolymers* **29**, 1847–1862.

Brünger, A. T., Kuriyan, J., and Karplus, M. (1987) *Science* **235**, 458–460.

Bryant, S. H., and Amzel, L. M. (1987) *Int. J. Peptide Protein Res.* **29**, 46–52.

Burgess, A., Scheraga, H. A. (1975) *Proc. Natl. Acad. Sci. U.S.A.* **72**, 1221–1225.

Chandrasekhar, J., Smith, S. F., and Jorgensen, W. L. (1985) *J. Amer. Chem. Soc.* **107**, 154–163.

Cherfils, J., Duquerroy, S., and Janin, J. (1991) *Proteins* **11**, 271–280.

Chiche, L., Gregoret, L. M., Cohen, F. E., and Kollman, P. A. (1990) *Proc. Natl. Acad. Sci. U.S.A.* **87**, 3240–3243.

Chou, P. Y., and Fasman, G. D. (1974a) *Biochemistry* 13, 211-222.

Chou, P. Y., and Fasman, G. D. (1974b) *Biochemistry* 13, 222-245.

Chothia, C. (1975) *Nature (London)* 254, 304-308.

Chothia, C. (1984) *Annu. Rev. Biochem.* 53, 537-572.

Chothia, C., and Lesk, A. M. (1987) *J. Mol. Biol.* 196, 901-917.

Chothia, C., Lesk, A. M., Levitt, M., Amit, A. G., Mariuzza, R. A., Phillips, S. E. V., and Poljak, R. J. (1986) *Science* 233, 755-758.

Chothia, C., Lesk, A. M., Tramontano, A., Levitt, M., Smith-Gill, S. J., Air G., Sheriff, S., Padlan, E. A., Davies, D., Tulip, W. R., Colman, P. M., Spinelli, S., Alzari, P., and Poljak, R. J. (1989) *Nature (London)* 342, 877-883.

Cohen, F. E., Abarbanel, R. M., Kuntz, I. D., and Fleterick, R. J. (1983) *Biochemistry* 22, 4894-4904.

Cohen, F. E., Richmond, T. J., and Richards, F. M. (1979) *J. Mol. Biol.* 132, 275-288.

Cohen, F. E., Sternberg, M. J. E., and Taylor, W. (1982) *J. Mol. Biol.* 156, 821-862.

Connolly, M. L. (1983) *Science* 221, 709-713.

Connolly, M. L. (1985) *J. Amer. Chem. Soc.* 107, 1118-1124.

Connolly, M. L. (1986) *Biopolymers* 25, 1229-1247.

Davis, M. E., and McCammon, J. A. (1989) *J. Comp. Chem.* 10, 386.

DeGrado, W., Wasserman, Z. R., and Lear, J. D. (1989) *Science* 243, 622-628.

Diamond, R. (1966) *Acta Crystallogr.* 21(2), 253-266.

Dill, K. A. (1990) *Biochemistry* 29, 7133-7155.

Eisenberg, D., and McLachlan (1986) *Nature (London)* 319, 4943-4952.

Epstein, C. J. (1964) *Nature (London)* 203, 1350-1352.

Epstein, C. J. (1966) *Nature (London)* 210, 25-28.

Fasman, G. D. (1989) In *Prediction of Protein Structure and Principle of Protein Conformation* (G. D. Fasman, Ed.), pp. 193-316. Plenum, New York.

Feldmann, R. J. (1976) *Annu. Rev. Biophys. Bioeng.* 5, 477-510.

Feldmann, R. J., Bind, D. H., Furie, B. C., and Furie, B. (1978) *Proc. Natl. Acad. Sci. U.S.A.* 75, 5409-5412.

Fine, R. M., Wand, H., Shenkin, P. S., Yarmush, D. L., and Levinthal, C. (1986) *Proteins* 1, 342-362.

Garnier, J., Osguthorpe, D. J., and Robson, B. (1978) *J. Mol. Biol.* 120, 97-120.

Gilson, M., Sharp, K. A., and Honig, B. (1988) *J. Comp. Chem.* 9, 327-335.

Go, N., and Scheraga, H. A. (1970) *Macromolecules* 3, 178-187.

Gregoret, L. M., and Cohen, F. E. (1990) *J. Mol. Biol.* 211, 959-974.

Gregoret, L. M., and Cohen, F. E. (1991) *J. Mol. Biol.* 219, 109-122.

Greer, J. (1980) *Proc. Natl. Acad. Sci. U.S.A.* 77, 3393-3397.

Greer, J. (1981) *J. Mol. Biol.* 153, 1027-1042.

Hagler, A. T., and Honig, B. (1978) *Proc. Natl. Acad. Sci. U.S.A.* 75, 554-558.

Hagler, A. T., Huler, E., and Lifson, S. (1974) *J. Amer. Chem. Soc.* 96, 5319-5327.

Harvey, S. C. (1989) *Proteins* 5, 78-92.

Hecht, M. H., Richardson, J. S., Richardson, D. C., and Ogden, R. C. (1990) *Science* 249, 884-891.

Hendlich, M., Lackner, P., Weitckus, S., Floeckner, H., Froschauer, R., Gottsbacher, K., Casari, G., and Sippl, M. J. (1990) *J. Mol. Biol.* 216, 167-180.

Hermans, J., and McQueen. (1974) *A. Cryst.* A30, 730-739.

Holley, L. H., and Karplus, M. (1989) *Proc. Natl. Acad. Sci. U.S.A.* 86, 152-156.

Holm, L., and Sander, C. (1992) *J. Mol. Biol.* 225, 93-105.

Horton, N., and Lewis, M. (1992) *Protein Sci.* 1, 169-181.

Jean-Charles, A., Nicholls, A., Sharp, K., Honig, B., Tempczyk, A., Hendrickson, T., and Still, C. (1990) *J. Amer. Chem. Soc.* 113, 1454-1455.

Jiang, F., and Kim, S. H. (1991) *J. Mol. Biol.* 219, 79-102.

Jin, L., Fendly, B., and Wells, J. A. (1992) *J. Mol. Biol.*, in press.

Johnson, C. K. (1965) *ORTEP: A Fortran Thermal-Elipsoid Plot Program for Crystal Structure Illustration.* ORNL-3794, Oak Ridge, Tennessee.

Jones, T. A. (1978) *Methods Enzymol.* **115**, 157–171.

Jones, T. A., and Thirup, S. (1986) *EMBO J.* **5**, 819–822.

Kabsh, W., and Sander, C. (1984) *Proc. Natl. Acad. Sci. U.S.A.* **81**, 1075–1078.

Kaptein, R., Zuiderweg, E. R. P., Scheek, R. M., Boelens, R., and van Gunsteren, W. F. (1985) *J. Mol. Biol.* **182**, 179–181.

Karplus, M., and Petsko, G. A. (1990) *Nature (London)* **347**, 631–639.

Katz, L., Levinthal, C. (1972) *Annu. Rev. Biophys. Bioeng.* **1**, 465–504.

Kauzmann, W. (1959) *Adv. Protein Chem.* **14**, 1–63.

Kirkpatrick, S., Gelatt, C. D., and Vecchi, M. P. (1983) *Science* **220**, 671–680.

Kneller, D. G., Cohen, F. E., and Langridge, R. (1990) *J. Mol. Biol.* **214**, 171–182.

Kuntz, I. E. (1972) *J. Amer. Chem. Soc.* **94**, 4009–4012.

Kuntz, I. D., and Crippen, G. M. (1979) *Int. J. Peptide Protein Res.* **13**, 223–228.

Kuntz, I.D., Crippen, G. M., Kollman, P. A., and Kimelman, D. (1976) *J. Mol. Biol.* **106**, 983–994.

Langone, J. J. (Ed.) (1991) *Methods Enzymol.* **202, 203**.

Langridge, R., and MacEwan, E. A. W. (1965) *IBM Symp. Computer-Aided Exp.*, pp. 133–143.

Leszczynski, J. F., and Rose, G. D. (1986) *Science* **234**, 845–855.

Lee, B. K., and Richards, F. M. (1971) *J. Mol. Biol.* **55**, 379–400.

Lee, C., and Levitt, M. (1991) *Nature (London)* **352**, 448–451.

Levinthal, C. (1966) *Scientific Amer.* **214(June)**, 42–52.

Levitt, M., and Chothia, C. (1976) *Nature (London)* **261**, 552–558.

Levitt, M., and Lifson, S. (1969) *J. Mol. Biol.* **46**, 269–279.

Levitt, M., and Sharon, R. (1988) *Proc. Natl. Acad. Sci. U.S.A.* **85**, 7557–7561.

Levitt, M., and Warshel, A. (1975) *Nature (London)* **253**, 694–698.

Lewis, R. A. (1989) *J. Computer-Aided Mol. Design* **3**, 133–147.

Lüthy, R., Bowie, J. W., and Eisenberg, D. (1992) *Nature (London)* **356**, 83–85.

Lybrand, T. P., Ghosh, I., and McCammon, J. A. (1985) *Amer. Chem. Soc.* **107**, 7793–7794.

Lybrand, T. P., McCammon, J. A., and Wipff, G. (1986) *Proc. Natl. Acad. Sci. U.S.A.* **83**, 833–835.

Maiorov, V. N., and Crippen, G. M. (1992) *J. Mol. Biol.* **227**, 876–888.

McCammon, J. A., Gelin, B., and Karplus, M. (1977) *Nature (London)* **267**, 585–590.

McCammon, J. A., and Harvey, S. C. (1987) *Dynamics of Proteins and Nucleic Acids.* Cambridge Univ. Press, Cambridge UK.

McLachlan, A. D. (1972). *J. Mol. Biol.* **64**, 417–437.

McLachlan, A. D., and Shotton, D. M. (1971) *Nature New Biol.* **229**, 202–205.

Meng, E. C., Shoichet, B. K., and Kunz, I. D. (1992) *J. Comput. Chem.* **13**, 505–524.

Mezei, M., Swaminathan, S., and Beveridge, D. L. (1978) *J. Am. Chem. Soc.* **100**, 3255–3256.

Moult, J., and James, M. N. G. (1986) *Proteins* **1**, 146–163.

Murphy, K. P., Privalov, P. L., and Gill, S. J. (1990) *Science* **247**, 559–561.

Nicholls, A., Sharp, K. A., and Honig, B. (1991) *Proteins* **11**, 281–296.

Novotny, J., and Auffray, C. (1984) *Nucleic Acids Res.* **12**, 243–255.

Novotny, J., Bruccoleri, R. E., and Karplus, M. (1984) *J. Mol. Biol.* **177**, 787–818.

Novotny, J., Bruccoleri, R. E., and Saul, F. A. (1989) *Biochemistry* **28**, 4735–4749.

Novotny, J., Handschumacher, M., Haber, E., Bruccoleri, R. E., Carlson, W. D., Fanning, D. W., Smith, J. A., and Rose, G. D. (1986) *Proc. Natl. Acad. Sci. U.S.A.* **83**, 226–230.

Novotny, J., Rashin, A. A., and Bruccoleri, R. E. (1988) *Proteins* **4**, 19–30.

Novotny, J., and Sharp, K. A. (1992) *Progress Biophys. Mol. Biol.* **58**, 203–204.

O'Donnell, M., and Olson, A. (1985) *Computer Graphics* **15**, 3.

Ooi, T., Oobatake, M., Nemethy, G., and Scheraga, H. A. (1987) *Proc. Natl. Acad. Sci. U.S.A.* **84,** 3086–3090.

Padlan, E. A., Davies, D. R., Pecht, I., Givol, D., and Wright, C. (1976) *Cold Spring Harbor Symp. Quant. Biol.* **41,** 627–637.

Ponder, J. W., and Richards, F. M. (1987) *J. Mol. Biol.* **193,** 775–791.

Presta, L. G., and Rose, G. D. (1988) *Science* **240,** 1632–1641.

Qian, N., and Sejnowski, T. J. (1988) *J. Mol. Biol.* **202,** 865–884.

Rao, B. G., and Singh, U. C. (1991) *J. Amer. Chem. Soc.* **113,** 6735–6750.

Rao, S. N., Singh, U. C., Bash, P. A., and Kollman, P. A. (1987) *Nature (London)* **328,** 551–554.

Rashin, A. A., and Namboodiri, K. (1987) *J. Chem. Phys.* **91,** 6003–6012.

Richards, F. M. (1974) *J. Mol. Biol.* **82,** 1–14.

Richards, F. M. (1977) *Annu. Rev. Biophys. Bioeng.* **6,** 151–176.

Richardson, J. S. (1981) *Adv. Protein Chem.* **34,** 167–339.

Richardson, J. S., and Richardson, D. C. (1988) *Science* **240,** 1648–1652.

Richardson, J. S., and Richardson, D. C. (1989) In *Prediction of Protein Structure and Principle of Protein Conformation* (G. D. Fasman, Ed.), pp. 1–98. Plenum, New York.

Richmond, T. J. (1984) *J. Mol. Biol.* **178,** 63–89.

Risler, J. L., Delorme, O., Delacroix, H., and Henaut, A. (1988) *J. Mol. Biol.* **204,** 1019–1029.

Robson, B. (1974) *Biochem. J.* **141,** 853–867.

Rooman, M. J., and Wodak, S. J. (1988) *Nature (London)* **335,** 45–49.

Rose, G., and Roy, S. (1980) *Proc. Natl. Acad. Sci. U.S.A.* **77,** 4643–4647.

Schiffer, C. A., Caldwell, J. W., Stroud, R. M., and Kollman, P. A. (1992) *Protein Sci.* **1,** 396–400.

Schulz, G. E. (1988) *Annu. Rev. Biophys. Biophys. Chem.* **17,** 1–21.

Schulz, G. E., Barry, C. D., Friedman, J., Chou, P. Y., Fasman, G. D., Finkelstein, A. V., Lim, V. I., Ptitsyn, O. B., Kabat, E. A., Wu, T. T., Levitt, M., Robson, B., and Nagano, K. (1974) *Nature (London)* **250,** 140–142.

Sharp, K. A. (1991) *Current Opinion Struct. Biol.* **1,** 171–174.

Sharp, K. A., Fine, R., and Honig, B. (1987) *Science* **236,** 1460–1462.

Sharp, K., and Honig, B. (1990a) *J. Phys. Chem.* **94,** 7684–7692.

Sharp, K., and Honig, B. (1990b) *Annu. Rev. Biophys. Biophys. Chem.* **19,** 301–332.

Sharp, K. A., Nicholls, A., Fine, R. M., and Honig, B. (1991a) *Science* **252,** 106–109.

Sharp, K. A., Nicholls, A., Friedman, R., and Honig, B. (1991b) *Biochemistry* **30,** 9686–9697.

Shire, S. G., Hanania, G. I. H., and Gurd, F. R. N. (1974) *Biochemistry* **14,** 2967–2974.

Shotton, D. M., and Hartley, B. S. (1970) *Nature (London)* **225,** 802–816.

Snow, M. E., and Amzel, M. (1986) *Proteins* **1,** 267–279.

Straatsma, T. P., and McCammon, J. A. (1992) *Annu. Rev. Phys. Chem.* **43,** 407–435.

Stouch, T. R., and Jurs, P. C. (1986) *J. Chem. Information Comput. Sci.* **26,** 4–12.

Summers, N. L., and Karplus, M. (1991) *Methods Enzymol.* **202,** 156–203.

Tanaka, S., and Scheraga, H. A. (1975) *Proc. Natl. Acad. Sci. U.S.A.* **72,** 3802–3806.

Tanford, C. (1979) *Proc. Natl. Acad. Sci. U.S.A.* **76,** 4175–4176.

Tramontano, A., Chothia, C., and Lesk, A. M. (1990) *J. Mol. Biol.* **215,** 175–182.

van Gunsteren, W. F., and Berendsen, H. J. C. (1987) *GROMOS, Groningen Molecular Simulation Computer Program Package.* University of Groningen, The Netherlands.

Vila, J., Williams, R. L., Vasquez, M., and Scheraga, H. A. (1991) *Proteins* **10,** 199–218.

Warme, H. K., and Scheraga, H. A. (1974) *Biochemistry* **13,** 757–767.

Warme, P. K., Momany, F. A., Rumball, S. V., Tuttle, R. W., and Scheraga, H. A. (1974) *Biochemistry* **13,** 768–782.

Warshel, A., and Levitt, M. (1976) *J. Mol. Biol.* **103,** 227–249.

Warshel, A., Russell, S. T., and Churg, A. K. (1984) *Proc. Natl. Acad. Sci. U.S.A.* **81,** 4785–4789.

Warwicker, J., and Watson, H. C. (1982) *J. Mol. Biol.* **157**, 671–679.

Weiner, S. J., Kollman, P. A., Nguyen, D. T., and Case, D. A. (1986) *J. Comp. Chem.* **7**, 230–252.

Wesson, L., and Eisenberg, D. (1992) *Protein Sci.* **1**, 227–235.

Williams, D. H., Cox, J. P. L., Doig, A. J., Gardner, M., Gerhard, U., Perry, T., Kaye, Lal A. R., Nicholls, I. A., Salter, C. J., and Mitchell, R. C. (1991) *J. Amer. Chem. Soc.* **113**, 7020–7030.

Wilson, C., Mace, J. E., and Agard, D. A. (1991) *J. Mol. Biol.* **220**, 495–506.

Zauhar, R. J., and Morgan, R. S. (1985) *J. Mol. Biol.* **186**, 815–820.

Zheng, C., Wong, C. F., McCammon, J. A., and Wolynes, P. G. (1988) *Nature (London)* **334**, 726–728.

Part II

PRODUCTION

3

The Art of Expression: Sites and Strategies for Heterologous Expression

JASON R. ROSÉ*

CHARLES S. CRAIK†

*University of California, San Francisco, Department of Pharmacology, Medical Sciences Building S1210, San Francisco, California 94143-0450; and †University of California, San Francisco, Department of Pharmaceutical Chemistry, Medical Sciences Building S926, San Francisco, California 94143-0446

I. WHY *ESCHERICHIA COLI?*

The choice of an expression system will largely depend on the desired use of the expressed protein. There is no universal system for expression of proteins, and any one of the several available systems may have specific advantages that make it more preferable. There are many reasons, however, for attempting the expression of a protein in *E. coli*. As the original organism in which recombinant DNA technology and extensive molecular genetics were developed, it represents an ideal subject for the manipulations required for protein expression. A large and ever-increasing body of information exists regarding the expression of a myriad of different proteins, along with different conditions for optimizing their expression. A large number of commercially available vectors developed specifically for the purpose of expression are now available, making the development of an expression system for a new target protein a more routine procedure. The ease with which *E. coli* can be grown allows a large number of experiments to be performed quickly during the process of expression optimization.

It should be noted here that the optimization of expression involves the consideration of a number of variables, any one of which may be rate-limiting for expression of a given gene. For instance, increasing promoter strength may not have an obvious effect if the gene of interest contains sequences that decrease the overall efficiency of translation. We have attempted to mention in this review those factors that have been observed to be important in maximizing gene expression. It is hoped that they will provide a list of options to be considered in the event of an initial expression failure.

Detailed work has been done in *E. coli* regarding genetic manipulation. Over 150 promoters have been identified in the course of studying prokaryotic expression (Hawley and McClure, 1983). The molecular mechanisms of regulation have been extensively characterized for several of them. These promoters and engineered variants of them will be discussed in this chapter. By exploiting past research in *E. coli* on promoter function, genetic background, and growth conditions, it is now possible to produce proteins at levels as high as 10–40% of the cell mass. These levels can be achieved for target proteins that represent a very small fraction of the total protein expressed in their natural hosts. Whether dealing with proteins expressed naturally in *E. coli* (homologous) or in other organisms (heterologous), this amplification greatly improves our ability to provide reagent quantities of material.

From the standpoint of an individual optimizing protein expression, *E. coli* has the advantage of being easy to engineer on the genetic level using the technology currently available. The ease with which recombinant technology is applied to this organism allows for the consideration and manipulation of a wide array of genetic factors influencing protein expression, most of which will be at least mentioned in the following discussion. Genetic and cell-biological studies in this organism are continually providing additional information about the processes that affect protein overexpression.

As an organism, *E. coli* itself has advantages that make it a preferred expression host. Not only are bacteria easy to grow, but their maintenance is very inexpensive relative to eukaryotic hosts. Its simple growth requirements also mean that scale-up to high level production is feasible. Many fermentation devices are available that can be used to maximize the growth and expression potential of the organism in high density fermentations. Since these allow for the production of very large quantities of bacteria in small volumes, even a low-expressing system can provide significant amounts of material for study. The architecture of a bacterium offers the possibility of expressing protein at different sites: the cytoplasm, periplasmic space, or the extracellular environment. Each of these environments differs, and the advantages of this variation will be discussed later in the review.

Although *E. coli* represents an ideal host for the production of many proteins, it has several shortcomings which may necessitate the use of an alternate host. Since bacteria are incapable of carrying out the extensive post-translational modifications found on many eukaryotic proteins, the expression of proteins that require these modifications for proper activity is best left to another system. On the other hand, bacterial expression may provide a source of nonmodified protein for X-ray crystallography, which is generally confounded by such modifications. Whether the chaperonins in *E. coli* will be able to carry out the functions needed for correct folding or association of large, multisubunit eukaryotic proteins is not yet fully resolved.

Biotechnologists are currently working on developing vectors that will direct expression in other strains of Gram-negative bacteria, which have advantages in protein stability, secretion, or overall adaptability to large-scale reactors (Rangwala *et al.*, 1991). These efforts are progressing but are often hampered by a lack of understanding of the difference in requirements for plasmid replication, transcription, or translation. Advanced systems already exist for expression in the Gram-positive bacterium *Bacillus subtilis*. This is generally the first-choice prokaryote for protein secretion (Henner, 1990), but suffers from poor recognition of the better-understood *E. coli* transcription and translation signals, as well as a poor transformation efficiency.

This chapter is designed to review what is currently known about the means of expression optimization in *E. coli*. It is divided into separate sections dealing with (1) genetic considerations involving both the host and the target gene to be expressed, (2) the form in which the protein is expressed, and (3) localization of the protein in the host. Our intention is to provide an introduction to the many factors that should be considered in expressing a recombinant gene, and consequently will avoid details that can be obtained from more in-depth reports. Reviews of the subject matter that we have found to be particularly useful are listed at the end of each section.

General Subjects/Reviews

Das, A. (1990) In *Methods in Enzymology* (M. P. Deutscher, Eds.), Vol. 182, pp. 93–112. Academic Press, New York.

Goeddel, D. V., Ed. (1990) *Methods in Enzymology*, Vol. 185. Academic Press, New York.

Reznikoff, W., and Gold, L., Eds. (1986) *Maximizing Gene Expression*. Butterworths, Boston.

Rodriguez, R. L., and Denhardt, D. T., Eds. (1988) *Vectors: A Survey of Molecular Cloning Vectors and Their Uses*. Butterworths, Boston.

Wu, R., and Grossman, L., Eds. (1987) *Methods in Enzymology*, Vol. 153. Academic Press, New York.

II. GENETIC CONSIDERATIONS IN EXPRESSION

Overexpression of a desired gene product can be accomplished by optimization of two major regulatory steps: transcription and translation. Transcriptional regulation primarily involves the interactions of RNA polymerase with the promoter region of the gene to be expressed. Translational (post-transcriptional) control of expression appears to be a far more complex pathway of regulation. As a result of extensive genetic analysis most of the determinants of transcriptional control have been identified and optimized to the point where post-transcriptional regulation will probably become the rate-limiting step in the production of a desired protein.

A. Regulation of Transcription

This section will discuss various strong promoters that are commonly used, and the means by which they can be used to modulate transcription for overexpression purposes. It will also mention genetic elements of the host that should be considered when expressing foreign genes.

1. Polymerase and Promoter Function

Transcription of prokaryotic genes by RNA polymerase involves several steps, each of which has the potential to be a regulatory step. The first step involves the formation of the "closed complex," in which the polymerase complex ($\alpha_2\beta\beta'\sigma$) associates with the promoter core region through interactions between the DNA and the sigma subunit. The second step involves the melting of the DNA to expose the transcription start site—referred to as the formation of the "open complex." The final steps promote the initiation of the mRNA chain, which involves both the loss of the sigma subunit and escape of the polymerase from the promoter. The elongation of the mRNA chain may be affected by pause sites or termination sites contained within the gene. The affinity of a promoter for the polymerase (K), rate of isomerization of the closed to open complex (k_2), escape from the promoter, and extent of pausing downstream of the promoter are all potential steps for regulation of transcription.

A diagram of the various control regions in *E. coli* promoters is shown in Fig. 1. Promoter strength is regulated by the nature of the nucleotide sequences at positions: -10 (Pribnow box), -35, and -45 (AT box). By footprinting and protection experiments it has been shown that the σ subunit of RNA polymerase interacts primarily with the core -10 and -35 sequences. The exact role of the AT box has not yet been determined.

The selectivity of a promoter is generated by the degree to which the core regions (-10 and -35) match the consensus shown above. These sequences and the spacing between them (generally 17 ± 1 base pairs) affect the values of K and k_2, but it is rare that the effects of mutations within either region can be shown to affect only one of these kinetic constants. Typically, strong promoters in *E. coli* have K and k_2 values of $10^9 - 10^8$ M^{-1} and $0.1 - 0.01$ s^{-1}, respectively (Reznikoff and McClure, 1986). Promoters may vary widely in the rate of polymerase escape—as seen in the difference between *lacUV5* (slow) and T7 (fast). The determinants of this effect are not clearly understood and consequently are not yet amenable to manipulation.

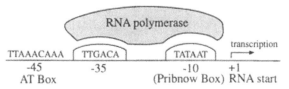

FIGURE 1 Schematic of the interaction of RNA polymerase with an *E. coli* promoter. Consensus sequences at −10 and −35 are those identified by Hawley and McClure (1983). The AT box sequence is that determined by Yamada *et al.*, from the *B. stearothermophilus nprM* promoter (Yamada *et al.*, 1991).

Following release from the promoter, the RNA polymerase elongation complex must be maintained in a highly processive state to reduce sensitivity to premature termination events. The study of bacteriophage transcription has led to the discovery of antitermination sequences (Nut sites) immediately downstream of some promoters. These cause accessory proteins to be added to the elongation complex and render it less sensitive to internal pause or termination sites. These sites are available in most vectors that utilize the bacteriophage λ leftward promoter (P_L).

The commercially available promoters that are most commonly used are *lac*, *trp*, *tac*, P_L, and more recently the T7 and *tac/att* hybrids. All of these promoters are considered strong promoters—where strong refers to the ability to direct the production of a given gene product to levels between 10 and 40% of the total cellular protein. The characteristics of these promoters vary in terms of host requirements, repression, and methods of induction—any of which may affect the choice of a promoter. A summary of these factors is shown in Table I.

2. Regulatable Promoter Systems

a. Repression/induction. The most common mode of promoter induction involves relief of repression (*lac/tac*, *phoA*, *trp*, P_L). Transcription from the *lac/tac* family of promoters is repressed by the *lac* repressor (*lacI*). In the absence of lactose, this repressor binds to an operator sequence immediately downstream of the transcription start site, preventing the isomerization of the closed to open polymerase complex (Straney and Crothers, 1987). This process is schematized in Fig. 2. The addition of lactose or the gratuitous chemical inducer isopropyl-β-D-thiogalactoside (IPTG) causes the dissociation of the repressor protein and activates transcription. The *trp* repressor (*trpR*) acts in a similar fashion, but is bound in the presence of tryptophan. Depletion of tryptophan or addition of 3-β-indole-acrylic acid (β−IAA) causes the loss of the repressor from the operator. In the case of *phoA*, repression is relieved by phosphate starvation, inducing transcription from the promoter. For cases in which regulated expression is not required, strains containing mutations in the *phoR* gene are available that allow con-

TABLE I Commonly Used Promoter Systems

Promoter	Repression requirements	Induction	Benefits	Drawbacks
lac/lacUV5[a]	lac I[q]	IPTG (lactose)	Strong promoter, easily induced	Leakage
trp	trp R	trp starvation or β-IAA addition	Strong promoter	Difficult to repress, induction toxicity
tac[b]	lac I[q]	IPTG (lactose)	Very strong promoter, easily induced	Leakage
P_L, P_R[c]	phage repressor (cI857)	Heat shock or nalidixic acid	Very strong promoter	Leakage, induction of SOS and heat shock responses, slow growth
tac or aac and phage att sites[d]	lysogen λxis⁻cI857	Short heat shock inverts promoter	Strong promoters, no leakage	Efficiency, timing of inversion may vary
T7[c]	T7 lysozyme (pLys) or F' for infection	IPTG induces T7 pol expression T7 phage infection	Strong promoter, easily induced No leakage	Lysozyme toxicity in cells, some leakage may occur, Limited to engineered strains
phoA[f]	phoR	Phosphate starvation	Inexpensive, growth regulated induction	May activate SOS response

[a] pSL301, Invitrogen.
[b] pKK223-3, Pharmacia; pPROK-1, Clontech.
[c] pPL-lambda, Pharmacia.
[d] pNH series, Stratagene.
[e] pET series, Novagen.
[f] pBAce, Stratagene.

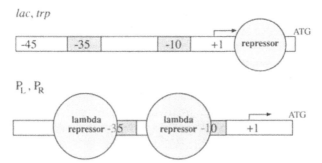

FIGURE 2 Schematic of repressor interaction with *E. coli* promoters. Repressors of the *lac/trp* type bind to operator sequences downstream of the RNA polymerase binding region. Lambda repressor binds to operator sequences in the RNA polymerase binding region.

stitutive (constantly active) transcription from the *phoA* promoter (Wanner, 1986).

The lambda repressor (*cI*) differs from the other repressors mentioned in that it binds to operators in the promoter and appears to interfere with the initial binding of the polymerase complex (Hawley *et al.*, 1985). Temperature sensitive (TS) mutants of the repressor (cI857) fail to bind to the operator at 42°C, resulting in a relief of repression.

Another mode of induction involves an increase in the copy number of the plasmid (runaway replication), since the amount of protein produced is linked to the copy number of the plasmid. Runaway replicons may produce up to 2000 copies of a plasmid per cell. The use of this system was first described in 1979 (Uhlin *et al.*, 1979), when a TS copy number mutant of the RI plasmid was shown to amplify protein production by 400-fold when shifted to the nonpermissive temperature. An updated version of this plasmid offering tighter control of copy number induction is now commercially available (Hasan and Szybalski, 1987). These vectors are better for rapid production of protein, as the runaway replication of the plasmid is invariably lethal to the host.

The T7 polymerase promoter is another very strong promoter that is gaining popularity for use in overexpression (Studier *et al.*, 1990). This promoter can be regulated in two ways. The first involves a strain that contains two plasmids: one coding for T7 polymerase under control of an inducible promoter (*lacUV5*), and a second plasmid that constitutively expresses T7 lysozyme. T7 lysozyme binds specifically to T7 polymerase and prevents transcription, thus inhibiting any T7 polymerase produced from leakage of the *lacUV5* promoter. Induction of T7 polymerase with IPTG titrates the T7 lysozyme and produces high levels of the polymerase, which, in turn, transcribes the gene of interest at high levels. T7 polymerase can also be provided by phage infection, which allows a total repression of the gene of interest until induction. A drawback of using the T7 system is that work is limited to a small set of engineered host strains.

The recent development of the *tac/att* hybrid promoter system provides another means of maintaining promoters in a totally inactive state prior to induction (Podhajska, *et al.*, 1985). In this system the promoter to be used (*lac, tac*) points away from the gene to be expressed and is located between two divergent λ *att* sites, which are used as recombination substrates by the λ integrase (Int) protein. The Int protein is produced from a cI857 prophage in the strain by means of a short heat pulse. Int then inverts the promoter, which can be further repressed or allowed to freely transcribe the gene of interest.

b. Usage of promoters.   Given the number of promoters available, which one should be used to express a given protein? All the promoters mentioned above are roughly equal in strength *in vivo* but some contain unique features that make them preferable in particular situations. These situations will often be defined by factors ranging from protein toxicity and method of induction to requirements for specific export functions. We hope the factors presented in Table I will serve as a brief guideline for promoter selection.

3. Promoter/Strain Interactions—Preferences for Expression

Upon overexpression proteins may become toxic to the *E. coli* host. The potential toxicity of the recombinant product is often a major factor in determining the host/vector combination to be used. In this context a leaky promoter makes the expression plasmid unstable by virtue of its toxicity to the host. In the case of promoters that are regulated by the *lac* repressor, extra repressor can be provided by a *lac* I^q on the plasmid or in the strain. A superexpressing *lac* repressor strain (I^{sq})—DH21—is also available (Hanahan, 1983). Maintaining the expression plasmid at a low copy number may also be useful in maintaining full repression. Most promoters, however, exhibit some basal level of transcription (leakage) even under heavy repression (Table I).

The T7 and *tac/att* hybrid promoter systems are probably the most ideal for controlling leaky expression. In the first system, as mentioned previously, the polymerase is not present in the cell until induction or phage infection. In the *tac/att* system the promoter is maintained in a reversed, and consequently inactive, orientation until induction. If the promoter placed between the *att* sites is sensitive to chemical inducers such as IPTG, its activity may be titrated following promoter inversion if a further level of control is desired. Few test cases using this promoter have been reported to date. A possible problem with it involves differing efficiency of inversion within a population, leading to heterogenous populations of cells expressing the protein of interest.

The level of full-length mRNA may be affected by the presence of internal transcription termination signals in the gene to be expressed. Promoter systems based on phage promoters contain antitermination sequences that can be used to offset this problem. Recently an antitermination sequence from the ribosomal RNA operon rrnB has been placed downstream of a modified tac promoter to provide the same function for nonphage promoters. Strong transcription terminators from bacteriophage have been cloned and used as transcription termination modules in many plasmids (Brosius, 1988).

The mode of promoter induction may also be a major consideration in determining which system to use. Many proteins have been shown to be unstable and prone to precipitation at 42°C, the temperature used to induce cI857 controlled vectors (P_L) or runaway replicons. This heat shock may therefore result in protein aggregation or increased sensitivity to proteolysis. The tac/att promoters avoid this problem by requiring only a short heat shock. Nalidixic acid, an alternate lambda promoter inducer activates the SOS responses in the cell, which may result in decreased protein synthesis and activation of the heat shock pathway (Neidhardt et al., 1984). The heat shock response in E. coli can be problematic in that it causes the induction of general degradative proteases.

Rampant proteolysis in E. coli is often a problem when expressing proteins. In 1985, Goff and Goldberg demonstrated that expression of abnormal proteins induces the lon-encoded protease La, and mutants in lon resulted in increased stability of the overproduced protein (Goff and Goldberg, 1985). lon⁻ strains have the phenotypic disadvantage of being highly sensitive to DNA-damaging agents and are prone to overproduction of mucoids, making them difficult to work with. To circumvent this problem, mutations have been made in the htpR gene, which codes for the σ_{32} polymerase subunit used to transcribe all heat shock genes. htpR⁻ strains are also protease deficient (Baker et al., 1984) but do not display the detrimental lon phenotypes (Goff et al., 1984).

A final means of inhibiting proteolysis is derived from the observation that bacteriophage T4 inhibits the degradation of protein fragments in infected cells (Simon et al., 1978). An expression vector (pDIP) has been constructed that allows a foreign gene under control of the T7 promoter/terminator system to be recombined into the T4 genome. Infection of E. coli containing T7 polymerase with "DIP phage" results in expression of proteins that may be stabilized against degradation as much as sixfold (Singer and Gold, 1991).

It may often arise that a protein of interest is not expressed in a strain for reasons that are not obvious. In this case, a rapid screen of other available strains may be used to identify one that produces the protein at acceptable levels (Browner *et al.*, 1991; Frorath *et al.*, 1992).

B. Regulation of Translation

A number of regulatory mechanisms can control the rate of translation of most genes. This section will discuss the post-transcriptional regulatory mechanisms that have been identified in prokaryotes and ways in which they can be manipulated to enhance production of a desired protein. It should be noted here that, for the most part, prokaryotic genes can be highly expressed in *E. coli* without much manipulation (McGrath *et al.*, 1991). A summary of possible targets for translation regulation is presented in Table II.

1. Post-transcriptional Regulation of mRNA Levels

One of the simplest means of regulating translation is by affecting the population of messenger RNAs in the cell. This seems to be done primarily by affecting the half-life of the transcript in question. The half-life of an average mRNA in *E. coli* is on the order of 2–6 min but genes have been identified whose transcripts have much longer half-lives (~15–20 min). Some messages exhibit cell growth control; having long half-lives under conditions where growth is rapid and short half-lives when growth is slow (Nilsson *et al.*, 1984). Messenger RNA degradation, shown in Fig. 3, occurs by exo- and endonucleolytic attack, which occurs at the 5′ and 3′ ends as well as at internal sites in the transcript.

a. 5′ Stem loops. By examination of the T4 gene 32 and *omp* A RNA transcripts that demonstrate high stability in *E. coli* it was demonstrated

---

TABLE II Manipulation of Post-transcriptional Regulation

Target for regulation	Downregulatory mechanisms	Strategies to alter regulation
mRNA	exonuclease degradation	Put stem loops at 5′ and 3′ ends, REP[a] sequences at 3′ end
	endonuclease degradation	Protect sites with ribosomes
Preinitiation complex	weak SD	Lengthen SD[b], add TP[c]
	stem-loops at SD/ATG	Reduce 2° structure around SD
	non-optimal SD-ATG spacer	Change length, sequence
Elongation complex	5′ codons (+4 to +30)	Maximize A:T usage
	ribosomal pausing	Reduce 2° structure within gene
	rare codon usage	Codon optimization

[a] REP, repetitive extragenic palindromic sequence.
[b] SD, Shine–Dalgarno sequence.
[c] TP, translation–initiation promoting site.

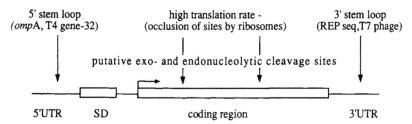

FIGURE 3 Sites of degradation of mRNA and mechanisms used to stabilize transcripts against nucleolytic attack. Examples of stem loops shown to confer stability are indicated and also referred to in the text.

that sequences upstream of the RBS played a role in affecting the rate of degradation (Emory *et al.*, 1992; Gorski *et al.*, 1985). Both of these transcripts have long extensions in the 5′ untranslated region (5′UTR) that are predicted to contain stable stem loops. In addition, these stem loops are formed so that there are only 1 or 2 unpaired nucleotides at the extreme 5′ end of the transcript. Emory *et al.* demonstrated that increasing the number of unpaired nucleotides at the extreme 5′ end increased the degradation rate of the mRNA, suggesting that a 5′ exonuclease or 5′-end-dependent endonuclease may be responsible for inactivating the 5′ ends of mRNAs (Emory and Belasco, 1990). Consistent with this idea, a synthetic stem loop placed at the 5′ end of the β-lactamase transcript increased its half-life approximately three- to fivefold (Belasco *et al.*, 1986).

b. 3′ Stem loops. The effect of stem loops at the 3′ end on mRNA stability was initially identified with transcripts from T7 phage. Processing of these transcripts by RNAse III leads to the formation of a stem loop structure with a fully base-paired 3′ terminus that exhibits increased stability relative to uncleaved transcripts (Dunn and Studier, 1983). In the phage lambda *int* transcript an RNAse III cleavage removes the entire stem loop and results in rapid inactivation of the transcript. A similar stabilization mechanism exists in bacterial genes in the form of repetitive extragenic palindromes (REP sequences). These highly conserved inverted repeats are found in regions between cistrons in many *E. coli* operons (Higgins *et al.*, 1982). The REP sequences are predicted to form high energy stem loops ($\Delta G = -24$ to -54 kcal) downstream of the coding region, and should provide a significant barrier to 3′-5′ exonucleases. Placing REPs downstream of coding regions has been shown to increase translation (Newbury *et al.*, 1987). These observations suggest that the placement of stem loop "cassettes" at both the 5′ and the 3′ UTRs of a cloned gene could be used to stabilize the mRNAs.

c. Internal cleavage sites. Messenger RNAs may also be degraded by endonucleolytic cleavage of sequences within the coding region, which lead to mRNA species with heterogenous 3′ ends. These are subjected to further rapid exonucleolytic degradation (Kennell, 1988). The exact nature of internal endonuclease cleavage sites is not well understood. The presence of these cleavage sites in a gene is ultimately a rate-limiting determinant for stability, since they will not be affected by sequences at the 5′ or 3′ UTR. The rate of cleavage at these sites, however, does appear to be regulated. Investigation of the effects of blocking translation or pausing ribosomes on mRNAs has provided strong evidence that the presence of ribosomes on transcripts can stabilize them against degradation. Kennell and Riezman have demonstrated that the rate of message degradation of the three cistrons of the *lac* operon is inversely related to the frequency of translation (Kennell and Riezman, 1977). From this, it seems that the best way to stabilize mRNAs that contain internal endonuclease cleavage sites is by an indirect method such as the maximization of ribosome binding and initiation.

2. Mechanism of Ribosome Binding

Translation is initiated by a series of binding events that assemble the ribosomal subunits on a messenger RNA. The mRNA initially binds to the 30S ribosomal subunit in the presence of the f-Met-tRNA to form the 30S ternary complex. This complex then associates with the 50S ribosomal subunit to form the 70S initiation complex upon the loss of various initiation factors. The 70S complex is competent to carry out the initiation and elongation steps of polypeptide synthesis.

Initial association of the mRNA with the 30S subunit was proposed to occur by means of a nucleotide sequence (5′-GAUCACCUCCUUA-OH-3′) on the 3′ end of the 16S rRNA (Shine and Dalgarno, 1974). This sequence has a complementary consensus sequence on the mRNA known as the Shine–Dalgarno (SD) sequence, or ribosome binding site (RBS). The SD sequence is located at approximately −10 relative to the start AUG and typically consists of 3–5 contiguous bases from the consensus sequence AGGAGGU.

In addition to the role of the SD sequence, it was demonstrated that a region from −53 to −30 could promote efficient rebinding of messenger RNA to ribosomes (Borisova, *et al.*, 1979). Examination of sequences upstream of the SD sequences in highly expressed *E. coli* genes has revealed a second consensus sequence—UGAUCC—referred to as the translation–initiation promoting site (TP) (Thanaraj and Pandit, 1989). This suggests an additional role for upstream nucleotide sequences in increasing the rate of translation initiation relative to less well-expressed genes. The multiple elements that regulate translation initiation are shown in Fig. 4.

a. Regulation of ribosome binding/preinitiation complexes.   Genes cloned from organisms other than *E. coli* will most likely not contain optimal Shine–Dalgarno sequences. Ribosome binding site optimization should

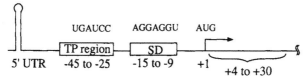

FIGURE 4 Schematic of the 5′ end of a prokaryotic mRNA. Consensus sequences are shown for the Shine–Dalgarno (SD) sequence and the translation–initiation promoting region (TP). The translation start site is indicated by +1 and nearby downstream codons by +4 to +30.

take into account the observation that even well-expressed *E. coli* genes do not contain long, highly complementary SD sequences. Ribosome binding in *E. coli* is optimized for both capture of preinitiation complexes and release of translation-competent ribosomes. The current perception in the literature is that very long SD sequences hinder ribosome release and therefore decrease the frequency of polypeptide chain initiation. In contrast, a system for expressing the HIV-1 protease using a maximally complementary sequence AGGAGGU directed the production of twofold more protease than the shorter sequence AGGAG (J. R. Rosé, unpublished results). This suggests that a long SD sequence is not always detrimental to translation.

Additional sequences around the ribosome binding site appear to be important regulators of expression. A systematic analysis of the nucleotide spacing between the SD sequence and the initiating AUG (Shepard *et al.*, 1982) indicates that the optimal spacing is 9 ± 2 nucleotides. Deviation from this spacing can affect expression as much as 200 to 500-fold. This might be expected, given the constrained structural environment of the ribosome–mRNA initiation complex. As discussed by Stormo, the sequence of the spacer region appears to be biased heavily in favor of A and U, perhaps in order to reduce self-complementarity and unambiguously define the SD sequence (Stormo *et al.*, 1982).

Ribosome binding to the SD sequence may also be affected by the formation of local secondary structures that occlude the site. A number of well-studied mRNAs contain predicted stem–loops that encompass the SD and/or initiator AUG and are, in general, poorly translated. Mutations that reduce the self-complementarity in the stem–loop can increase the translation of these genes by as much as 20 to 100-fold (Hall *et al.*, 1982). Examination of regulatory mechanisms in the *atp* operon has led to the discovery of a 30 nucleotide "intragenic enhancer" upstream of the *atpE* gene that displays little or no predicted secondary structure. Expression vectors that incorporate this enhancer region and the *atpE* SD sequence have demonstrated enhanced translational efficiency with a number of heterologous genes (McCarthy *et al.*, 1986). Consequently, genes cloned into expression vectors along with their own upstream sequences should be examined for possible secondary structures that may be formed by nucleotide sequences around the SD site and start codon, which could inhibit ribosome binding. Alternatively, heterologous genes may be cloned downstream of a SD sequence that has been optimized with respect to the factors mentioned previously, with the caveat that translation will still be inhibited by secondary structure in the 5′ end of the cloned gene.

In the past making optimized SD constructs has been hindered by the limited restriction sites available for placement of the 5′ end of a target gene relative to the SD sequence. With the advent of the polymerase chain reaction (PCR), strategies have been developed to circumvent this constraint. The PCR primers can be chosen such that any restriction site can be incorporated into the 5′ end of a gene and used to clone the gene into the available restriction sites. Restriction sites such as *Nco*I, *Sph*I, or *Nde*I are commonly used for cloning next to SD sequences because they contain internal ATGs that are used as the start codon for the target gene.

Using *Nco*I and *Sph*I sites leads to some restrictions since the sequence of the restriction site itself dictates the first base of the codon following the ATG. As described by Remaut *et al.*, these sites can be tailored to leave blunt ATGs, to which genes left blunt-ended by PCR can be ligated (Remaut *et al.*, 1989). *Nde*I sites, on the other hand, terminate with the ATG, allowing for fusion to genes without concern for the nature of the nucleotide sequences downstream of the site.

b. Regulation during polypeptide elongation.   Once a ribosome initiates a polypeptide chain there are a number of factors that appear to affect its ability to complete efficiently the elongation process. The first of these is the nucleotide bias in codons immediately downstream of the initiator AUG. Taniguchi and Weissmann originally demonstrated that mutation of the second codon from GCA to ACA had a dramatic effect on translation (Taniguchi and Weissmann, 1978), suggesting that the codons immediately downstream of the start codon could affect translation efficiency. In keeping with this suggestion, it has been observed that G:C-rich genes, such as those cloned from parasites, display very poor expression in *E. coli* (Button *et al.*, 1991). Maximizing the first four codons after the AUG for A:T usage without affecting the amino acid sequence resulted in a 20 to 30-fold increase in expression. Although the presence of G:C-rich codons may promote the formation of secondary structures that could block elongating ribosomes, no satisfactory correlation has been made between the effect of A:T substitutions in the 5′ region on protein expression and disruption of predicted stem loops. Whatever the mechanism, it is generally beneficial to increase the A:T content of the 5′ end of the gene to be expressed whenever possible. If the amino terminal sequence of the gene cannot be optimized due to sequence constraints, fusion of the gene with the first 10–12 amino acids of a highly expressed bacterial gene may produce the same effect. This, of course, requires the ability to process this amino terminal extension away from the fusion product, a topic that will be discussed later.

Additional elements of secondary structure in the body of a gene may affect expression levels. The presence of stem loops in an mRNA creates a barrier to elongating ribosomes that may cause pausing and an overall decrease in translational efficiency as high as 50% (Kubo and Imanaka, 1989). Pausing may also be followed by discontinuation of the polypeptide chain. The observation of discrete polypeptides of less than full length *in vivo,* or monitoring *in vitro* transcription may indicate the presence of such an event (Varenne *et al.,* 1982).

A good deal of information has been published with respect to the role of rare codons in affecting translational efficiency. In many cases, changing codon usage of a foreign gene to match that of the host leads to an improvement in expression levels of at least 2 to 20-fold (Kotula and Curtis, 1991; Williams *et al.,* 1988). Certain regulatory genes of *E. coli* have provided evidence that rare codons are used to modulate expression levels (Konigsberg and Godson, 1983). In many other cases, however, heterologous genes with high percentages of rare codons (15–25%) are able to be expressed to very high levels (Devlin *et al.,* 1988; Nassal *et al.,* 1987). It is likely that rare codons modulate gene expression in a case-by-case manner and that this represents a level of regulation that is subordinate to the others mentioned above. It has been suggested that rare codons primarily affect the rate, not the overall level, of translation from a given gene (Andersson and Kurland, 1990). Furthermore, evidence exists that suggests rare codon usage becomes rate-limiting when translation is occurring at very high rates, providing a growth-phase-dependent type of regulation (Robinson *et al.,* 1984). Consequently, we recommend optimization of the aforementioned conditions prior to the extensive replacement of rare codons with codons optimized for *E. coli.*

Transcription

McClure, W. R. (1985) Mechanism and control of transcription initiation in prokaryotes. *Annu. Rev. Biochem.* 54, 171–204.

Translation

Belasco, J. G., and Higgins, C. F. (1988) Mechanisms of mRNA decay in bacteria: A Perspective. *Gene* 72, 15–23.

McCarthy, J. E. G., and Gualerzi, C. (1990) Translational control of prokaryotic gene expression. *Trends Genet.* 6(3), 78-85.

Codon Usage

deBoer, H. A., and Kastelein, R. A. (1986) In *Maximizing Gene Expression* (W. Reznikoff and L. Gold, Eds.), Vol. 9, pp. 225–285. Butterworths, Boston.

Kurland, C. G. (1991) Codon bias and gene expression. *FEBS Lett.* 285, 165–169.

III. POST-TRANSLATIONAL MANIPULATIONS: FORM OF THE EXPRESSED PROTEIN

A major concern in the expression of heterologous proteins is the form in which the protein is expressed. A protein may be expressed as the native polypeptide or fused to one of a number of standard partner polypeptides. In addition, there is the additional concern whether the protein is expressed in a soluble or insoluble form. These two topics will be addressed in this section.

A. To Fuse or Not to Fuse

1. Uses for Fusion Proteins

The use of polypeptide fusions in expression systems has become an important method of recovering the final gene products from an expression system. There are currently at least five commercial vectors that allow the fusion of the gene of interest to partners such as glutathione *S*-transferase or the maltose binding protein, expressly for the purpose of providing a "handle" for use in affinity purification. Smaller peptide sequences are also available (FLAG, poly-His) for immuno- or metal affinity purification. A list of commercially available fusion-purification systems is shown in Table III. These fusions are very useful for high-yield purification of proteins that represent a small fraction of all polypeptides in the cell or for which no purification protocols are available. Furthermore, removal of the fusion partner by proteolysis or other cleavage methods usually generates a recombinant product with homogenous or native-like termini.

Fusion partners may be useful in stabilizing proteins that are normally unstable or rapidly degraded in *E. coli*. Many polypeptide hormones, which tend to be small and basic, exhibit very rapid kinetics of degradation and are difficult to express. Saito *et al.* demonstrated that fusion of these proteins to more acidic polypeptides could result in levels of accumulation as high as 32% of the cellular protein (Saito *et al.*, 1987). A similar approach was taken to express even smaller peptides through fusions with protein A (Löwenadler *et al.*, 1987). This not only stabilized the peptides, but allowed for easy purification and generation of antipeptide antibodies. The use of fusions to target proteins to regions of the cell will be discussed later.

2. The Case for Native Expression

Although fusion proteins have many potentially useful properties, there are many cases where it is desirable to produce a protein in its native state. Logically, native, or direct, expression is preferred whenever possible to guarantee the authenticity of the protein with regard to structure and function. The presence of a fusion partner polypeptide may have serious effects on the folding or activity of the protein of interest. By interfering with correct folding, a fusion construct may also affect the solubility and proteolytic stability of the recombinant product. For example, when partial fragments of a polypeptide domain were fused to a target protein, misfolding and extensive proteolytic degradation were observed (Bishai *et al.*, 1987).

Most importantly, direct expression has the advantage that no further proteolytic processing is necessary following purification. When the protein is expressed as a fusion, the partner polypeptide can be removed by recombinant and highly specific proteases such as those mentioned in Table III, but this removal may be plagued by both poor cleavage efficiency and extreme expense of the processing enzyme upon scale-up. In the case in which the recombinant fusion must be recovered from an insoluble pellet, the presence of denaturing agents needed to maintain solubility of the fusion protein may also prevent the processing enzymes from functioning efficiently.

Although fusions can be useful for increasing translation, as mentioned previously, it may be desirable not to have to process the fusion product to obtain the protein in its desired state. A method has been described that relieves the requirement for post-translational processing of such a translation-enhancing amino terminal extension by creating a two-cistron expression system (Schoner *et al.*, 1986). The first cistron is designed to maximize translation initiation and has a termination codon and a new Shine–Dalgarno sequence engineered in its C-terminal sequences. The second cistron consists of the gene for the target protein, appropriately placed to utilize the SD site (SD2) in the C-terminus of the primary cistron. This sort of construction, shown in Fig. 5, allows for coupling of the translation of the downstream gene to that of the primary cistron without producing a polypeptide fusion. Such a system may also prove useful for the coordinated production of the subunits of a multi-subunit enzyme or structural protein.

B. Soluble and Insoluble Systems

In the course of expressing a protein in *E. coli* the final recombinant product may be found in a soluble or an insoluble form. Many bacterial proteins remain soluble even when expressed at high levels, but foreign proteins show a greater tendency to form insoluble, refractile aggregates known as inclusion bodies. These are typically associated with high levels of expression (15–50% of the cell mass) but can occur even when proteins are expressed at low levels (Schoemaker *et al.*, 1985). Several possible causes of aggregation include: (1) size, (2) the presence of large hydrophobic regions, (3) misfolding or instability of the folded product, (4) incorrect disulfide formation, and (5) high concentrations of the recombinant product. This section will discuss the rationale for attempting to produce a protein in a soluble or insoluble form, and describe ways to bias an expression system toward one or the other form.

TABLE III Commercially Available Fusion Expression/Purification Systems

Fusion partner	Site of fusion	Localization	Purification	Cleavage site	Company
FLAG peptide	N-term/C-term	Periplasm/media	Immuno-affinity	Enterokinase	IBI (Kodak)
Poly-His	N-term/C-term	Cytoplasm	Ni-affinity	Not provided	Qiagen
Poly-His	N-term	Cytoplasm	Ni-affinity	Enterokinase	Invitrogen
Glutathione-S-transferase	N-terminal	Cytoplasm	Glutathione Sepharose	Thrombin, factor Xa	Pharmacia
Protein A IgG binding domains	N-terminal	Cytoplasm	IgG sepharose	Not provided	Pharmacia
Z domain[a]	N-terminal	Media	IgG Sepharose	Not provided	Pharmacia
Maltose binding protein	N-terminal	Cytoplasm/periplasm	Amylose resin	Factor Xa	NEB

[a] Modeled on the IgG binding domain of protein A. (Nilsson et al., 1987).

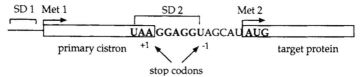

FIGURE 5   Diagram of a bicistronic construction. Two stop codons are present in different reading frames (+1 and −1) to prevent read-through translation. The sequence of SD2 has been maximized for complementarity to the 16S rRNA to increase reinitiation efficiency. The construction contains an *Nde*I site (5′-CAUAUG-3′) at the end of SD2 for cloning of target proteins.

1. Soluble Expression

The most obvious reason for preferring the soluble expression of a protein is to ensure that the physical characteristics of the final product represent authentic, native protein. Proteins that are purified from inclusion bodies and refolded *in vitro* may undergo chemical modifications as a result of the solubilization process. This process often involves the use of strong denaturants and extremes of pH. Use of such reagents may lead to artifacts such as modification of functional groups, thereby affecting activity or the homogeneity of a sample to be used for applications such as X-ray crystallography and mass spectrometry. Refolded proteins may represent nonnative conformations of the protein that are only partially active or may re-precipitate at a later date. Furthermore, the *in vitro* refolding of proteins is currently a very inefficient process, and may result in only a 1–10% recovery of active protein from insoluble material.

In terms of monitoring the presence of a protein by means of its activity, it is far more preferable to express soluble protein. One case where soluble protein is mandatory is the development of genetic selections to screen for mutant proteins with altered functional properties. Evnin *et al.* developed such a selection using the proteolytic activity of trypsin to provide arginine from a nonnutritive source to *E. coli* that had been made auxotrophic for that amino acid (Evnin *et al.*, 1990). Mutants that retained catalytic activity could be selected by their ability to grow on minimal media supplemented with the nonnutritional arginine source. Sufficient soluble protein must be present in the cell to generate the activity around which the selection is based. The requirement for soluble, and presumably active, protein also holds true for most of the fusion systems mentioned previously, where the success of an affinity purification rests on the ability of the fusion partner to interact with its ligand. For the most part this requires a native and soluble conformation. The development of a polyhistidine fusion partner that can interact with immobilized metal ions even when denatured, however, has removed this requirement in certain cases (Henco, 1991).

2. Insoluble Expression

The expression of proteins as insoluble inclusion bodies has a number of advantages. Since the recombinant protein tends to precipitate exclusive of other cellular proteins, the inclusion body itself represents a good source of relatively pure protein (~90%). Furthermore, these inclusion bodies are fairly easy to purify by differential centrifugation. In a situation where antibodies to the protein are desired, inclusion bodies can be rapidly purified and used as a primary immunogen. Since the production level is not limited by the intrinsic solubility of the protein in the cell, and as inclusion bodies can represent a very large fraction of the cell, it is possible to produce large amounts of protein with very little effort. Several recombinant polypeptides of therapeutic interest have been produced in this fashion (Marston, 1986).

Inclusion of a recombinant product is also a means of protecting the bacterium and the protein of interest from one another. Proteins that are toxic by virtue of activity or that are rapidly degraded by *E. coli* may be ultimately easier to work with when inactivated or sequestered from cytoplasmic proteases by precipitation. In situations where proteolysis is a major problem and the genetic mechanisms for avoiding proteolysis described earlier are not available, this may be one of the only means of protecting the protein from degradation. It should be noted, however, that even material obtained from inclusion bodies can contain proteolysis products (Schein and Noteborn, 1988), probably as a consequence of slow precipitation following denaturation.

3. Tilting the Balance

More often than not, a protein that is expressed in bacteria ends up in an inclusion body. The successful refolding of a protein is unpredictable and usually inefficient; hence it may be preferable to settle for a lower yield of soluble material. There are a few means of changing an insoluble system to a soluble one, although most of them have been identified empirically for individual proteins and may not be generally applicable.

The factors that dictate the solubility of a protein are not yet understood. Since aggregation is likely a function of intrinsic solubility, and the concentration of both mature and nascent (nonfolded) protein, it may be possible to shift toward soluble material by limiting the activity of the promoter. This is particularly true in the case of proteins targeted for secretion, where overexpression often leads to overloading of the secretion system, followed by aggregation. The use of titratable or weaker promoters is recommended in this case. Promoters that are de-repressed by the addition of chemical agents (*lac, tac*) are more amenable to titration of this sort, although the concentrations of inducer involved must be empirically obtained for each system being studied.

An additional genetic approach involves increasing the expression of proteins from the *groE* and *dnaK* operons, which serve a heat-shock-induced chaperonin function in *E. coli*. These proteins often increase the fraction of the total recombinant protein that is properly folded and soluble in the cytoplasm of the cell (Lee and Olins, 1992). They have also proven useful in maintaining proteins in secretion-competent forms prior to extracytoplasmic transport (Phillips and Silhavy, 1990).

The effect of temperature on the solubility of recombinant proteins can be fairly dramatic. For example, when diphtheria toxin is expressed cytoplasmically, up to 90% of the protein being produced is insoluble, an effect shown to be independent of the length of the protein or the host strain. Upon shifting from 37 to 30°C, however, up to 50% of the material could be recovered as a soluble form (Bishai *et al.*, 1987). It might be expected that large heterologous proteins would be more likely to require chaperonin functions to assist in folding, thus rendering them more prone to aggregation in *E. coli*. Despite this, more than 90% of glycogen phosphorylase (842 amino acids) can be recovered in a soluble form when produced at 22°C, whereas it is totally insoluble at 37°C (Browner *et al.*, 1991). These observations may represent a barrier to using promoter systems that require heat shock for activation.

Other, less rational methods may help in the attempt to bias the form of the protein expressed. Levels of inclusion body formation may be dramatically altered by switching host strain. In other cases, altering the conditions of growth from standard media (Luria broth) to rich or minimal media has a dramatic effect, perhaps by reducing the stress on the organism or changing the rate at which the protein is produced relative to its rate of folding. Because the factors that dictate these effects are not understood, these represent methods that must be tested empirically for each protein and expression system.

A final means of affecting the distribution of a protein between soluble and insoluble forms is to target it to different compartments in the host. This will be the topic of discussion in the next section.

The factors that contribute to protein insolubility, both in the producing organism and *in vitro*, are not clearly understood. Further research may eventually yield methods to increase the solubility of a recombinant product, but the solution is likely to be complex.

Fusion Strategies

Uhlén, M., and Moks, T. (1990) In *Methods in Enzymology* (D. V. Goeddel, Ed.), Vol. 185, p. 129. Academic Press, New York.

Solubility

Marston, F.A.O. (1986) The purification of eukaryotic polypeptides synthesized in *Escherichia coli*. *Biochem. J.* **240**, 1–12.
Schein, C. H. (1989) The production of soluble recombinant proteins in bacteria. *Bio/Technology* **7**, 1141–1148.

IV. SITES OF EXPRESSION

Many eukaryotic proteins are specifically targeted to cellular organelles, and it is reasonable to assume that they have been optimized to the environment in those compartments. Although *E. coli* is not nearly as compartmentalized as a eukaryotic cell, it does maintain regions with differing environments which, for the purposes of expression, can be considered separate compartments. These are schematized in Figure 6.

The various benefits of each of these compartments, and the means of getting a protein to a particular location, are an important consideration. *E. coli* is not a highly efficient host for secretion when compared to the Gram-positive bacteria *Bacillus* or *Streptomyces*. These organisms can efficiently secrete proteins into the extracellular space and have found a primary use in biotechnology for polypeptide secretion. Genetic analysis has helped to identify many of the pathways responsible for secretion, however, and may eventually allow for manipulation of the *E. coli* pathways to provide more efficient secretion. This section will discuss both the machineries that are currently available and others that, through engineering, may become more generally useful.

A. Cytoplasmic Expression

The cytoplasm represents the default location of recombinant proteins expressed without the signals required for export or secretion. The cytoplasm is a protein-rich, reducing environment that is probably a fair approx-

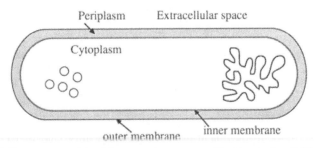

FIGURE 6   Diagram of the different environments and barriers in *E. coli*. The circles and irregular figure represent plasmid and genomic DNAs, respectively.

imation of the eukaryotic cytoplasmic space. Consequently it may represent the best site for expression of eukaryotic cytosolic proteins to maintain their native conformations or activities.

As a general rule, production of proteins is much more efficient in the cytoplasm and results in the accumulation of larger amounts of the recombinant protein. As described earlier, however, there is a tendency for the protein to aggregate and form inclusion bodies in many cases of overexpression.

Cytoplasmic expression in *E. coli* may affect the amino terminal sequence of a heterologous protein. The generation of a free (native) amino terminus is a major requirement for proteins whose biological activities are sensitive to the presence of the initiator methionine. The enzyme methionine aminopeptidase (MAP), which has been cloned and overproduced, carries out efficient removal of the methoinine when the second amino acid is small (Ala, Gly, Pro, Ser, Thr). Processing is variable, however, when the amino acid is larger or charged. When the recombinant protein is being produced at very high levels not every amino terminus may be processed by the limited amount of MAP normally present in the cell, leading to a heterogenous mix of processed and unprocessed material. The cloning of the gene and the development of MAP hyperproducing strains have been useful in overcoming this problem (Ben-Bassat *et al.*, 1987).

Cytoplasmic expression is not ideal for proteins that require the formation of disulfide bonds, since the cytoplasm is a reducing environment. In the event that the protein being produced is toxic to the cell and must be maintained in the native state, the only option available may be to remove it from the cytoplasmic compartment. There are two possible alternatives: periplasmic expression (export) or targeting of the protein to the extracellular space (secretion).

B. Periplasmic Expression (Export)

The first and simplest compartment a protein may be directed to is the periplasmic space. The periplasm is a considerably different environment than the cytoplasm; it has a higher oxidizing potential and a relatively small number of proteins present. As a result, the probability of correct disulfide formation increases, and fewer contaminating proteins are present during purification. It should be noted, however, that inclusion bodies can be formed in the periplasm as well as the cytoplasm (Georgiou *et al.*, 1986). Osmotic shock or lysozyme treatment can be used to release specifically the contents of the periplasm, resulting in a significant purification away from cytoplasmic proteins (Vasquez *et al.*, 1989). As mentioned previously, it is a good place to localize proteins that are toxic or subject to extensive proteolysis when cytoplasmically expressed. Although a "housekeeping protease" (DegP) has been identified that may present a threat to translocated proteins, mutants are available that do not compromise the viability of the host at 30 or 37°C (Strauch *et al.*, 1989).

Periplasmic signal peptidases (*lep* or *lsp*) process signal peptides to liberate mature forms of an exported protein. For instance, this processing event has been exploited in the periplasmic expression of trypsin. By fusing the *his* J signal peptide directly to mature trypsin, a native amino terminus could be generated without the addition of enteropeptidase, which is normally required for maturation of pro-forms of the enzyme (Higaki *et al.*, 1989). The signal peptidase activity may also be used to generate proteins with free amino termini, which may be important in the production of pharmaceutically active polypeptides. Furthermore, these processing enzymes display a little more flexibility in cleavage sites than the MAP.

Proteins are transported into the periplasm through the interaction of a signal peptide with the *sec/prl* machinery. This event is schematized in Fig. 7 as a single-step translocation event, although the actual mechanics are far more complex and not fully understood.

The signal sequences from bacterial proteins such as β-lactamase, phoA, and ompA are widely used for periplasmic targeting. It should be noted here that the amino terminus of the protein to which a signal peptide is fused can have a major impact on the efficiency with which the protein is secreted. For instance, the presence of a positively charged residue within three amino acids of the phoA signal sequence reduces export by a factor of 50 (Li *et al.*, 1988). In some cases the β-lactamase signal peptide alone is insufficient for efficient export and requires the fusion of the target gene to a larger amino terminal fragment (≥12 aa) of β-lactamase (Summers and Knowles, 1989). By comparison to β-lactamase, the ompA signal peptide appears to be much more efficient and has no requirement for an additional amino acid spacer before the target protein. The ompA peptide has proven very effective at exporting both prokaryotic and eukaryotic cytoplasmic proteins (Takahara *et al.*, 1988).

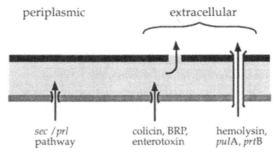

FIGURE 7   Schematic of the three major secretion pathways available in *E. coli*. The *sec/prl* pathway represents the periplasmic export system native to *E. coli*. The colicin and hemolysin pathways represent machinery that has been engineered into *E. coli* from plasmid genes (colicin, enterotoxin, BRP) or other organisms (hemolysin, *pulA*, *prtB*).

The road to developing a periplasmic expression system, however, is fraught with difficulties. The *sec/prl* pathway seems to be highly sensitive to size, folded state, and the presence of hydrophobic regions in the protein. Any of these factors may be rate-limiting for secretion. Attempts to optimize the export of a recombinant protein should consider all of these as possible problems. When expressing large, multisubunit proteins, the presence of most of these factors may confound the use of the *sec/prl* pathway. It should be stressed that fusing a signal peptide to a target gene is not guaranteed to result in efficient export, and the reasons behind a failure are unclear due to a lack of understanding about the machinery involved.

The classic example of export failure is β-galactosidase (*lacZ*), which was shown to cause blockage of the transport machinery (Emr *et al.*, 1980). This is invariably lethal to the cell, as it prevents the secretion of proteins that are required for normal function. The secretion machinery may also be overloaded by high levels of expression of proteins that are transported slowly, which results in the cytoplasmic accumulation of immature precursors. The presence of these precursors in the insoluble fraction, along with immature or partly processed polypeptides associated with the membrane, is usually a good indication of translocation block.

C. Extracellular Expression (Secretion)

The secretion of heterologous proteins into the media offers benefits similar to secretion into the periplasm, with the added incentive that it is more likely that large quantities of the protein may be produced without fear of exceeding its solubility. Secretion is also useful for production of proteins in continuous reactors, since the recombinant product can be recovered without any direct manipulations being performed on the bacteria producing it. Until recently, the requirements for a protein to be secreted by *E. coli* were poorly understood. The recent successes with genetic manipulation of secretion that will be mentioned have increased the probability that more proteins will be able to be secreted.

Evidence has accumulated to suggest that some proteins are secreted by hyperexpression into the periplasm, followed by leakage of the exported protein into the extracellular media. For example, PhoS directed to the periplasm by its own signal peptide was processed and released at a constant rate into the media following a lag phase in which high concentrations of PhoS accumulated in the periplasm (Pages *et al.*, 1987). The requirement for massive overproduction prior to secretion suggested that leakiness of the outer membrane was primarily responsible for the secretion event.

From this have come expression systems that take advantage of the release mechanisms of bacteriocin release protein (BRP) and the product of the ColE1 plasmid *kil* gene. The general mechanism is shown in Fig. 7 as a two-step secretion process. The BRP is a 28 residue polypeptide that is inserted into the outer membrane and activates phospholipase A_2 (deGraaf and Oudega, 1986). This causes release of periplasmic proteins into the media. By producing a low level of this protein in conjunction with a high-level recombinant expression plasmid, target proteins can be loaded and released fairly selectively from the periplasm (Hsiung *et al.*, 1989). The colicins and the product of the *kil* gene perform similar functions and have also been engineered into a secretion system for controlled release of peri-plasmic proteins (Kato *et al.*, 1987). It is interesting to note that cells can be stimulated to secrete periplasmic proteins by colicin expression in growth media containing 1% Triton, which appears to be sufficient to perturb the outer membrane (Pugsley and Schwartz, 1984).

Secretion mechanisms relying on outer membrane perturbation, how-ever, suffer in that they require the loading of the target protein into the periplasmic space prior to secretion. Additionally, unless the membrane-disrupting protein is kept under tight regulation, there is a high probability of cell lysis during long growth periods.

In the past 5 years a number of operons have been identified in other Gram-negative bacteria that are specifically involved in the secretion of a single gene product. The most well studied of these is the hemolysin operon (*hly A, B, C, D*) in which hemolysin (*hlyA*) is secreted by a transpor-ter/ATPase composed of a transmembrane protein (*hlyD*) and an inner-membrane-associated ATPase (*hlyB*). The details of this machinery, shown schematically on the far right in Fig. 7, were recently reviewed (Holland *et al.*, 1990a). The export signal has been localized to an 80–100 amino acid polypeptide at the C-terminus of *hlyA*, which appears to be both necessary and sufficient for recognition by the secretion system. To date at least three proteins have been successfully secreted by a system involving expression of fusions with the C-terminal domain of *hlyA* in *E. coli* containing the *hlyB* and -*D* genes (Holland *et al.*, 1990b).

Several other operons have been identified from *Klebsiella pneumoniae* (*pulA*) (d'Enfert *et al.*, 1987) and *Erwinia chrysanthemi* (*prtB*) (Delepelaire, 1990). The transport signals for these operons are very similar to that from *hlyA* and appear to function efficiently in an *hly*+ background. These oper-on secretion systems have the advantage that they appear to be able to secrete efficiently proteins as large as 140 kDa without killing the host cell. Consequently, they may serve as a means to secrete the larger multimeric or eukaryotic proteins that cannot be transported by the *sec/prl* pathways. They do not, however, remove the C-terminal transport signal and therefore require postproduction processing with a recombinant protease.

Secretion

Hirst, T. R., and Welch, R. A. (1988) Mechanisms for secretion of extracellular proteins by Gram-negative bacteria. *TIBS* 13, 265–269.

Wickner, W., Driessen, A. J., and Hartl, F. U. (1991) The enzymology of protein translocation across the *Escherichia coli* plasma membrane. *Annu. Rev. Biochem.* 60, 101–24.

V. CONCLUSION

By virtue of its routine use in molecular genetics, *E. coli* represents a readily available, inexpensive, and well-characterized organism for expression trials. Because *E. coli* may be manipulated easily and rapidly, we consider it the best system for performing the wide variety of experiments that may be necessary for optimizing the expression of a heterologous gene.

In terms of deciding which course to pursue in expressing a protein, a few basic guidelines should be followed. The intended use of the expressed protein should dictate the design of an expression system. For example, massive expression levels are not always desirable. They are frequently associated with the production of insoluble material that is useful for antibody production but not for protein crystallization trials or a genetic selection that requires active protein. Similarly, low levels of expression are not necessarily bad. An expression system that makes less protein, all of which is soluble and active, may still be useful if an efficient purification scheme is available and high density fermentation can be used to scale-up the amount of starting material.

As far as promoter selection is concerned, since most of the promoters mentioned in this review are roughly equal in strength we recommend choosing a promoter as a function of which strain and/or induction requirements are best for your purpose. For instance, you may choose a *tac* promoter over T7 to give greater flexibility in choice of host strains, or phoA over P_L to avoid heat shock responses after induction.

In our opinion, the manipulation of mRNA interactions with degredative and translational machinery will have the most significant impact on expression level. Given that each step in translation is affected by the one preceding it, attempts to maximize translation should proceed from the 5′ to 3′ end of the transcript. Thus, the most important concern is maximization of the Shine–Dalgarno sequence and SD–ATG spacing, followed in order of importance by elimination of stem loops involving the SD, maximization of A:T usage in the 5′ translated region, reduction of internal stem loops, and biasing codon usage for *E. coli*. If initial efforts along this pathway are unsuccessful, it is important to examine mRNA half-life and consider the use of 5′ or 3′ stem loops to stabilize mRNAs.

Many biotechnology support companies are developing and marketing vectors to address the rising need for broadly applicable *E. coli* expression systems. Most of these seek to optimize as many variables as possible in the areas of transcription and translation, as well as export and secretion signals. These plasmids serve as good starting points in developing an expression system. In most cases, however, some degree of fine tuning regarding the gene of interest is necessary. We hope that further development of vectors and the *E. coli* host itself, coupled with continued research on bacterial expression, will provide us with prokaryotic systems for the biosynthesis of any protein at reagent levels.

ACKNOWLEDGMENTS

We thank Drs. Michelle Browner, David Corey, David Sloane, and Jill Winter for comments and discussions regarding this manuscript. This work was supported by NIH Grant GM 39552 and National Science Foundation Grant DMB8904956.

REFERENCES

Andersson, S. G. E., and Kurland, C. G. (1990) *Microbiol. Rev.* **54**, 198–210.
Baker, T. A., Grossman, A. D., and Gross, C. A. (1984) *Proc. Natl. Acad. Sci. U.S.A.* **81**, 6779–6783.
Belasco, J. G., Nilsson, G., vonGabain, A., and Cohen, S. N. (1986) *Cell* **46**, 245–251.
Ben-Bassat, A., Bauer, K., Chang, S.-Y., Myambo, K., Boosman, A., and Chang, S. (1987) *J. Bacteriol.* **169**, 751–757.
Bishai, W. R., Rappuoli, R., and Murphy, J. R. (1987) *J. Bacteriol.* **169**, 5140–5151.
Borisova, G. P., Volkova, T. M., Berzin, V., Rosenthal, G., and Gren, E. J. (1979) *Nucleic Acids Res.* **6**, 1761–1774.
Brosius, J. (1988) In *Vectors: A Survey of Molecular Cloning Vectors and Their Uses* (R. Rodriguez and D. Denhardt, Eds.), Vol. 10, pp. 179–203. Butterworths, Boston.
Browner, M. F., Rasor, P., Tugendreich, S., and Fletterick, R. J. (1991) *Protein Engineering* **4**, 351–357.
Button, L. L., Reiner, N. E., and McMaster, R. W. (1991) *Mol. Biochem. Parasitol.* **44**, 213–224.
d'Enfert, C., Ryter, A., and Pugsley, A. P. (1987) *EMBO J.* **6**, 3531–3538.
deGraaf, F. K., and Oudega, B. (1986) *Curr. Topics Microbiol. Immunol.* **125**, 183–205.
Delepelaire, P. a. W., C. (1990) *J. Biol. Chem.* **265**, 17118–17125.
Devlin, P. E., Drummond, R. J., Toy, P., Mark, D. F., Watt, K. W. K., and Devlin, J. J. (1988) *Gene* **65**, 13–22.
Dunn, J. J., and Studier, F. W. (1983) *J. Mol. Biol.* **166**, 477–535.
Emory, S. A., and Belasco, J. G. (1990) *J. Bacteriol.* **172**, 4472–4481.
Emory, S. A., Bouvet, P., and Belasco, J. G. (1992) *Genes Dev.* **6**, 135–148.
Emr, S. D., Hedgpeth, J., Clément, J.-M., Silhavy, T. J., and Hofnung, M. (1980) *Nature (London)* **285**, 82–85.
Evnin, L. B., Vásquez, J. R., and Craik, C. S. (1990) *Proc. Natl. Acad. Sci. U.S.A.* **87**, 6659–6663.
Frorath, B., Abney, C., Berthold, H., Scanarini, M., and Northemann, W. (1992) *BioTechniques* **12**, 558–563.
Georgiou, G., Telford, J. N., Shuler, M. L., and Wilson, D. B. (1986) *Appl. Environ. Microbiol.* **52**, 1157–1161.
Goff, S. A., Casson, L. P., and Goldberg, A. L. (1984) *Proc. Natl. Acad. Sci. U.S.A.* **81**, 6647–6651.
Goff, S. A., and Goldberg, A. L. (1985) *Cell* **41**, 587–595.
Gorski, K., Roch, J.-M., and Krisch, H. M. (1985) *Cell* **43**, 461–469.

Hall, M. N., Gabay, J., Débarbouillé, and Schwartz, M. (1982) *Nature (London)* **285**, 616–618.

Hanahan, D. (1983) *J. Mol. Biol.* **166**, 557–580.

Hasan, N., and Szybalski, W. (1987) *Gene* **56**, 145–151.

Hawley, D. K., Johnson, A. D., and McClure, W. R. (1985) *J. Biol. Chem.* **260**, 8618–8626.

Hawley, D. K., and McClure, W. R. (1983) *Nucleic Acids Res.* **11**, 2237–2255.

Henco, K. (1991) *The QIAexpressionist.* DIAGEN GmbH, Düsseldorf.

Henner, D. J. (1990) In *Methods in Enzymology* (D. V. Goeddel, Ed.), Vol. 185, pp. 199–201. Academic Press, New York.

Higaki, J. N., Evnin, L. B., and Craik, C. S. (1989) *Biochemistry* **28**, 9256–9263.

Higgins, C. F., Ames, G. F.-L., Barnes, W. M., Clement, J. M., and Hofnung, M. (1982) *Nature (London)* **298**, 760–762.

Holland, I. B., Blight, M. A., and Kenny, B. (1990a) *J. Bioenerg. Biomembr.* **22**, 473–491.

Holland, I. B., Kenny, B., Steipe, B., and Plückthun, A. (1990b) In *Methods in Enzymology* (M. P. Deutscher, Eds.), Vol. 182, pp. 132–143. Academic Press, New York.

Hsiung, H. M., Cantrell, A., Luirink, J., Oudega, B., Veros, A. J., and Becker, G. W. (1989) *Bio/Technol.* **7**, 267–271.

Kato, C., Kobayashi, T., Kudo, T., Furusato, T., Murakami, Y., Tanaka, T., Baba, H., Oishi, T., Ohtsuka, E., and Ikehara, M. (1987) *Gene* **54**, 197–202.

Kennell, D., and Riezman, H. (1977) *J. Mol. Biol.* **114**, 1–21.

Kennell, D. E. (1986) In *Maximizing Gene Expression* (W. Reznikoff and L. Gold, Eds.), Vol. 9, pp. 101–142. Butterworths, Boston.

Konigsberg, W., and Godson, G. N. (1983) *Proc. Natl. Acad. Sci. U.S.A.* **80**, 687–691.

Kotula, L., and Curtis, P. J. (1991) *Bio/Technol.* **9**, 1386–1389.

Kubo, M., and Imanaka, T. (1989) *J. Bacteriol.* **171**, 4080–4082.

Lee, S. C., and Olins, P. O. (1992) *J. Biol. Chem.* **267**, 2849–2852.

Li, P., Beckwith, J., and Inouye, H. (1988) *Proc. Natl. Acad. Sci. U.S.A.* **85**, 7685–7689.

Löwenadler, B., Jansson, B., Paleus, S., Holmgren, E., Nilsson, B., Moks, T., Palm, G., Josephson, S., Philipson, L., and Uhlén, M. (1987) *Gene* **58**, 87–97.

Marston, F. A. O. (1986) *Biochem. J.* **240**, 1–12.

McCarthy, J. E. G., Sebald, W., Gross, G., and Lammers, R. (1986) *Gene* **41**, 201–206.

McGrath, M. E., Erpel, T., Browner, M. F., and Fletterick, R. J. (1991) *J. Mol. Biol.* **222**, 129–142.

Nassal, M., Mogi, T., Karnik, S. S., and Khorana, H. G. (1987) *J. Biol. Chem.* **262**, 9264–9270.

Neidhardt, F. C., VanBogelen, R. A., and Vaughn, V. (1984) *Annu. Rev. Genet.* **18**, 295–329.

Newbury, S. F., Smith, N. H., and Higgins, C. F. (1987) *Cell* **51**, 1131–1143.

Nilsson, B., Moks, T., Jansson, B., Abrahmsén, L., Elmblad, A., Holmgren, E., Henrichson, C., Jones, T. A., and Uhlen, M. (1987) *Prot. Engineering* **1**, 107–113.

Nilsson, G., Belasco, J. G., Cohen, S. N., and von Gabain, A. (1984) *Nature* **312**, 75–77.

Pages, J.-M., Anba, J., and Lazdunski, C. (1987) *J. Bacteriol.* **169**, 1386–1390.

Phillips, G. J., and Silhavy, T. J. (1990) *Nature (London)* **341**, 882–884.

Podhajska, A. J., Hasan, N., and Szybalski, W. (1985) *Gene* **40**, 163–168.

Pugsley, A. P., and Schwartz, M. (1984) *EMBO J.* **3**, 2393–2397.

Rangwala, S. H., Fuchs, R. L., Drahos, D. J., and Olins, P. (1991) *Bio/Technol.* **9**, 477–479.

Remaut, E., Marmenout, A., Simons, G., and Fiers, W. (1989) In *Recombinant DNA Methodology* (R. Wu, L. Grossman, and K. Moldave, Eds.), pp. 445–460. Academic Press, San Diego.

Reznikoff, W. S., and McClure, W. R. (1986) In *Maximizing Gene Expression* (W. Reznikoff and L. Gold, Eds.), Vol. 9, pp. 1–33. Butterworths, Boston.

Robinson, M., Lilley, R., Little, S., Emtage, J. S., Yarranton, G., Stephens, P., Millican, A., Eaton, M., and Humphreys, G. (1984) *Nucleic Acids Res.* **12**, 6663–6671.

Saito, Y., Ishii, Y., Koyama, S., Tsuji, K., Yamada, H., Terai, T., Kobayashi, M., Ono, T., Niwa, M., and Ueda, I. (1987) *J. Biochem.* **102**, 111–122.

Schein, C. H., and Noteborn, M. H. M. (1988) *Bio/Technol.* **6**, 291–294.
Schoemaker, J. M., Brasnett, A. H., and Marston, F. A. O. (1985) *EMBO J.* **4**, 775–780.
Schoner, B. E., Belagaje, R. M., and Schoner, R. G. (1986) *Proc. Natl. Acad. Sci. U.S.A.* **83**, 8506–8510.
Shepard, H. M., Yelverton, E., and Goeddel, D. V. (1982) *DNA* **1**, 125–131.
Shine, J., and Dalgarno, L. (1974) *Proc. Natl. Acad. Sci. U.S.A.* **71**, 1342–1346.
Simon, L. D., Tomczak, K., and St. John, A. C. (1978) *Nature (London)* **275**, 424–428.
Singer, B. S., and Gold, L. (1991) *Gene* **106**, 1–6.
Stormo, G. D., Schneider, T. D., and Gold, L. M. (1982) *Nucleic Acids Res.* **10**, 2971–2996.
Straney, S. B., and Crothers, D. M. (1987) *Biochemistry* **26**, 5063–5070.
Strauch, K. L., Johnson, K., and Bechwith, J. (1989). *J. Bacteriol.* **171**, 2689–2696.
Studier, F. W., Rosenberg, A. H., Dunn, J. J., and Dubendorff, J. W. (1990) In *Methods in Enzymology* (D. V. Goeddel, Ed.), Vol. 185, pp. 60–89. Academic Press, New York.
Summers, R. G., and Knowles, J. R. (1989) *J. Biol. Chem.* **264**, 20074–20081.
Takahara, M., Sagai, H., Inouye, S., and Inouye, M. (1988) *Bio/Technol.* **6**, 195–198.
Taniguchi, T., and Weissmann, C. (1978) *J. Mol. Biol.* **118**, 533–565.
Thanaraj, T. A., and Pandit, M. W. (1989) *Nucleic Acids Res.* **17**, 2973–2985.
Uhlin, B. E., Molin, S., Gustafsson, P., and Nordström, K. (1979) *Gene* **6**, 91–106.
Varenne, S., Knibiehler, M., Cavard, D., Morlon, J., and Lazdunski, C. (1982) *J. Mol. Biol.* **159**, 57–70.
Vásquez, J. R., Evnin, L. B., Higaki, J. N., and Craik, C. S. (1989) *J. Cell. Biochem.* **39**, 265–276.
Wanner, B. L. (1986) *J. Bacteriol.* **168**, 1366–1371.
Williams, D. P., Regier, D., Akiyoshi, D., Genbauffe, F., and Murphy, J. R. (1988) *Nucleic Acids. Res.* **16**, 10453–10467.
Yamada, M., Kubo, M., Miyake, T., Sakaguchi, R., Higo, Y., and Imanaka, T. (1991) *Gene* **99**, 109–14.

4

Methods for Expressing Recombinant Proteins in Yeast

VIVIAN L. MACKAY*

THOMAS KELLEHER†

*ZymoGenetics, Inc., 1201 Eastlake Ave. E, Seattle, Washington 98102
and †ZymoGenetics, Inc., 4225 Roosevelt Way NE, Seattle, Washington
98105

I. INTRODUCTION

A. Choice of Expression Systems

The production of recombinant heterologous proteins is generally undertaken to achieve one of several possible goals: (1) expression of a cellular protein in a suitable biological system to assess its function in cell biology; (2) bench-scale production of a protein of research interest; and (3) the large-scale production and purification of therapeutic or industrial proteins for commercial purposes. Our objective in this chapter is to enable academic and industry investigators who have little or no previous experience with yeast to express successfully their favorite recombinant protein in this organism. Therefore, we will focus on the first two goals, since large-scale production is often a further empirical development of bench-scale results, conducted by scientists and engineers experienced in scale-up issues. We will attempt to describe briefly the basic components and factors that facilitate or influence yeast expression; more detail for each of them can be found in the references cited and in a recent thorough review of yeast expression by Romanos *et al.* (1992).

The choice of expression systems for any recombinant protein is dictated by the experimental goals and the characteristics of the protein (often a mammalian protein). Although bacteria (mainly *Escherichia coli*) have a strong track record, these systems are much less successful for proteins with disulfide bonds, usually requiring cumbersome denaturation–renaturation cycles to obtain product with proper conformation (Kohno *et al.*, 1990). Moreover, bacterial cell biology is sufficiently different from mammalian biology that functional questions are rarely addressed. Conversely, complex proteins can usually be readily expressed in native conformation in mammalian cells, but bench-scale production can be more expensive and time-consuming than in microbial systems. The recombinant protein can also be deleterious to mammalian cell growth or alter its physiology, and its functional role may be difficult to assess in the mammalian cell background. Therefore, expression in lower eukaryotes, i.e., yeast and filamentous fungi, provides alternative strategies to circumvent these difficulties.

Although many yeast and fungal species have been employed for production of heterologous proteins, the technology developed for the first one, *Saccharomyces cerevisiae* (bakers' yeast), is the most diverse and the principles apply to the other systems. In addition, the extensive genetic, molecular, and cell biological characterization of this yeast has led to its frequent use in recent years as a model eukaryote to understand the function and properties of many mammalian proteins. For these reasons (and for brevity), in this chapter we will discuss only expression in *S. cerevisiae*. For information about expression of recombinant proteins in other yeasts, the reader is directed to recent extensive reviews by Buckholz and Gleeson (1991), Tschopp and Cregg (1991), and Romanos *et al.* (1992).

B. Use of *Saccharomyces cerevisiae* for Protein Production

1. Advantages as an Expression System (Why Yeast?)

Hundreds of recombinant proteins from other yeasts, fungi, higher eukaryotes, mammals, and humans have now been expressed in *S. cerevisiae;* Table I demonstrates the wide variety of active, functional human proteins made at high enough levels to permit purification and characterization of the product. Protein expression in yeast can be accomplished by one of two general approaches. Proteins lacking disulfide bonds, such as α-1-antitrypsin, superoxide dismutase, lipocortin V, or Factor XIII (see Table I), are generally expressed cytoplasmically, often at high levels ($\geq$ 20% of soluble protein or grams of product per liter of culture). The development of inclusion bodies is rare (but see Cousens *et al.*, 1987). In contrast, proteins that require disulfide bond formation to attain native, active conformation are usually expressed with secretion leaders that direct them through the secretory pathway (which is quite similar to the secretory system of mammalian cells; Schekman, 1985; Cleves and Bankaitis, 1992; Bennett and Scheller,

TABLE I   Partial List of Human Proteins Produced
in *Saccharomyces cerevisiae* at High Levels[a,b]

Protein (molecular weight)	Reference/source
Cytoplasmic expression	
Cu, Zn superoxide dismutase (15.7 kDa)	Hallewell *et al.*, 1987, 1991
γ-Interferon (16.8 kDa)	Derynck *et al.*, 1983; Fieschko *et al.*, 1987
Fibroblast growth factor (18 kDa)	Barr *et al.*, 1988; G. McKnight, pers. comm.
Hepatitis B surface antigen (22 kDa)	Hitzeman *et al.*, 1983a; Miyanohara *et al.*, 1983; McAleer *et al.*, 1984; Bitter and Egan, 1984; Bitter *et al.*, 1988; Jacobs *et al.*, 1989
Hepatitis B core antigen (22 kDa)	Kniskern *et al.*, 1986
Lipocortin V (35 kDa)	M. Irani, pers. comm.
Gα,1 subunit (41 kDa)	T. Jones and V. L. MacKay, unpublished
α-1-Antitrypsin (44.7 kDa)	Cabezón *et al.*, 1984; Casolaro *et al.*, 1987
Platelet-derived endothelial cell growth factor (45 kDa)	Finnis *et al.*, 1992
Hemoglobin (62 kDa)	Wagenbach *et al.*, 1991; Coghlan *et al.*, 1992; Ogden *et al.*, 1991, 1992
cAMP phosphodiesterase isozyme IV (77 kDa)	McHale *et al.*, 1991
Coagulation factor XIIIa (83 kDa)	Bishop *et al.*, 1990
HIV-1 reverse transcriptase (117 kDa)	Bathurst *et al.*, 1990
Secretory expression	
Epidermal growth factor (5.5 kDa)	Brake *et al.*, 1984; George-Nascimento *et al.*, 1988
Insulin precursors (6 kDa)	Thim *et al.*, 1986
Insulin-like growth factor I (7.5 kDa)	Bayne *et al.*, 1988; Steube *et al.*, 1991
Parathyroid hormone (9.4 kDa)	Gabrielsen *et al.*, 1990
Granulocyte-macrophage colony-stimulating factor (14 kDa)	Miyajima *et al.*, 1986; Ernst, 1988
Lysozyme (14.7 kDa)	Jigami *et al.*, 1986; Taniyama *et al.*, 1988; Ichikawa *et al.*, 1989
Interleukin-1α, β (17 kDa)	Baldari *et al.*, 1987; Ernst, 1988; Livi *et al.*, 1990
α-Interferon (20 kDa)	Hitzeman *et al.*, 1983b; Singh *et al.*, 1984; Bitter *et al.*, 1984; Mellor *et al.*, 1985; Chang *et al.*, 1986; Zsebo *et al.*, 1986
Growth hormone (22 kDa)	Hitzeman *et al.*, 1984; Tokunaga *et al.*, 1985; Hiramatsu *et al.*, 1990
Interleukin-6 (22 kDa)	Guisez *et al.*, 1991
Erythropoietin (24 kDa)	Elliott *et al.*, 1989
Platelet-derived growth factor (30 kDa)	Kelly *et al.*, 1985; Östman *et al.*, 1989
Thrombin zymogens (36–69 kDa)	H. Han and V. L. MacKay, unpublished
Single-chain urokinase (47 kDa)	Melnick *et al.*, 1990
β1–4 Galactosyltransferase (48 kDa)	Krezdorn *et al.*, 1993
Chimeric L6 antibody Fab (48 kDa)	Horwitz *et al.*, 1988; Better and Horwitz, 1989

(Continues)

TABLE I *(Continued)*

Protein (molecular weight)	Reference/source
Microsomal epoxide hydrolase (52.9 kDa)	Eugster *et al.*, 1991
Serum albumin (65 kDa)	Chisholm *et al.*, 1990; Hitzeman *et al.*, 1990; Kingsman *et al.*, 1990; Sleep *et al.*, 1990
Chimeric L6 antibody (150 kDa)	Horwitz *et al.*, 1988; Better and Horwitz, 1989
Epstein–Barr virus major envelope glycoprotein (350 kDa)	Schultz *et al.*, 1987

[a] Proteins listed are those expressed at high enough levels (approximately 1 mg/liter or higher) for protein purification and/or characterization.
[b] Our apologies to investigators whose work was not included.

1993), and often into the extracellular medium (Table I). *S. cerevisiae* is a GRAS (Generally Regarded As Safe) species that is as easily manipulated as *E. coli* (although yeast smells better) and that can be grown in high yield, low cost fermentations. Because of the decades of classical and molecular genetic research with *S. cerevisiae,* suitable host strains exist (or can be easily generated) bearing alterations that enhance protein production by eliminating major proteases, by increasing transcription, expression, or secretion levels, or by modifying post-translational events, such as hyperglycosylation. Similarly, a variety of plasmids, promoters, secretion leaders, etc., have been described and are generally available from the large number of investigators who usually share materials and information freely. In most studies, autonomously replicating plasmids are used, thereby avoiding the problems of integration site effects on expression and of complex multiple integrations that complicate mammalian and fungal transfection/transformation. These materials will be described in more detail in later sections.

2. Disadvantages as an Expression System (Why Not Yeast?)

Many proteins of interest for production in yeast contain multiple disulfide bonds that generally do not form properly in the reducing environment of the cytoplasm. These require expression via the secretion pathway to attain native conformation. However, the success of secretion is unpredictable; some proteins are efficiently exported to the medium (see Table I), whereas others are detected only at low levels (< mg/liter levels). Either the polypeptide is poorly expressed or the majority (50–90%) of the secreted recombinant protein remains cell-associated, particularly at high expression levels (utilizing strong promoters, high copy number vectors, and complex growth media). (Throughout this chapter, we will distinguish between "secretion" and "export." For example, yeast invertase is rapidly and efficiently secreted to the periplasmic space, but remains there as a cell-associated protein; i.e., it is not exported to the medium.) Although, in general, small nonglycosylated proteins are exported fairly efficiently by yeast whereas larger glycosylated ones are not, there are enough exceptions to this pattern to indicate that protein conformation and/or hydrophobicity is probably a major factor and therefore potential success cannot be predicted. A well-known failure at many companies was the production of human tissue plasminogen activator (tPA) in yeast. The two Kringle domains in tPA were shown to be at least partly responsible for the secretion/export block (B. L. A. Carter, pers. comm.), but other homologous serine proteases with Kringle domains have been successfully exported (Zaworski *et al.*, 1989; Melnick *et al.*, 1990; H. Han and V. L. MacKay, unpublished).

A second major disadvantage of yeast expression is the absence of some complex mammalian post-translational modifications, such as the γ-carboxylation and hydroxymethylation that are characteristic of many blood proteins or the synthesis of complex oligosaccharides. Although yeast does have a glycosylation system that recognizes asparagine-linked glycosylation sites in mammalian proteins with fidelity, the core oligosaccharide (identical in yeast and mammalian cells) that is transferred to the polypeptide in the endoplasmic reticulum is usually modified and extended in the yeast Golgi to become a large, heterogeneous polymer that is highly immunogenic (Ballou, 1982, 1990) and that can interfere with protein activity, purification, and characterization. In addition, some secreted recombinant proteins may receive a low level of serine/threonine-linked glycosylation (O-linked; Settineri *et al.*, 1990) that is not characteristic of the native molecule. (See Section II,B,5 for strategies to minimize hyperglycosylation.) As noted by Romanos *et al.* (1992), *S. cerevisiae* can, however, carry out many other eukaryotic post-translational modifications of its own or recombinant proteins correctly, including amino-terminal acetylation, phosphorylation, myristylation, and isoprenylation.

Although a few yeast genes have introns, it is necessary to use cDNAs (or genes without introns) for heterologous expression of mammalian proteins, since yeast uses different signals for RNA processing than mammalian cells. Since cDNAs are generally more available than genes, this requirement presents no serious limitations.

Finally, there are usually a few minor technical obstacles to overcome before undertaking a yeast expression project. Most of the useful biological materials are not commercially available and must be obtained from the individual investigators; some new small equipment (25–30°C incubator and shaker), media, and supplies are generally required. Either an effective collaboration or some understanding of the genetics, molecular biology, and/or cellular biology of yeast is also usually necessary, although the recent publication of methods volumes may obviate this limitation for basic expression efforts (Wu *et al.*, 1983; Hinnen *et al.*, 1989; Goeddel, 1990; Guthrie and Fink, 1991); many of the references cited in this chapter are included in these volumes. Finally, anyone using yeast must be prepared for the occasional false accusation that this laboratory yeast is the source of tissue culture contamination!

C. Use of *Saccharomyces cerevisiae* for Cell Biology Research

Of major interest recently is the use of *S. cerevisiae* for eukaryotic cell biology research; results from the past 5 to 10 years have demonstrated numerous parallels between yeast and mammalian cells, e.g., the cell division cycle (including oncogenes, kinases, and phosphatases), signal transduction and G proteins, the secretory pathway, multiple drug resistance, metabolic enzymes, transcription factors, and others. The available yeast mutants defective in specific steps of these pathways provide suitable hosts to ask whether or how expression of cDNAs for specific mammalian proteins affects the yeast cell. The goal of these experiments is not the production of large quantities of protein for characterization but the understanding of the cellular function for the protein of interest or the identification of other interacting proteins. Therefore, choices of the promoter (constitutive or regulated), the plasmid system (single-copy vs multi- or high-copy vector), and characteristics of the host strain must be considered. In particular, the host strain should contain appropriate regulatory and "reporter" genes to promote and detect functional expression.

II. MATERIALS AND METHODOLOGY FOR PROTEIN EXPRESSION IN *SACCHAROMYCES CEREVISIAE*

High-level expression can be achieved by optimizing at three levels: plasmid components including the vector, transcription elements, and secretion leaders; host strains with appropriate mutations and alterations; culture conditions and media for optimizing protein production. These will be discussed individually in the following sections. For an excellent introduction to the characteristics of *S. cerevisiae* and its genetics, growth, and medium requirements, see "Getting Started with Yeast" by Sherman (1991).

A. Plasmid Elements

1. Yeast Vectors

S. cerevisiae vectors are available that provide a range of copy number from one to >200/cell (Table II and text below; see also Schneider and Guarente, 1991). For cytosolically produced proteins, expression level is generally correlated with copy number: the higher the copy number of the plasmid (vector + expression cassette), the higher the product level. How-

TABLE II   Examples of Vectors for Yeast Expression

Plasmid type	Copy number	Selectable marker[a]	Comments
Integrative (YIp)	One	URA3, LEU2	Stable integration into chromosome
Multi-integrative	One–five	URA3, LEU2	Stable integration into chromosome
Centromere (YCp)	One–two	TRP1, URA3, LEU2, HIS3	Stable autonomously replicating from ARS element or 2-μm origin
ARS (YRp)	Moderate	TRP1	Autonomously replicating but unstable
2 μm (YEp)	Moderate	LEU2, URA3, HIS3	Autonomously replicating but fairly stable
2 μm (YEp)	High	LEU2-d, POT	Autonomously replicating, stable

[a] Examples of the common selectable markers available for each vector type; other combinations of selectable markers and vector type have been developed in individual laboratories.

ever, for some secreted proteins, maximal expression via the secretory pathway may not be achieved with the highest copy number plasmid (or the strongest promoter; see Ernst, 1986, and Section II,A,2). For example, for both tissue plasminogen activator and porcine urokinase, cells transformed with moderate copy number plasmids (i.e., YEp13) grew better and produced higher product levels than those transformed with a higher copy number plasmid (POT) (V. L. MacKay, unpublished). Similarly, Elliott *et al.* (1989) reported that levels of exported erythropoietin expressed from centromere plasmids (1–2/cell) was higher with a stronger constitutive promoter. The use of a strong constitutive promoter on a multicopy 2-μm plasmid, however, did not result in higher exported levels, but instead led to slower cell growth and a reduction in the number of plasmid-containing cells and average copy number per cell. Gabrielsen *et al.* (1990) have also reported a lack of correlation between plasmid copy number and levels of exported human parathyroid hormone.

All yeast vectors used for protein expression include a bacterial origin of replication and an antibiotic-resistance gene (e.g., resistance to ampicillin or tetracycline) for selection of *E. coli* transformants to facilitate plasmid construction or preparation. As shown in Table II, selection of yeast transformants is nearly always achieved by complementation of an auxotrophic or nutritional mutation in the host strain with a plasmid gene (although successes with selection for dominant resistance to G418, hygromycin B, tunicamycin, and copper have been reported) (Webster and Dickson, 1983; Gritz and Davies, 1983; Rine *et al.*, 1983; Kaster *et al.*, 1984; Henderson *et al.*, 1985). It is therefore normally essential to match the plasmid with a host strain bearing the appropriate mutation to permit selection (see Section II,B,1 below). For example, *leu2* mutants cannot grow on synthetic medium lacking leucine, but *leu2* cells that have been transformed with a plasmid bearing the wild-type *LEU2* gene can be selected on -leucine medium. The *URA3* gene presents a particular advantage in that transformants containing the gene can be selected on medium lacking uracil and cells that have lost the plasmid or have acquired a mutation or alteration (e.g., a disruption or deletion) of the *URA3* gene can be counter-selected on medium containing the toxic inhibitor 5-fluoro-orotic acid (Boeke *et al.*, 1984). Similarly, α-aminoadipic acid can be used as a counter-selection for *lys2* mutants (Chattoo *et al.*, 1979), although few vectors carry the *LYS2* gene. These selection/counter-selection strategies are especially beneficial in developing host strains with specific alterations.

Chromosomal integration through targeted, site-specific homologous recombination is used to achieve stable modifications of the host strain, such as the following examples: the disruption of a deleterious yeast gene or the creation of null or specific mutations into a gene of interest; the generation of fusion proteins, e.g., for "reporter" genes; the introduction of a second or more highly expressed copy of a chromosomal gene; the insertion of a mammalian cDNA to be expressed from a yeast promoter. A number of strategies have been developed to facilitate such targeted integration events and several excellent methods chapters have been devoted to the design and use of integration plasmids and linear fragments (Orr-Weaver *et al.*, 1983; Rothstein, 1983, 1991; Winston *et al.*, 1983; Stearns *et al.*, 1990).

Multiple integration strategies have also been occasionally used to generate high level production strains for commercial products (Smith *et al.*, 1985; Moir and Davidow, 1991); after the integrations have been selected, it is no longer necessary to maintain selection and the transformants can be grown in any medium that maximizes expression. Integration vectors have been developed that facilitate site-specific integration into the 140 tandem repeat copies at the rDNA locus (Lopes *et al.*, 1989, 1991) or into the δ sequences (Shuster *et al.*, 1990), which are frequently present at 100 copies per haploid genome, existing either independently or as part of a *Ty* element (a retroviral-like transposon found in many yeast strains) (Boeke and Sandmeyer, 1991). Integration vectors based on *Ty* elements have also been used to generate strains with multiple copies of the integrated expression cassette (Jacobs *et al.*, 1989). Because the construction of strains with multiple integrations of an expression cassette can be laborious, it is expedient to optimize other expression components, such as transcription elements and secretion leaders, and characteristics of the host strain before undertaking the multiple integration approach. For bench-scale expression, use of autonomous plasmids is less cumbersome and time-consuming.

Most expression projects employ autonomously replicating vectors (see Table II), in which a chromosomal *ARS* (autonomously replicating sequence) element or the origin of replication from the endogenous *S. cerevisiae* 2-μm plasmid promote replication. The *ARS* (YRp) vectors are maintained at a high copy number per cell but are "unstable" due to inefficient partitioning into daughter cells (loss rate of 1–10% per cell per generation; Broach and Volkert, 1991). This instability is largely overcome by the addition of a chromosomal CEN (centromere) fragment that, however, reduces the copy number to 1–2/cell. Most investigators employ more stable moderate copy number vectors that are derived from the endogenous S. cerevisiae 2-μm plasmid (Rose and Broach, 1990). (Nearly all S. cerevisiae strains have the endogenous plasmid, i.e., are [cir+]. See Broach and Volkert, 1991, for a review of the 2-μm and other naturally occurring yeast plasmids.) For replication in yeast, these recombinant YEp vectors utilize the 2-μm origin and cis-acting REP3 locus that is needed for partitioning in cell division; most of the vectors (e.g., YEp13 or YEp24; Broach, 1983) lack the trans-acting REP1, REP2, and FLP1 functions of the endogenous 2-μm plasmid and therefore require the presence of the endogenous plasmid for stable maintenance and partitioning. Although the YEp-type plasmids can be stably maintained at 30 copies/cell (Broach, 1983), we typically find a somewhat lower copy number (10–20) that can decrease to 2–3/cell, depending on the heterologous protein being expressed (H. Kato and V. L. MacKay, unpublished).

Higher copy number, 50–400/cell, of these vectors can be achieved by using partially defective selectable markers (although as above the copy number generally decreases with expression of a heterologous product). For example, the leu2-d allele (Beggs, 1981) has an intact coding sequence but a truncated promoter (Erhart and Hollenberg, 1983); therefore, high copy number of the leu2-d gene is required to generate enough gene product (β-isopropylmalate dehydrogenase) for the transformants to grow in the absence of leucine. (We have observed that, at least with our host strains, selection of Leu+ transformants with plasmids bearing the leu2-d allele cannot be done at 30°C; the plates must be incubated at a lower temperature [25°C]. Perhaps the lower growth rate of the cells at 25°C allows the establishment of the high copy number needed for growth in the absence of leucine. Once colonies have developed, the transformants can then be grown and maintained at 30°C). If the vector contains all of the 2-μm functions and the leu2-d gene, it can be transformed into leu2 mutants that have been cured of the endogenous 2-μm plasmid. In these strains, copy numbers of 100–400/cell have been observed (Rose and Broach, 1990; Price et al., 1990; Schneider and Guarente, 1991).

An elegant use of a partially defective gene was developed by Glenn Kawasaki at ZymoGenetics in the generation of the POT vectors (Thim et al., 1986; Egel-Mitani et al., 1988; MacKay et al., 1990; Bishop et al., 1990). S. cerevisiae strains with mutations in the TPI1 gene (encoding triose phosphate isomerase) grow very poorly (if at all) with glucose, sucrose, or other fermentable carbon sources, but can be maintained on glycerol + ethanol (with some galactose added for cell wall synthesis). This defect can be complemented by the POT gene (encoding the triose phosphate isomerase from the unrelated yeast *Schizosaccharomyces pombe*), the promoter of which is poorly utilized by *S. cerevisiae*. Thus, transformed cells with high copy number (50–60/cell of the POT vector without insert) are continuously selected. Because growth on glucose (or sucrose) is used for selection, the investigator or production engineer is not limited to the synthetic, chemically defined growth medium that lacks leucine or uracil, but has greater flexibility for the development of optimal growth media. The vectors also contain the *leu2-d* gene, so that cells with even higher copy number of the plasmid can be selected. This plasmid system is extremely stable, even after weeks in continuous culture, and is being used for the commercial production of a secreted human insulin precursor, platelet-derived growth factor, coagulation factor XIII, and others.

Although high copy number plasmids have obvious advantages for protein production, they can also be useful in expression efforts directed toward cell biology studies. For example, using strong promoters, we have recently expressed mammalian cDNAs encoding subunits from heterotrimeric G proteins involved in signal transduction, specifically the transducin β_1 and $G\alpha_i 1$ subunits. For the β_1 subunit, expression from YEp13 (moderate copy number) gave no detectable phenotype or polypeptide visible on Western blots, while expression on a POT plasmid (higher copy number) gave detectable polypeptide and even higher copy number (through the *leu2-d* allele on the same plasmid) led to phenotypic effects, i.e., growth inhibition (V. L. MacKay, unpublished). Similarly, expression of the $G\alpha_i 1$ subunit on a POT plasmid gave detectable polypeptide and a complex phenotype, whereas YEp13 expression was apparently too low (V. L. MacKay and T. Jones, unpublished). Why use of a high copy vector coupled with a strong promoter is required for detectable expression of these proteins is unclear.

2. Promoters

a. Constitutively transcribed promoters. The most straightforward approach for high-level protein production is to select the strongest "constitutive" promoter available. For yeast, these are the promoters from genes for the abundant glycolytic enzymes, such as glyceraldehyde-3-phosphate dehydrogenase (*TDH3;* Holland and Holland, 1980; Rosenberg *et al.*, 1990), triose phosphate isomerase (*TPI1;* Alber and Kawasaki, 1982), or phosphoglycerate kinase (*PGK1;* Hitzeman *et al.*, 1982; Kingsman *et al.*, 1990). Although these promoters are not strictly constitutive since transcription from them does decrease somewhat as cultures shift from early glycolytic growth to late-phase gluconeogenic growth, this variation is much less than that seen with regulated promoters. Most recombinant proteins are produced using one of these workhorses. Moderately strong constitutive promoters include the *ADH1* (previously denoted *ADC1*) promoter from the gene encoding alcohol dehydrogenase isozyme I, which is available with several useful modifications at its 3' end (Ammerer, 1983), and the promoter from the *MFα1* gene (encoding the mating pheromone α-factor) that is constitutive in *MATα* cells and nonfunctional in *MATa* cells and a/α diploids (Kurjan and Herskowitz, 1982). However, at least for some secreted heterologous proteins, e.g., somatomedin-C (Ernst, 1986) and porcine urokinase (Sledziewski *et al.*, 1988), higher product levels were obtained with weaker promoters, suggesting that efficient secretion of these recombinant proteins is probably hindered by high precursor levels entering the secretory pathway.

b. Regulated promoters. Use of a regulated promoter is indicated when either the product itself or its synthesis and/or secretion is deleterious to the cells or acceptable growth rates or, in cell biology studies, when one wants to investigate the impact on cellular physiology of a rapid, high level induction of the recombinant protein. The glucose-repressed strong promoter from the *ADH2* gene (encoding alcohol dehydrogenase isozyme II) has been used for the production of several human therapeutic proteins (Price *et al.*, 1990); the promoter is tightly repressed during growth on glucose, but after glucose depletion (<0.5% w/v) it is activated by the Adr1 regulatory protein and drives high level protein production. In shake flask cultures with medium containing 1% yeast extract, 2% peptone, 1% glucose, and adenine and uracil supplements, the glucose was depleted by 8–12 h after inoculation (Price *et al.*, 1990). In controllable fermentors, the promoter can be made constitutive by careful adjustment of the glucose feed rate. A mutant form of the promoter, *ADH2-4c*, is partially constitutive in glucose-containing medium and shows higher levels of derepression than the wild-type *ADH2* promoter after exhaustion of glucose (Russell *et al.*, 1983).

We have superimposed temperature regulation onto both the constitutive *TPI1* promoter and the glucose-repressible *ADH2* promoter (Sledziewski *et al.*, 1988, 1990) by inserting mating-type regulatory elements into the 5′ noncoding regions of these promoters that will cause them to be repressed by the *MATα2* gene product. Plasmids with these promoters are then transformed into *sir3-8* mutant host strains in which synthesis of the Matα2 protein occurs at 35°C, but is blocked at 25°C. Genes/cDNAs under the control of the hybrid, temperature-regulated promoters are therefore expressed at 25°C, very poorly at 35°C, and partially at intermediate temperatures, apparently because of a low level expression of the Matα2 protein in each cell. Thus, the promoter is subject to a "rheostat" type control in which promoter strength can be adjusted by temperature. For porcine urokinase, secreted activity was actually highest at 35°C where promoter strength is lowest (C. Yip and V. L. MacKay, unpublished). Temperature regulation of these hybrid promoters has not yet been tested in fermentors.

Another hybrid regulated promoter has been recently described (Picard *et al.*, 1990; Schena *et al.*, 1991) in which three 26-base pair mammalian glucocorticoid response elements (GREs) were inserted in the *CYC1* promoter joined to a reporter gene. Cells co-transformed with this plasmid and a second one for expression of the glucocorticoid receptor exhibited 50- to 100-fold increased expression of the reporter gene after induction with deoxycorticosterone with no apparent effects on endogenous gene expression. Stable chromosomal integration of the expression cassette for the glucocorticoid receptor might improve the utility of the system. Although this regulated promoter system may not be feasible for large-scale production, it could be quite useful for cell biology research.

Regulated promoters that have been used extensively, particularly for bench-scale protein production and cell biology research, are those of the galactose-utilization pathway, particularly the *GAL1* and *GAL10* promoters, which are both glucose-repressible and galactose-inducible (Johnston, 1987; Schneider and Guarente, 1991). Although these promoters are not as strong as the glycolytic and *ADH2* promoters, the *GAL1* or *GAL10* promoter is induced 1000-fold by galactose in the absence of glucose. (The regulatory region from these promoters can be introduced into the strong constitutive glycolytic promoters, conferring galactose regulation on them) (Bitter and Egan, 1988; Wahleithner *et al.*, 1991; Demolder *et al.*, 1992). Thus, cultures can be grown to glucose depletion and induced by addition of galactose (or harvested and resuspended in glucose-free galactose medium) and rapidly induced for the gene/cDNA under control of the *GAL* promoter. Transformants harboring expression plasmids for toxic or growth-inhibitory proteins (e.g., those that interact with the cell cycle or signal transduction) can be maintained on glucose. Similarly, the function of essential gene products can be investigated by constructing strains containing the essential gene under the control of the *GAL* promoter. Cells are maintained on galactose and growth on glucose is used to produce defective mutants (Deshaies *et al.*, 1988; Han *et al.*, 1988). One potential drawback in use of the *GAL* promoters to achieve regulated expression is that many laboratory yeast strains carry *gal2* mutations that inactivate the galactose permease. (S288c, the progenitor of most strains, carries this mutation.) However, either the mutation is leaky or there is an alternative uptake system for galactose since the strains do grow with galactose as sole carbon source and do exhibit galactose regulation of the *GAL* promoters. The requirement for galactose addition (and its cost) would seem to limit use of this regulated promoter system for large-scale production, although use of a strong galactose-regulated promoter to produce γ-interferon in fermentation optimization studies has been reported (Fieschko *et al.*, 1987).

Additional regulated promoters include the strong MT promoter (Etcheverry, 1990) from the yeast *CUP1* gene, whose gene product is the Cu-chelatin (or Cu-metallothionein) protein, and the *PHO5* promoter (Kramer *et al.*, 1984) from the gene encoding acid phosphatase. The latter is induced by phosphate limitation, which can be easily achieved at laboratory scale by changing the growth medium (Kramer *et al.*, 1984; Eugster *et al.*,

1991), but is cumbersome for large-scale production. The MT promoter is rapidly induced by copper sulfate (0.01 to 0.5 mM, depending on the copper resistance of the host strain), although a low level of mRNA can be detected in the absence of exogenous copper (Etcheverry, 1990). As discussed later (Section II,B,2), natural resistance to copper and levels necessary for induction vary among laboratory strains.

3. Transcription Terminators

Termination of RNA transcripts in yeast is poorly understood and appears not to have precise identifiable signals (Donahue and Cigan, 1990). For example, the terminator region from the *CYC1* gene will function when inserted in either orientation at the 3' end of a heterologous coding sequence, although this fragment may contain a terminator from another gene that is transcribed in the opposite direction (G. McKnight, pers. comm.) Similarly, attempts to discern specific regions by deletion analysis have been unsuccessful. For expression of heterologous genes or cDNAs, the terminator from any yeast gene would appear to be sufficient; those from the *CYC1*, *MFα1*, *TPI1*, *PGK1*, and others have been used. Failure to include a terminator in the expression cassette might not preclude satisfactory expression levels (Goff *et al.*, 1984; although see Hitzeman *et al.*, 1984), but could cause interference with expression of other plasmid genes. It is generally advisable to place the yeast terminator immediately 3' to the coding region with a minimal length of heterologous DNA between them. Expression of human coagulation factor XIII was depressed 50-fold by the inclusion of 119 base pairs of 3' noncoding cDNA sequence before the yeast *TPI1* terminator (Bishop *et al.*, 1990). Similarly, Demolder *et al.* (1992) reported a 10-fold improvement in expression of murine interleukin-2 (and an increase in steady-state mRNA) when the cDNA's 3' untranslated sequence was truncated.

4. Secretion Leaders

An appropriate combination of the vectors and transcription elements described above will usually lead to efficient cytoplasmic production of heterologous proteins. However, secretory expression may be preferred for recombinant proteins containing disulfide bonds, to obtain properly processed amino termini (lacking the initiator methionine), or to facilitate protein isolation and purification.

In both yeast and mammalian cells, there are several types of secretion signals. Entry into the secretory pathway requires a cleavable signal peptide or presequence that is removed by signal peptidase in the endoplasmic reticulum; many proteins then complete the secretory pathway and are targeted to their final destination without additional processing events. Some proteins, however, also contain a prosequence that is cleaved later in the pathway, generally late in the Golgi, while proteins destined for the mammalian lysosome or its equivalent in yeast, the vacuole, contain specific targeting sequences that are removed during processing/activation in the vacuole. (For a review of the yeast secretory pathway, see Cleves and Bankaitis, 1992.) The pre- or pro-secretion signals of some heterologous proteins are recognized by yeast to yield detectable or higher product levels; these include interferon α (Hitzeman *et al.*, 1983b), human growth hormone (Hitzeman *et al.*, 1984), calf prochymosin (Goff *et al.*, 1984), influenza viral hemagglutinin (Jabbar *et al.*, 1985), human antithrombin III (Bröker *et al.*, 1987), and human serum albumin (Etcheverry *et al.*, 1986). Where analyzed, the signal peptides appear to be cleaved at the correct (native) site. However, for some of these proteins and for many others, such as preproinsulin, tissue plasminogen activator, porcine urokinase, and human urokinase, detectable or higher level expression has required substituting yeast secretion signals for the endogenous ones.

For some heterologous proteins, use of a yeast signal peptide alone is sufficient to achieve export of the product to the culture medium; for others, however, the recombinant protein tends to accumulate in the endoplasmic reticulum or be degraded if there is not a suitable endogenous or yeast prosequence on the protein. There are few published studies that have compared the abilities of an endogenous signal peptide or prepro sequence, yeast signal peptide, and yeast prepro sequence to facilitate secretion/export of the heterologous protein (see II,A,4,c. below). When beginning expression efforts for a new protein, it is therefore generally advisable to join the cDNA for the mature protein or the desired secreted form to a prepro sequence (secretion leader) that functions in yeast or, if possible, both a yeast signal peptide and a pro sequence.

a. Signal peptides (presequences).   The signal peptides from the yeast *SUC2* gene, encoding invertase, and the *PHO5* gene, specifying acid phosphatase, have been the most widely used yeast signal peptides (Meyhack *et al.*, 1982; Smith *et al.*, 1985; Hitzeman *et al.*, 1990; Moir and Davidow, 1991). Signal peptide cleavage appears to be efficient and largely unaffected by the nature of the amino acid immediately after the cleavage site. Examples of heterologous proteins expressed efficiently with yeast signal peptides include calf prochymosin (Smith *et al.*, 1985), human granulocyte-macrophage colony stimulating factor and interleukin-1β (Ernst, 1988; Baldari *et al.*, 1987), α2-interferon (Chang *et al.*, 1986), full-length and Fab chimeric antibodies (Horwitz *et al.*, 1988; Better and Horwitz, 1989), and mouse immunoglobulin kappa chain (Kotula and Curtis, 1991).

b. Secretion leaders (prepro sequences).   In contrast to the examples cited above, signal peptides are insufficient to facilitate secretion and export of other heterologous proteins; these proteins require the addition of a pro sequence to progress through the secretory pathway (Chaudhuri *et al.*,

1992; C. Yip and V. L. MacKay, unpublished). The most widely used yeast secretion leader is the 85 or 89 amino acid prepro region from the *MFα1* gene (Kurjan and Herskowitz, 1982; Brake, 1990; Hitzeman *et al.*, 1990). As diagrammed in Fig. 1A, the prepro includes a 19 amino acid signal peptide that is cleaved in the endoplasmic reticulum, three asparagine-linked glycosylation sites, a dibasic amino acid pair (Lys-Arg) for cleavage by the Kex2 protease localized in the late Golgi (Redding *et al.*, 1991), and two Glu-Ala pairs that are removed from the amino terminus of the heterologous protein through the action of the dipeptidylaminopeptidase encoded by the *STE13* gene. As discussed by Brake (1990), the Glu-Ala pairs can be deleted to ensure a homogeneous amino terminus on the secreted protein, but their deletion can reduce Kex2 processing efficiency, depending on the amino terminal sequence of the heterologous protein. Kex2 processing can be less than 50% efficient, particularly at high expression levels (Zsebo *et al.*, 1986; Elliott *et al.*, 1989; Gabrielsen *et al.*, 1990; Kotula and Curtis, 1991; Calderón-Cacia *et al.*, 1992). Unprocessed *MFα1* prepro-product fusions can accumulate in the cell, in the culture medium, or both. Inefficient processing may be improved either by modifying the processing site (Brake *et al.*, 1984; Guisez *et al.*, 1991) or by increasing the expression of the Kex2 protease (Barr *et al.*, 1987). In a few cases where Kex2 protease cleavage is blocked by

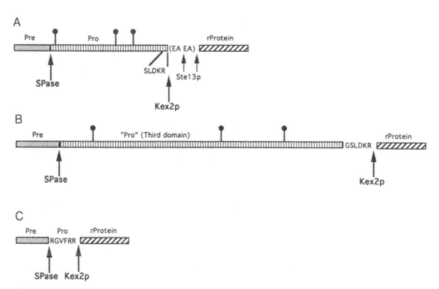

FIGURE 1 Diagrammatic sketches of three yeast secretion leaders. (A) the *MFα1* prepro, (B) the Bar leader, and (C) the HSA leader. Amino acid sequences immediately before the Kex2 protease cleavage site are shown in the one-letter code. The (EA EA) sequence in the *MFα1* prepro can be omitted, although Kex2 protease cleavage may be less efficient. SPase denotes the signal peptidase cleavage site and closed circles with stems indicate potential N-linked glycosylation sites.

mutation or by apparent inaccessibility of the protease to the cleavage site, another protease has been detected that cleaves either at the Kex2 site or upstream in the MFα1 prepro sequence (Egel-Mitani *et al.*, 1988; Bourbonnais *et al.*, 1991, 1993).

We have developed another secretion leader (Fig. 1B; MacKay *et al.*, 1990) by joining fragments of the yeast *BAR1* gene that encodes an extracellular, heavily glycosylated aspartyl protease (MacKay *et al.*, 1988, 1991). The leader includes the signal peptide, the first 10 amino acids of the first catalytic domain, the last 6 amino acids of the second catalytic domain, part of the third domain which is required for export of Bar protease itself, and a Kex2 protease cleavage site. In addition to the three asparagine-linked glycosylation sites shown, the third domain fragment contains substantial serine/threonine-linked glycosylation (M. Jars, pers. comm.), which may enhance its passage through the secretory pathway (S. Welch, M. Zavortink, and V. L. MacKay, unpublished). We have shown that the Bar secretion leader will direct efficient secretion of a precursor of human insulin, transforming growth factor-α, epidermal growth factor, platelet-derived growth factor (PDGF), and a precursor of human thrombin. In addition, a truncated version of the Bar leader is nearly as efficient for PDGF secretion as the original leader.

Human serum albumin (HSA) is efficiently exported to the culture medium when expressed with its own signal peptide and short (6 amino acid) prosequence (Etcheverry *et al.*, 1986); the prosequence is probably cleaved by the endogenous yeast Kex2 protease (Hitzeman *et al.*, 1990). Moreover, the HSA prepro will also direct secretion of other heterologous proteins or HSA fusion proteins (Fig. 1C; Hitzeman *et al.*, 1990). Unlike the MFα1 and Bar leaders described above, neither the HSA prepro nor HSA itself is glycosylated, although glycosylation is thought to be an important factor for the efficiency of the other two leaders (Caplan *et al.*, 1991; S. Welch, M. Zavortink, and V. L. MacKay, unpublished).

All of these prepro secretion leaders depend on the Kex2 protease for processing of the precursor to the mature heterologous protein. For some proteins, however, Kex2 also cleaves within the product; strategies for eliminating or reducing these cleavages will be discussed in Sections II,A,6; II,B,3; and II,C,2 below.

c. Comparison of secretion leaders. For two heterologous proteins (porcine urokinase [uPA] and tissue plasminogen activator [tPA]), we have made direct comparisons of different secretion signals, in which the transcription elements, plasmid constructions, host strains, growth conditions, etc., have been constant (Table III). For both proteins, the endogenous signal peptide (uPA) or prepro sequence (tPA) was inefficient for secretion in yeast. Replacement of these sequences with yeast signal peptides led to better expression levels, but the proteins appeared to be blocked in transit through

TABLE III Comparison of Signal Peptides and Prepro Sequences for the Secretion/Export of Human Tissue Plasminogen Activator (tPA) and Porcine Urokinase (uPA)[a]

A. Tissue plasminogen activator (activity in cell extracts)

Signal peptide/prepro	tPA activity, relative units
None (cytoplasmic)	0.05
tPA (endogenous)	1.0
PHO5 signal peptide	3.8[b]
SUC2 signal peptide	6.3[b]
MFα1 prepro [+(EA)$_2$]	2.5[b]
MFα1 prepro [−(EA)$_2$]	3.2[b]

B. Porcine urokinase (activity in cell extracts and exported to medium)

Signal peptide/prepro	Total uPA activity, relative units	% activity exported
uPA (endogenous)	1.0	2
PHO5 signal peptide	5.2	1
MFα1 prepro [+(EA)$_2$]	17.3	18
MFα1 prepro [−(EA)$_2$]	1.9	<4.3

[a] All expression units used the same promotor (TPI1) inserted in the same orientation in plasmid YEp13 (LEU2-bearing moderate copy number plasmid). The same host strain was used for all growth curves, which were conducted simultaneously for the set of tPA and uPA plasmid, respectively.

[b] tPA expressed with signal peptides appeared to have only core N-liked glycosylation, indicating its retention in the endoplasmic reticulum, whereas tPA expressed with the MFα1 prepro was heavily glycosylated and lacked the prepro peptide, indicating its passage through the Golgi.

the secretary pathway (i.e., retained in the endoplasmic reticulum). Use of the MFα1 leader facilitated passage through the secretory pathway, either to the extracellular medium or at least past the Kex2 cleavage step, the last known Golgi event in yeast (Redding et al., 1991). However, similar experiments with prochymosin (Smith et al., 1985) showed that the MFα1 leader was no more efficient for prochymosin secretion than the SUC2 signal peptide, possibly because the prochymosin prosequence promoted transit from the endoplasmic reticulum through the Golgi. Similarly, Ernst (1988) demonstrated that the signal peptide (pre sequence) of the MFα1 gene facilitates secretion and export of human granulocyte-macrophage stimulating factor and interleukin-1β as well as the full-length MFα1 prepro, whereas Chaudhuri et al. (1992) reported that the MFα1 pro region (coupled to any one of three signal peptides) is essential for secretion of insulin-like growth factor-I. In summary, as stated at the beginning of this section, it is advisable to try both a yeast signal peptide and a prepro secretion leader for initial attempts at secretion/export of a new heterologous protein.

5. Features Affecting Translation

As in mammalian cells, translation in yeast seems to follow the ribosome scanning model with few structural or sequence requirements. (See Donahue and Cigan, 1990, for a review of translation requirements.) Yeast mRNAs exhibit a bias (75%) for A at the -3 position 5' to the AUG initiator, but even this feature is not required and other conserved sequences around the AUG can be mutated with little or no effect on expression (Cigan et al., 1988). However, translation in yeast appears to be more sensitive to secondary structure in the 5' nonflanking region than in mammalian cells; thus, promoter–cDNA junctions in expression plasmids should be constructed to minimize the length of 5' noncoding sequence from the heterologous cDNA.

A frequent concern for high-level expression of mammalian cDNAs in yeast is codon bias, since yeast genes encoding abundant proteins have preferred codons that are generally more AT-rich than preferred codons in higher eukaryotes (Bennetzen and Hall, 1982; Maruyama et al., 1986; Sharp et al., 1986). The few studies comparing expression levels from natural and codon-optimized cDNAs are difficult to interpret since secondary structure of either the natural or the synthetic sequence could have a strong influence on translation efficiency or unknown sequences within the cDNA could affect transcription or mRNA stability (Chen et al., 1984; Mellor et al., 1985). (See reviews by Donahue and Cigan, 1990, and Romanos et al., 1992, for references and additional discussion.) For example, Bitter and Egan (1984) modified the 5' nontranslated sequence of the hepatitis B surface antigen gene (HBsAg) to the untranslated leader of the highly expressed glyceraldehyde-3-phosphate dehydrogenase gene (the promoter of which was used for expression) and used optimal yeast codons for the first 30 amino acids; HBsAg expression increased 10- to 15-fold, but the relative effects of the optimized 5' leader and the optimized codons could not be discriminated. Using an expression competition assay, Egel-Mitani et al. (1988) reported that a codon-optimized gene for an insulin precursor competed favorably with a non-optimized gene. Kotula and Curtis (1991) recently reported that yeast expression of an immunoglobulin kappa chain from a codon-optimized cDNA was 50-fold higher than from the natural cDNA. In contrast, Ernst and Kawashima (1988) found no correlation between codon usage bias and expression level for somatomedin-C in either S. cerevisiae or E. coli, while substitution of 12 consecutive CAG codons (infrequent in yeast) in murine interleukin-2 with the preferred CAA codons had no effect on expression levels (Demolder et al., 1992). In addition, hepatitis B virus core antigen has a very low codon bias for S. cerevisiae, yet is expressed cytoplasmically at 40% of soluble protein (Kniskern et al., 1986). Mutating codons of a cDNA to the preferred yeast codons might thus be a futile, or perhaps even deleterious, effort to improve expression levels. In general, expression of the natural cDNA is reasonable for initial efforts, but codon optimization should be considered if recombinant protein levels are low (due to inefficient translation) but its mRNA is abundant.

6. Modifications of the cDNA

There are two major post-translational events that could alter a heterologous secreted protein: Kex2 protease cleavage at an internal Lys-Arg or Arg-Arg pair (Julius et al., 1984; Bitter et al., 1984; Elliott et al., 1989; Gabrielsen et al., 1990) and hyperglycosylation at an Asn-X-Ser/Thr (N-linked) glycosylation site (Ballou, 1982; see Section II,B,5 below). Both of these modifications can be blocked by site-specific mutation of the cDNA, although the secreted product would then be slightly different than the native protein. Kex2 cleavage can be avoided by mutating the dibasic pair to Lys-Lys, which seems at least in vitro to be poorly cleaved by the Kex2 protease (Fuller et al., 1985). Similarly, addition of core oligosaccharide in the endoplasmic reticulum can be prevented by mutation of the Asn-X-Ser/Thr recognition site. Miyajima et al. (1986) mutated one or both Asn-X-Thr glycosylation sites in murine granulocyte-macrophage colony-stimulating factor (GM-CSF) to Asn-X-Val without loss of activity and obtained higher levels of the unglycosylated species in the medium. (It is of interest that human GM-CSF also contains two Asn-linked glycosylation sites, but most of the mature protein exported by yeast appeared to be unglycosylated [Miyajima et al., 1986].) We have mutated the three N-linked sites in tissue plasminogen activator to Gln-X-Ser/Thr and recovered higher levels of ELISA-detectable polypeptide and of tPA activity (MacKay, 1987); however, a similar mutation at the single N-linked site in a thrombin precursor severely reduced levels of exported protein (A. Bell and A. Sledziewski, pers. comm.). Since Gln is probably not a conservative change from a glycosylated Asn, other mutations, such as Asn to Asp or Ser/Thr to Ala, might be better choices. Recently, Hasnain et al. (1992) reported that mutation of the Asn-Gly-Ser glycosylation site in rat cathepsin B to Asn-Gly-Ala was still exported by yeast and was functionally identical with the native enzyme (and more active than the glycosylated native protein made in yeast). In an analog of tissue plasminogen activator, mutation of one N-linked glycosylation site from Asn to Gln had no effect on export, whereas a similar mutation at the other site led to a three- to five-fold decrease in exported activity (Gill et al., 1990), although it is not clear if the mutation affects activity or export.

7. Fusion Proteins

High level cytoplasmic expression of heterologous proteins can be thwarted by instability and degradation of the protein product; in many cases, degradation can be prevented by using host strains with mutations in one or more protease genes (see Section II,B,3 below), but expression levels may still be lower than expected. Two groups have achieved high level expression of heterologous proteins by fusing their cDNAs to either human superoxide dismutase (SOD) (Cousens *et al.*, 1987) or ubiquitin (Ecker *et al.*, 1989). Human SOD is a highly stable, soluble protein even when expressed at 20% of soluble protein. However, although the SOD–proinsulin fusion protein was expressed at approximately equivalent levels, it appeared to be contained predominantly within inclusion body-like structures, presumably because of misfolding due to the proinsulin moiety. Misfolded proinsulin was released from SOD by cyanogen bromide treatment and renatured to native proinsulin. Ubiquitin fusions with the mammalian $G_s\alpha$ subunit, sCD4 (a soluble fragment of the T cell receptor protein), and the human urokinase protease domain were produced at lower levels than the SOD–proinsulin protein, but were all processed *in vivo* by the endogenous yeast ubiquitin endoprotease to yield correctly cleaved product. Although the urokinase protease was inactive, the $G_s\alpha$ subunit (which does not contain disulfide bonds) could be reconstituted functionally in mammalian cell membranes. Thus, both of these fusion protein systems can be used to generate moderate to high levels of polypeptide, but correct disulfide bond formation depends on *in vitro* renaturation.

The fusion protein strategy could also be used for secreted heterologous proteins. For example, in at least some cases, unprocessed secretion leader–product fusions are exported efficiently to the culture medium (Kotula and Curtis, 1991; Calderón-Cacia *et al.*, 1992; T. Karwaki and V. L. MacKay, unpublished). These fusions can be processed *in vitro* at the Kex2 site with trypsin if the heterologous protein lacks trypsin-sensitive sites (Calderón-Cacia *et al.*, 1992; V. L. MacKay, unpublished) or the Kex2 site can be mutated to a recognition site for another protease (e.g., Factor X).

B. Host Strain Characteristics

For any expression plasmid, it is necessary to select a complementary yeast host strain that will permit selection and maintenance of the plasmid, optimal transcription and processing of the heterologous cDNA or fusion, cytoplasmic accumulation or secretion of intact product, and the presence or absence of post-translational modifications. Perhaps most importantly, after transformation the host strain should exhibit vigorous growth under selective conditions or in production medium; levels of both cytoplasmic and secreted/exported heterologous products are generally correlated with cell mass. Although in the early development stage we usually work with haploid strains with multiple markers to identify characteristics that affect product levels, we prefer to use (nearly) prototrophic diploids for later development and production. These generally grow more rapidly and are more stable (i.e., less likely to be outgrown by mutants that produce less product).

In cell biology studies, to determine whether the expressed protein functions in yeast participate in or alter part of the cell's physiology, "reporter" genes that permit rapid assay and quantitation of the effect should be included. Host strains with the desired combination of mutations/modifications can be generated by standard genetic crosses (Sherman and Hicks, 1991) or by homologous recombination of linear fragments or integrating plasmids (see references cited in Section II,A,1,a above) into strains with suitable backgrounds. Yeast strains bearing a variety of appropriate mutations are available from individual investigators or the following sources (from Sherman, 1991):

- The Yeast Genetics Stock Center: MCB/Biophysics and Cell Physiology, 102 Donner Laboratory, University of California, Berkeley, California 94720
- The American Type Culture Collection: 12301 Parklawn Drive, Rockville, Maryland 20852
- The National Collection of Yeast Cultures: Food Research Institute, Colney Lane, Norwich NR4 7UA, United Kingdom
- The Centraalbureau voor Schimmelcultures: Yeast Division, Julianalaan 67a, 2628 BC Delft, The Netherlands
- The Czechoslovak Collection of Yeasts: Institute of Chemistry, Slovak Academy of Sciences, Dubravaska cesta, 809 33 Bratislava, Czech Republic.

1. Vector Selection and Maintenance

As discussed in Section II,A,1 above and Table II, most yeast vectors have a functional copy of a yeast gene (e.g., *LEU2, URA3, POT*) that is defective in the host strain, e.g., *leu2, ura3, tpi1*). For large-scale protein production, it is essential that the host mutation be completely defective, nonsuppressible, and nonreverting, so that host-cell revertants (that can grow without the plasmid) do not outgrow the transformants. It is also preferable that the host chromosomal mutation be either a complete deletion of the sequence borne on the plasmid, or a severely rearranged gene, to circumvent integration of the plasmid into the chromosome by homologous recombination. If vectors with dominant selectable markers are used, such as resistance to hygromycin or tunicamycin, potential host strains should be tested for their inherent sensitivity to the drug before and after transformation.

YEp-type vectors that contain the 2-μm *REP3* and *FLP*, but not *REP1* or *REP2*, genes can be stably maintained only in strains that contain the endogenous 2-μm plasmid and thereby the Rep1 and Rep2 proteins. If all of these genes are present on the vector, then strains that lack the endogenous plasmid (denoted [cir°]) might be better hosts to achieve higher copy number and to prevent homologous recombination between the expression plasmid and the 2-μm plasmid that, depending on the configuration of the expression plasmid, can lead to deletion of the expression cassette.

2. Transformation Methods

As discussed earlier (Section II,A,1), all yeast transformation techniques depend on having a dominant selectable marker on the plasmid or integrating fragment. Untransformed cells can be grown in any complex or chemically defined medium (Sherman, 1991) that supplies their required nutrients. Transformants are selected on "omission" or "drop-out" medium (e.g., medium lacking leucine or uracil) or on medium containing inhibitors such as G418 or copper. Transformants with stably integrated plasmids or fragments can subsequently be grown in any medium. Although transformants with autonomous plasmids are generally maintained on selective medium, those with highly stable plasmids can be grown for short periods (e.g., batch fermentations) with reduced or no selection without significant loss of plasmid or product expression.

There are now several methods available for transforming yeast cells, the choice depending on the goals of the experiment, taste of the investigator, and perhaps the genetic background of the yeast strain, which may influence efficiency of a specific method. The original procedure (Hinnen et al., 1978; Beggs, 1978; slightly modified in MacKay, 1983) involved removing or at least partially digesting the yeast cell wall enzymatically to form osmotic-sensitive spheroplasts. These are incubated with plasmid DNA ($\pm$ carrier DNA) in 10 mM CaCl$_2$, which is taken up during the subsequent treatment with 20% polyethylene glycol (PEG). The transformed spheroplasts are regenerated to form colonies within a top agar layer plated on selective media (see below) at frequencies of ca. 10^4 transformants/μg DNA. This method is relatively rapid (2–3 h) but has two major drawbacks. Many of the transformant colonies are buried within the top agar layer, requiring picking or resuspension and replating before secondary screening. More seriously, as many as 30% of the transformants may have undergone spheroplast (protoplast) fusion during the PEG step and are now polyploid relative to the original host strain. For example, transformants from a MATα haploid strain may now be homozygous α/α diploids, which are difficult to identify without genetic or DNA content analysis and which may be unsuitable for further strain development either by classical genetic crosses or by targeted homologous recombination to disrupt or modify chromosomal genes.

The disadvantages of spheroplast transformation are avoided in the lithium acetate method (Ito et al., 1983; Hill et al., 1991; Gietz et al., 1992) in which intact cells (not spheroplasts) are transformed by incubation with DNA in 0.1 M lithium acetate in Tris–EDTA and the subsequent addition of PEG. After a 15-min heat shock, the cells are spread directly on the surface of the selective plate. The method is rapid and extremely efficient (reported as 10^6 transformants/μg DNA; Gietz et al., 1992), but as with all methods efficiencies may vary with genetic background (and investigator).

Several other transformation methods have recently been described but not yet widely used: conjugation directly from bacteria into yeast (Heinemann and Sprague, 1991); electroporation (Becker and Guarente, 1991); dimethyl sulfoxide–ethylene glycol-induced DNA uptake (Dohmen et al., 1991); high-velocity bombardment with tungsten microprojectiles from a particle gun is the only method reported to transform DNA into mitochondria (Fox *et al.*, 1991).

3. Transcription

Most of the promoters described above will function efficiently in standard laboratory strains, although small differences due to undefined genetic background effects can be seen. Transcription from some of the regulated promoters, however, requires or can be enhanced by modifying key regulatory genes. For example, the *sir3-8* mutation is obligatory for use of the temperature-regulated hybrid promoter as a regulated promoter (Sledziewski *et al.*, 1988). For use of the copper-inducible MT (*CUP1*) promoter, potential host strains must be screened for their inherent copper resistance and number of endogenous *CUP1* genes, since this gene is amplified in many laboratory strains (Etcheverry, 1990). For the *ADH2* promoter, increased copy number of its activating protein Adr1 was shown to boost *ADH2-lacZ* expression 4- to 10-fold after derepression (Price *et al.*, 1990). Similarly, strains with enhanced expression of the Gal4 activating protein produce higher expression levels from the galactose-inducible *GAL1*, *GAL7*, and *GAL10* (Mylin *et al.*, 1990; Demolder *et al.*, 1992). For most laboratory experiments, however, expression without these transcription-enhancing modifications is adequate.

4. Protease Deficiency

Wild-type *S. cerevisiae* cells contain a group of proteases that are localized to the vacuole (equivalent to the mammalian lysosome); these include an acid endoproteinase (protease A or PrA), a serine protease (protease B or PrB), two carboxypeptidases (CpY and CpS), and two aminopeptidases (API and a Co^{2+}-dependent aminopeptidase, ApCo). Although other proteases and peptidases are located in other cellular compartments, these vacuolar enzymes, particularly PrB and to a lesser extent PrA and the carboxypeptidases, appear to be responsible for most of the degradation of endogenous yeast and heterologous proteins. Since two excellent methods chapters have been published (Jones, 1990, 1991) to which the reader is referred, this chapter will provide only a brief summary of these recent ones, describing the problem and the most useful solutions.

a. The problem. Cytoplasmically produced heterologous proteins are most susceptible to degradation by yeast proteases, although secreted proteins can also be affected (see below). The vacuolar proteases appear to be glucose-repressed and are therefore expressed late in culture growth after glucose depletion from the medium. Since heterologous protein levels also are maximal at high cell density, production cultures are generally grown into stationary phase. When cells (and their vacuoles) are broken open, usually with glass beads, PrA, PrB, and CpY each complex with specific polypeptide inhibitors that are in the cytoplasm. During incubation (or long-term storage, even at $-20°C$) of the extracts at pH 4-5, however, the inhibitors appear to be hydrolyzed, freeing the active proteases. Moreover, preparation of samples for gel electrophoresis by addition of SDS in a higher pH buffer and boiling activates PrB. In our first attempts to express human α-1-antitrypsin cytoplasmically in wild-type yeast, high levels of the protein were detectable by electrophoresis of cell extracts until approximately 16 hours of growth in a rich medium; at this time, the protein almost completely disappeared. Use of a pep4-3 mutant strain (see below) prevented degradation of the protein at later time points and facilitated purification of recombinant α-1-antitrypsin (J. Forstrom, pers. comm.).

Although one would expect vacuolar proteases to pose difficulties for the production and/or purification of proteins in the cytoplasm, loss of secreted heterologous proteins may also occur, due either to a low level of vacuolar protease release to the culture medium or to other uncharacterized proteases. Approximately 40% of secreted murine granulocyte-macrophage colony-stimulating factor lacked the first two amino acids of the mature protein (Price *et al.*, 1987). Internal cleavages in secreted human parathyroid hormone were significantly reduced by use of a host strain lacking PrA and PrB and by alterations in the culture medium (Gabrielsen *et al.*, 1990). In the production of a secreted thrombin precursor (even from a strain with a deletion in the gene encoding protease A; see below), rapid loss of activatable product occurred when glucose was depleted from the medium; higher initial glucose concentrations (6%) or a continuous glucose feed obviated this problem (H. Han and V. L. MacKay, unpublished). Since similar losses have not been seen for some other heterologous secreted proteins, this thrombin precursor may be particularly susceptible to proteolysis.

b. Prevention of proteolysis.   The easiest solution to most proteolysis problems is to use *S. cerevisiae* strains that are deficient in the major vacuolar proteases. These enzymes are targeted to the vacuole as zymogens, where PrA is required for maturation of PrB, CpY, ApI, and other vacuolar enzymes. Null mutations in the *PEP4* gene, which encodes PrA, block maturation of these other enzymes and are therefore pleiotropic. Most of the original null mutations, such as *pep4-3*, are, however, nonsense mutations; even a low level of suppression can result in enough active PrA to generate low levels of mature PrB, which can itself activate more of the PrB zymogen, so that degradation of heterologous proteins still occurs. Availability of the cloned gene (Ammerer *et al.*, 1986; Woolford *et al.*, 1986) has permitted the construction of new strains with nonreverting, nonsuppressible *pep4* disruptions (Jones, 1991). These have been sufficient for production of many heterologous proteins. Other products still exhibit degradation before or during purification, however, although generally at a much lower level than in wild-type *PEP4* strains. This loss is usually due to the presence of PrB that has been activated by an unknown *PEP4*-independent mechanism (Jones, 1991). Therefore, use of *pep4 prb1* double mutants with nonreverting mutations in both genes is advised; many yeast strains bearing one or more mutations in genes for vacuolar proteases have been generously deposited by Jones (1991) at the Yeast Genetics Stock Center.

Limited C-terminal proteolysis has also been reported for epidermal growth factor (George-Nascimento *et al.*, 1988), porcine ribonuclease inhibitor (Vicentini *et al.*, 1990), and hirudin exported from yeast (Takabayashi *et al.*, 1990). For hirudin, carboxypeptidase Y and the Kex1 carboxypeptidase were determined to be responsible for the degradation.

Although PrA, PrB, and the carboxypeptidases constitute the major degrading enzymes for heterologous proteins expressed in yeast, other uncharacterized proteases clearly exist. Their activity can be blocked or greatly reduced by the inclusion of a protease inhibitor cocktail in lysis and purification buffers; several of these cocktails are described in Jones (1991) and generally contain broad inhibitors for each of the major protease families, as well as more specific inhibitors.

In addition to potential degradation by vacuolar proteases, heterologous proteins expressed cytoplasmically may also undergo cotranslational aminoterminal processing or be recognized as abnormal and thus subject to ubiquitinylation and subsequent degradation by an ATP-dependent multimeric protease complex (Wilkinson, 1990; Kendall *et al.*, 1990). Although there have not been reports of heterologous proteins degraded by this enzyme system, it could be responsible for low product levels of certain proteins. Wilkinson (1990) has described methods for the detection of polypeptide–ubiquitin conjugates and ubiquitin-dependent proteolysis and the use of mutants to reduce degradation.

5. Enhancement of Secretion Efficiency

The efficiency of heterologous protein secretion and export from yeast is unpredictable and probably mostly dependent upon the characteristics of the heterologous protein. Heterologous proteins can accumulate at several block points (Shuster, 1991): translocation into the endoplasmic reticulum and/or signal peptide cleavage, exit from the endoplasmic reticulum to the Golgi, transit through the Golgi, trafficking from the post-Golgi to the cell surface, export through the cell wall to the culture medium. For a number of secreted recombinant proteins, particularly at high expression levels, a significant portion of the protein remains cell-associated. Since much of this cell-associated protein lacks the *MFα1* prepro secretion leader used (Elliott *et al.*, 1989; Steube *et al.*, 1991), these polypeptides have progressed beyond the Kex2 protease cleavage step, the last known event in the yeast Golgi. For a few heterologous proteins, the majority of the cell-associated protein has been shown to be localized to the vacuole by immunofluorescence in *pep4* mutant strains and by the absence of the protein in cell extracts of *PEP4* wild-type strains (C. Raymond and T. Karwaki, unpublished).

Although protein progress through the secretory pathway can be tracked to identify the bottlenecks (Franzusoff *et al.*, 1991), investigators who simply wish to produce large amounts of their recombinant protein will usually not be interested in such studies. Use of a few existing "supersecreting" mutants or undertaking the isolation of new mutants can significantly improve levels of secreted active product.

At least three mutants exist that have been shown to enhance secretion and export of heterologous proteins from yeast. The first two were described by Smith *et al.* (1985) for their effects on prochymosin export; one of these, *ssc1* (super-secreting), was subsequently reported to increase export of several other recombinant products (Rudolph *et al.*, 1989). The cloned and sequenced *SSC1* gene is identical to the *PMR1* gene that encodes a protein with homology to Ca^{2+} ATPases (Rudolph *et al.*, 1989), but the mechanism by which *ssc1* mutations enhance secretion remains unknown (Antebi and Fink, 1992). Characterization of the second mutant, *ssc2*, has not been reported. The third mutation, *vps17* (previously *vpt3*; Raymond *et al.*, 1992), was one of many isolated for the mislocalization of vacuolar proteins to the cell surface (Bankaitis *et al.*, 1986; Rothman and Stevens, 1986).The different *vps* mutations exhibit specificity for which vacuolar proteins are misdirected. Secretion of active porcine urokinase (uPA) was increased approximately sevenfold in this mutant (relative to an isogenic *VPS17* wild-type parent strain); levels of cell-associated active tissue plasminogen activator (tPA) were also enhanced in the *vps17* host (C. Yip and V. L. MacKay, unpublished). Both uPA and tPA have Kringle domains that are found on many blood proteins. We have recently found that secretion of thrombin precursors that include one or both of the two Kringle domains from prothrombin is also increased in a *vps17* host, whereas the mutation had no effect on secretion of a thrombin precursor lacking Kringle domains (H. Han and V. L. MacKay, unpublished). Similarly, there was no significant effect of the mutation on the secretion of several other unrelated recombinant proteins that lack Kringle domains (C. Yip and V. L. MacKay, unpublished). Thus, the specificity of *vps* mutations for different vacuolar proteins may extend to structural features of heterologous proteins; if so, one or more of the existing *vps* mutants could enhance secretion of a particular recombinant protein.

The *ssc* mutations described above were isolated as part of a screen for mutants with increased secretion of bovine prochymosin (Smith *et al.*, 1985). Frequently, starting with a production strain in which plasmid and host characteristics have already been optimized, it is easier to mutagenize and screen for new mutants that secrete more of the product than it is to incorporate existing mutations (e.g., *ssc* or *vps*) into the strain. Several recent articles describe selection and screening strategies for improved production strains (Chisholm *et al.*, 1990; Gill *et al.*, 1990; Moir and Davidow, 1991); such approaches are generally used when there is a long-term commitment to production of the protein.

6. Glycosylation

Although the steps required for the addition of core oligosaccharide to asparagine-linked (N-linked) glycosylation sites in secreted proteins appear to be very similar in yeast and mammalian cells (Kukuruzinska *et al.*, 1987; Fig. 2A), subsequent modification in the Golgi is quite diverse. Yeast lacks the enzymology to modify the core by removing several of the mannoses and replacing them with the other monosaccharide units characteristic of the complex type of mammalian glycosylation. Instead, the core oligosaccharides (at least those that are accessible) (Ziegler *et al.*, 1988) are extended with an $\alpha 1 \rightarrow 6$ mannose backbone that is further modified with $\alpha 1 \rightarrow 2$ and $\alpha 1 \rightarrow 3$ mannose bonds. The resulting "hyperglycosylated" structure (Fig. 2D) is heterogeneous and quite immunogenic, particularly the $\alpha 1 \rightarrow 3$ linkages (Ballou, 1990). Besides being unsuitable for therapeutic uses, these hyperglycosylated proteins are more difficult to detect on stained polyacrylamide gels or by immunological methods (see Section II,C,2) and to purify and characterize. Moreover, the yeast carbohydrates can interfere with folding and/or activity of the protein, e.g., tissue plasminogen activator (MacKay, 1987). An *in vitro* strategy for (partially) removing the N-linked hyperglycosylation is discussed in Section II,C,2. It should be noted that not all heterologous proteins with N-linked glycosylation sites become hyperglycosylated (Miyajima *et al.*, 1986; Livi *et al.*, 1990, 1991; Guisez *et al.*, 1991). We have recently purified a secreted thrombin precursor with a single N-linked site, which is not hyperglycosylated but has an apparently homogeneous oligosaccharide structure of as yet unknown structure (probably $\leq$ $Man_{12}GlcNAc_2$).

Hyperglycosylation can be avoided through the use of a double mutant *mnn1 mnn9* host strain (Tsai *et al.*, 1984), in which only a homogeneous $Man_{10}GlcNAc_2$ oligosaccharide (Fig. 2B) is present at N-linked sites in cell wall mannoproteins or secreted glycosylated proteins from the mutants. We have cloned and sequenced both the *MNN1* and the *MNN9* genes and used these to construct strains with nonreverting gene deletion/disruptions in both genes (MacKay *et al.*, 1990, 1991). Because cell wall mannoproteins apparently contribute significantly to the integrity and strength of the wall, *mnn9* (or *mnn1 mnn9*) mutants grow more slowly than wild-type strains and are osmotic sensitive, thereby requiring an osmotic support in the medium (usually 0.5–0.75 *M* sorbitol). Whereas these requirements may limit the use of the strains for large-scale protein production, we have grown up to 60-liter cultures in controlled fermentations and purified the secreted yeast Bar protease from the culture medium (MacKay *et al.*, 1991), although the

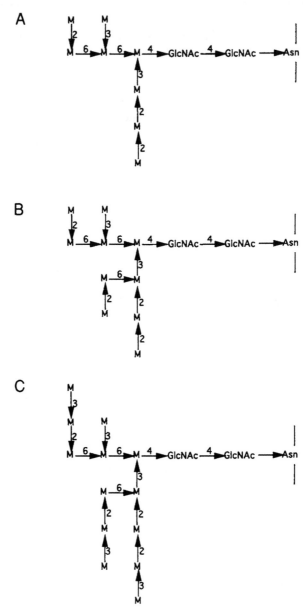

FIGURE 2 Structures of S. cerevisiae carbohydrates. (A) Core oligosaccharide; (B) *mnn1 mnn9* oligosaccharide; (C) *mnn9* oligosaccharide; (D) wild-type "hyper"glycosylated mannan; (E) O-linked mannose moieties.

yield of protease activity is lower than from *MNN9* wild-type strains.

The super-secreting *pmr1* (=*ssc1*) mutant described above (Section II,B,4) secretes yeast invertase, human α-1-antitrypsin, and other glycosy-

D

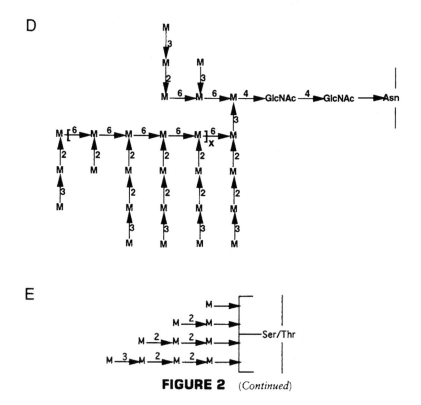

E

FIGURE 2   (*Continued*)

lated proteins with homogeneous truncated oligosaccharides (Moir and Dumais, 1987; Rudolph *et al.,* 1989; Antebi and Fink, 1992); since these oligosaccharides react with antisera against $\alpha 1 \rightarrow 6$ and $\alpha 1 \rightarrow 3$ mannose linkages (Antebi and Fink, 1992), they apparently represent a yet unknown modification in the Golgi of the core oligosaccharide added in the endoplasmic reticulum (Fig. 2A). It is also unclear whether *pmr1* affects the structure of N-linked carbohydrate on all secreted proteins. In contrast to N-linked glycosylation, the enzymology and site selection for O-linked carbohydrate on secreted proteins is still relatively poorly understood in any eukaryotes and appears to differ significantly between yeast (and other fungi) and mammalian cells. The oligosaccharide structure in yeast (Fig. 2E) is quite simple and lacks the GalNAc or GlcNAc that is characteristic of mammalian O-linked carbohydrate. Haselbeck and Tanner (1983) have proposed a pathway in which the first mannose is added to the serine or threonine in the endoplasmic reticulum by the enzyme dolichyl-phosphate-D-mannose:protein O-D-mannosyltransferase and subsequently modified in the Golgi. At least one heterologous protein (platelet-derived growth factor, BB homodimer) secreted from yeast has been shown to have one or more O-linked mannose in a fraction of the population (Settineri *et al.,* 1990).

Unfortunately, at present, it is not possible to predict which serine or threonine residues in a polypeptide are candidates for O-mannosylation (Gooley *et al.*, 1991; Strahl-Bolsinger and Tanner, 1991), there are no glycosidases that can remove the O-linked mannoses (Orlean *et al.*, 1991), and viable yeast mutants that are defective in O-linked carbohydrate synthesis or addition are not available. A recent report (Strahl-Bolsinger and Tanner, 1992) described the cloning of a gene encoding an endoplasmic reticulum Dol-P-Man:protein O-D-mannosyltransferase; cells carrying a gene disruption were viable and retained approximately 50% of the enzyme activity, suggesting the presence of a second gene encoding the same activity. If double mutants are found to be viable, they could be useful for secretion of those heterologous proteins that are subject to O-linked oligosaccharide addition in yeast.

7. Reporter Genes

Particularly in cell biology studies with a goal to assess the effect of a heterologous protein in a well-characterized area of yeast cell physiology, it is advantageous to have a rapid, simple test for the protein's function. Although endogenous yeast proteins can be assayed, fusions of yeast promoters or truncated genes with "reporter" genes are commonly used, particularly *E. coli lacZ* for nonsecreted proteins and the yeast *SUC2* gene (encoding invertase) for secreted ones. The reporter gene fusion can be either the direct target of investigation, for example, dissecting functional elements in a promoter joined to *lacZ* (e.g., see Guarente, 1983; Kronstad *et al.*, 1987; Sledziewski *et al.*, 1988, 1990; Price *et al.*, 1990; Schena *et al.*, 1991), or an indirect result of the heterologous protein's function. An example of the latter is the *FUS1-lacZ* fusion (Trueheart *et al.*, 1987) in which transcription from the *FUS1* promoter is strongly induced by the mating pheromone α-factor through a receptor-mediated multistep signal transduction pathway; expression and function of mammalian subunit homologs of the heterotrimeric G protein complex can be assayed by their indirect effect on β-galactosidase levels (King *et al.*, 1990; V. L. MacKay and T. Jones, unpublished). Fusions of the *SUC2* gene with portions of the *PEP4* and *PRC1* genes (encoding vacuolar enzymes protease A and carboxypeptidase Y) have been used to isolate and characterize the *vps* (vacuolar protein sorting) mutants and other features of the secretory system (Emr *et al.*, 1983; Bankaitis *et al.*, 1986; Rothman and Stevens, 1986), whereas fusions of the human serum albumin prepro sequence with a partially defective *SUC2* gene were used to select mutants able to utilize sucrose as sole carbon source due to enhanced secretion of the fusion protein (Chisholm *et al.*, 1990).

β-Galactosidase synthesis in colonies can be easily detected on plates or filters or assayed quantitatively in cell extracts (Miller, 1972; Rose and Botstein, 1983; Guarente, 1983; Ruby *et al.*, 1983; Sledziewski *et al.*, 1990). Transformants secreting *SUC2*-encoded invertase can be selected or identified by their ability to grow on sucrose as sole carbon source (Bankaitis *et al.*, 1986; Chisholm *et al.*, 1990) and the invertase activity analyzed by a gel electrophoresis assay (Gabriel and Wang, 1969; Esmon *et al.*, 1981) or quantitative assays in cell extracts (Goldstein and Lampen, 1975). Recently, a new sensitive cytoplasmic reporter system has been described that utilizes the β-glucuronidase gene from the yeast *Yarrowia lipolytica* (Bauer *et al.*, 1992).

C. Preliminary Analysis of Yeast Transformants Expressing Heterologous Proteins

Sensitive assays are generally required for initial detection and characterization of the recombinant protein; until expression is optimized, it is rare to be able to visualize the product in stained polyacrylamide gel electrophoresis of total cell extracts or unfractionated culture supernatants. A combination of assays that detect both total recombinant polypeptide, such as immunoblots or ELISAs, and properly processed and folded protein, such as activity assays or analytical HPLC systems, provides the most information about the product and the expression system. However, nearly all of these assay techniques are subject to interference by the presence of yeast proteins (at 90–95% contaminating levels) (and are generally too time-consuming to be useful for on-line monitoring of shake-flask cultures and fermentations). Therefore, before an expression project is initiated, some time and effort should be devoted to assay development and the assurance that the recombinant protein can be detected at least semiquantitatively in unfractionated cell extracts or culture supernatants.

1. Colony Blots

As discussed above (Section II,A,1), autonomously replicating plasmids are used for most expression efforts in yeast, so that potential chromosomal position effects on transcription of integrated expression cassettes (that are common in fungal transformations and mammalian cell transfections) are avoided. Transformants are therefore analyzed directly for protein expression, rather than for mRNA levels. (However, if protein expression levels are low or undetectable, Northern blot analysis of total or poly(A)$^+$ RNA [Köhrer and Domdey, 1991] should indicate poor transcription or mRNA degradation.) Also, with the exception of the YRp(*ARS*) plasmids, the plasmid copy number in independent transformant colonies is fairly constant, so that it is generally not necessary to characterize a large number of transformants for expression levels. Initially, we usually analyze four to eight transformants for each host/plasmid combination. Transformants are picked with sterile toothpicks from the transformation plates to selective media and regrown; in our hands, single-colony isolation has not been necessary. These plates serve as master plates for further work.

We usually use a quick, semiquantitative colony screen to determine if (or which) transformants are producing the heterologous protein before undertaking more time-consuming studies. Antibody-based colony blots are the most widely applicable (Fig. 3). Colonies on the master plate are replica-plated or spotted with a pronged applicator apparatus (Sherman, 1991) onto nitrocellulose filters placed on selective medium plates and regrown for 1–3 days. The filters are washed three times, each for 5 min, in transfer buffer (used for electrophoretic transfer of proteins from polyacrylamide gels to nitrocellulose for immunoblots; Towbin *et al.*, 1979) to remove the cells. The filters are then blocked, incubated with primary antibody, washed, incubated with second antibody, washed, and developed as for immunoblots (Fig. 3). This method will detect mainly only secreted protein. A similar method to detect intracellular expression has recently been published (Donovan *et al.*, 1992) in which the colonies on the filters are treated with lytic enzymes to remove the cell walls and the resulting spheroplasts are lysed. The filters can be used for enzymatic assays or for immunoblots as described above. It should be noted that most rabbits and many mice have circulating antibodies against yeast (mainly the carbohydrate structures, but also intracellular proteins) before immunization with antigen. If unfractionated antisera are used, it is essential that control transformants (host strain with the vector alone) be included on the filter to assess nonspecific background. Even monoclonal and affinity-purified antibodies may still generate some nonspecific reactions.

Colony screens employing enzymatic activity assays for the heterologous protein have also been used for initial evaluation of transformants producing prochymosin, human urokinase, porcine urokinase, and tissue

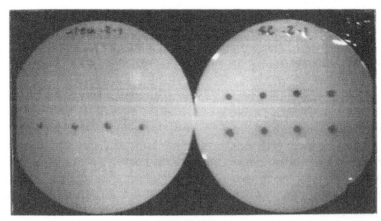

Row 1

Row 2

Row 3

FIGURE 3 Immunoblots of yeast colonies. Row 1, control vector transformants; row 2, transformants expressing an insulin precursor from a POT-*leu2-d* plasmid; row 3, a different host strain expressing the same insulin precursor from a POT-*leu2-d* plasmid.

plasminogen activator (Gill *et al.*, 1990; Moir and Davidow, 1991; V. L. MacKay, unpublished) or "reporter" gene products, such as β-galactosidase (Sledziewski *et al.*, 1990).

2. Polyacrylamide Gel Electrophoresis and Western Blots

A more informative analysis of the heterologous protein made by transformants is generally done by polyacrylamide gel electrophoresis of cell extracts and/or culture supernatants from small cultures (5–50 ml) of transformants grown in selective media. With constitutive promoters, we generally use medium containing 6% glucose to obtain higher yields and to minimize product degradation after glucose depletion. Yeast grown in unbuffered chemically defined yeast medium (Sherman, 1991) will produce enough acid during growth into stationary phase to lower the pH to approximately 2.7. Since low pH can irreversibly denature many heterologous proteins, it may be necessary to buffer the medium; yeast will grow well up to pH 6.5–7. (We usually use 0.1 M citrate-phosphate to buffer the medium at pH 5–6, but other buffers are also effective.) Rich media containing yeast extract and peptone have an initial pH of approximately 5.5 and drop only to approximately 4.5. Medium improvements are discussed in Section II,C,3 below.

After harvest by centrifugation at 4°C, cell pellets are washed with water or buffer and can be frozen at −20°C with no apparent loss of cell-associated protein. Yeast cells have thick cell walls and cannot be broken efficiently by freeze–thaw or sonication methods, but can be disrupted by vortexing with glass beads. In our protocol for small samples, cell extracts are prepared by resuspending the cells in extraction buffer (± protease inhibitors) at ≥ 10-fold the original concentration. An equal volume of acid-washed sterile glass beads (0.45 mm) is added and the mixture is vortexed. With our strains and vortex platform, three to five 1-min agitations (cooling between pulses) are necessary to obtain maximal protein release. Larger samples can be disrupted in a Braun homogenizer (Braun–Melsungen, West Germany) or a DynoMill (Willy A. Bachofen AG Maschinenfabrik, Basel, Switzerland). Soluble protein can be extracted with 5 mM EDTA in phosphate-buffered saline, assay buffer for the recombinant protein, or others. For extraction of membrane-associated proteins, we prefer TNEN (20 mM Tris–HCl, pH 8, 100 mM NaCl, 1 mM EDTA, 0.5% NP-40, a nonionic detergent) over harsher methods employing SDS, urea, heat treatment, etc. As noted in Section II,B,3, inclusion of a cocktail of protease inhibitors in the extraction buffer can minimize proteolytic degradation. After disruption, the extracts are clarified by centrifugation.

For the characterization of secreted heterologous proteins, the culture supernatants after centrifugation are usually filtered through 0.45-μm filters to remove residual cells or debris and the pH is adjusted if desired. If the concentration of the secreted protein is expected to be low or difficult to detect (see below), the supernatant proteins can be concentrated by ultrafiltration centrifugation through commercially available spin concentrators (Amicon) or frequently by precipitation of the proteins with 2 vol of 95% ethanol at $-20°C$ for ≥ 30 min (MacKay *et al.*, 1988).

Electrophoresis of cell extract and/or supernatant samples through native or denaturing (with or without reduction) polyacrylamide gels is followed by standard methods of detection, including staining of the gel itself or transfer to nitrocellulose for immunoblots (Towbin *et al.*, 1979). In cell extracts from transformants expressing the recombinant protein at high levels ($\geq 5\%$ solubilized protein), the recombinant protein band can usually be visualized by staining with reagents such as Coomassie Brilliant Blue G. For secreted proteins, however, staining with this reagent can give very misleading results. Most proteins in spent yeast medium are heavily glycosylated (Section II,B,5) and therefore stain poorly, if at all, with this reagent, even after removal of asparagine-linked carbohydrate with endoglycosidases (Fig. 4A), possibly because of large amounts of O-linked carbohydrate that is not cleaved by the endoglycosidases. If the recombinant protein is not hyperglycosylated, it will appear to be the major or only protein in the supernatant, leading to false estimates of purity and yield. Pretreatment of the gel with 1% periodic acid (Darbre, 1986) followed by silver staining (Merril, 1990) is the only method we have found to visualize certain glycosylated yeast proteins (Figs. 4B, 4C), although the results are not quantitative. If the heterologous protein is also hyperglycosylated, it may not be detectable in either stained gels or immunoblots, because the extent of glycosylation can be quite heterogeneous, leading to electrophoretic migration of the protein as a diffuse smear, and can block antibody binding to the polypeptide (see examples in Kotula and Curtis, 1991; Steube *et al.*, 1991; Calderón-Cacia *et al.*, 1992). Removal of asparagine-linked glycosylation by treatment with an endoglycosidase (Endo H, Endo F, or PNGase F) generally permits visualization on immunoblots and in silver-stained gels (see examples in Bröker *et al.*, 1987; Elliott *et al.*, 1989; Calderón-Cacia *et al.*, 1992).

The hyperglycosylation often seen on yeast-produced recombinant proteins containing N-linked glycosylation sites may also interfere with the activity of the heterologous protein (MacKay, 1987; Calderón-Cacia *et al.*, 1992). Some of the carbohydrate can be removed, and the activity improved, by treating either crude extracts/unfractionated supernatants or purified protein with an endoglycosidase under nondenaturing–nonreducing conditions (MacKay, 1987). Since the core oligosaccharides accessible on the surface of the protein are proposed to be the ones that are hyperglycosylated in the Golgi (Ziegler *et al.*, 1988), these are likely to be removed by endoglycosidase treatment.

3. Radiolabeling and Immunoprecipitation of Heterologous Proteins

For the expression of most recombinant proteins, immunoblots, ELISAs, and activity assays are sufficient methods to assess the quality and quantity

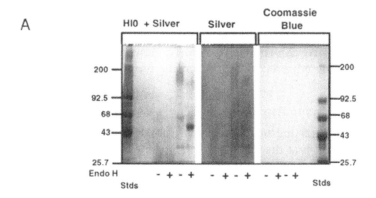

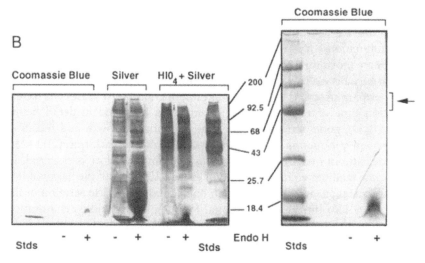

FIGURE 4 SDS–PAGE of secreted endogenous yeast proteins, stained with Coomassie brilliant blue G, silver, or silver after periodate pretreatment. Concentrated samples of culture medium or cell-free broth were incubated overnight at 37°C ± Endo H (Ziegler *et al.*, 1988) and fractionated on 10–20% SDS reducing gels. Silver staining was done according to the suppliers' directions (Daiichi, Integrated Separation Systems), with or without pretreatment with 1% (w/v) periodic acid. (A) In each panel, the first two samples are uninoculated complex medium, concentrated 10-fold by Centriprep 10 spin concentrators (Amicon), then incubated ± Endo H; the second two samples are cell-free culture broth from strain JG134 transformed with a POT vector and treated similarly. (B) Samples from strain ZM118 transformed with a POT vector and grown in complex medium; the cell-free culture broth was concentrated 5-fold and incubated ± Endo H as above. The panel to the right in B is a longer exposure of the Coomassie-stained gel to make visible the two weak bands in the Endo H-treated sample.

of the heterologous protein. However, in some cases, it may be of interest to determine the kinetics of its secretion or its interaction with yeast proteins. Conditions for pulse–chase labeling of extracellular or cell-associated proteins with radioisotopes, cell disruption, and immune precipitation are described in detail by Franzusoff *et al.* (1991); steady-state labeling (usually with [^{35}S]sulfate, methionine, or cysteine) can be achieved simply by incubating the cells overnight (10–15 h) in the selective medium containing the isotope, approximately 200 μCi in 0.5 ml (C. Raymond, pers. comm.). Modifications of these protocols depend on the properties of the specific antibodies to be used.

4. Growth Curves to Assess Culture Conditions

Although synthesis and secretion of heterologous proteins are usually correlated with growth of the culture, different transformant strains under different conditions grow and reach their maxima at different rates. In addition, export of some heterologous proteins continues after growth (increase in culture density) has effectively stopped, whereas others reach a maximum and then decrease. To determine optimal culture medium, temperature, or harvest time for maximal product levels or to compare different host/plasmid combinations, it is often necessary to do quantitative growth curves. Cultures are inoculated at a constant initial density and samples are taken at intervals (usually every 4 to 12 h) to monitor culture growth, product levels, and other parameters. (Under optimal growth conditions, yeast cells have a generation time of approximately 90 min.) As discussed in detail below (Section II,D), good aeration is essential for optimal growth and recombinant protein production. In shake flasks on rotary platforms (200–250 rpm), the ratio of medium volume to flask volume therefore is essential. In general, the medium volume should not exceed 20–25% of the flask volume to ensure adequate aeration. Although baffled flasks increase aeration of the culture, they also introduce substantial foaming and possible denaturation of the recombinant protein (V. L. MacKay, unpublished). For initial shake flask experiments, we avoid both baffled flasks and antifoams, which could affect the product. Results from these experiments are used to design conditions for large shake flask or fermenter cultures for protein production and purification.

The chemically defined culture media (0.67% Bacto yeast nitrogen base without amino acids, 2% glucose, supplemented with amino acids but lacking leucine or uracil, for example) used to select and maintain strains transformed with the most common vectors (Table II) promote relatively low final culture densities and therefore low product levels. These can be readily modified by the addition of supplements and then their effect on recombinant protein expression can be assessed in growth curves. For example, secreted human lysozyme levels were enhanced sevenfold by growth of the transformants bearing moderate copy number *LEU2* plasmids in standard -leucine medium with 10% glucose and additional mineral salts (Ichikawa *et al.*, 1989). We (MacKay *et al.*, 1991) have expressed the exported Bar proteinase of *S. cerevisiae* from the strong *TPI1* promoter on a moderate copy *LEU2* vector in a *mnn1 mnn9* glycosylation defective strain (see Section II,B,6). Transformants grown in a medium containing 1% yeast extract,

1.4% ammonium sulfate, mineral salts, trace elements, vitamins, and 0.5 M sorbitol (to support the osmotically sensitive mutant strain) exported 100-fold more Bar proteinase than the same strain grown in standard -leucine medium containing sorbitol (V. L. MacKay, unpublished). Apparently, yeast extract is deficient in leucine so that selection for plasmid-bearing cells is maintained. Similarly, the addition of 0.5–1% acid-hydrolyzed casamino acids to -uracil or -tryptophan medium significantly enhances growth and expression of recombinant proteins while maintaining plasmid selection (Fieschko *et al.*, 1987; Gabrielsen *et al.*, 1990). With these enriched media, the culture inocula are generally grown to high density in chemically defined medium for plasmid selection before transfer to the enriched medium. When cells transformed with POT plasmids (for which selection is growth on fermentable carbon sources) or stable integrative plasmids (which do not require selection) (Section II,A,1) are used for protein production, a much wider variety of culture media are available to achieve maximal expression levels, including complex media containing peptone.

D. Optimization of Protein Production

In the preparation of recombinant proteins, a process for the fermentation, recovery, and purification is critical if provision of a protein for characterization and experimentation is to be achieved. For bench-scale basic research or initial studies of a recombinant protein, cultures grown in shake flasks under conditions established in growth curves (see Section II,C,3 above) may produce sufficient levels of recombinant protein. However, for research requiring large amounts of recombinant protein or protein made under reproducible conditions (e.g., for *in vivo* animal or toxicology studies), fermentation optimization is often a major effort, since the efficiency and quality of the recovery and purification are frequently dependent upon the quality and productivity of the cells during the fermentation (Bailey and Ollis, 1986). To push forward with optimization of recovery or purification prematurely will often result in an unnecessary loss of time and in repeated efforts once the fermentation optimization is completed.

In the following sections we will discuss factors pertaining to optimization of recombinant protein production in large shake-flask cultures and small fermentors that are generally available to academic scientists and investigators in basic research departments of companies.

1. Establishment of a Master Cell Bank

A key first step is the creation of a master cell bank for the one or several transformant strains identified by preliminary analysis to yield the highest levels of the recombinant protein. Such a bank provides assurance that the same generational culture can be obtained and that the results of fermentation experimentation can be repeated. Typically, at least one vial per year is prepared for as many years as production from the strain is liberally estimated to be required. From a vial of the master cell bank, a production seed bank is made that should contain more than enough vials for a year of expected fermentations. Although establishment of master and working seed banks is essential for commercial production and for regulatory approval, it is also a highly recommended step for noncommercial production of recombinant proteins for basic research. Probably the most convenient method for storage of yeast transformants is in 15–50% glycerol at −80°C (Sherman, 1991), although yeast stocks can also be preserved by lyophilization or storage under oil (Martens et al., 1987).

2. Fermentation Conditions

In the production of recombinant proteins, expression levels, culture media, and fermentation conditions can markedly influence the structure, activity, localization, and modifications of the protein product (Benito et al., 1992; Clare et al., 1991; Van der Aar et al., 1992). In addition, as discussed previously (Section II,B,4), fermentation conditions may influence the expression of yeast proteases (Gabrielsen et al., 1990).

In general, high level production of recombinant proteins is strongly correlated with high biomass; i.e., it is closely tied to the primary metabolic physiology of the transformed strain (Sardonini and DiBasio, 1987; Wittrup et al., 1990). A mathematical description of recombinant yeast metabolism has been published using specific growth data during the expression of heterologous proteins in yeast (Coppella and Dhurjati, 1990). The fermentation conditions will be specifically dependent upon the energy and nutrient requirements of each transformant strain, as well as the expression vector, the constitutive or regulated promoter, properties of the recombinant product, etc. (Fieschko et al., 1987). Once these considerations are met, S. cerevisiae is capable of achieving relatively high cell densities, often ≥120 g/liter dry weight. Recombinant protein expression levels can be from 0.5 to 40% of the soluble protein fraction, i.e., from a few milligrams per liter to several grams per liter (Alberghina et al., 1991; Van der Aar et al., 1992).

The production of high levels of recombinant protein in yeast is a very aerobic, energy intensive process that is optimized by carefully regulated glucose levels (through continuous feeding of medium) and by appropriate dissolved oxygen levels (Oura, 1973; Furukawa et al., 1983; Wang et al., 1988). These conditions can be established in fermentors by monitoring culture parameters such as turbidity (Iijima et al., 1987), ion consumption (Garrido-Sanchez et al., 1988), glucose and ethanol concentrations (Kim and Lee, 1988; Fieschko et al., 1987), pH (Gibbons and Westby, 1986), dissolved oxygen (Phillips and Johnson, 1961), and off-gas content (carbon dioxide, oxygen, and nitrogen; Coppella and Dhurjati, 1987; Fieschko et al., 1987). Ideally, such fed batch fermentations are maintained by computer control to achieve reproducible high cell density (Nyiri et al., 1974; Alberghina *et al.*, 1991; O'Connor *et al.*, 1992). Insufficient dissolved oxygen levels can be avoided through the use of improved fermenter designs, oxygenation enhancers, and supplemental oxygen (Kawase and Moo-Young, 1987; Wang *et al.*, 1988; Rols *et al.*, 1990). In some cases, fermenters are being engineered to withstand hyperbaric oxygen (Taniguchi *et al.*, 1992), which increases the absolute pressure of oxygen in the gas phase, thereby increasing the formation of dissolved oxygen available to the yeast.

The production of high levels of recombinant protein requires the efficient ATP production afforded by aerobic growth, which in turn requires high dissolved oxygen and appropriate nutrient feed rates. The consequence of overfeeding the yeast culture with easily metabolized sugars, such as glucose, is that an internal anaerobic state can be created within the yeast. Glucose concentrations greater than 5% repress the oxidative pathway so that ethanol is produced under aerobic growth (i.e., the Crabtree effect; Suomalainen, 1969; Fieschko *et al.*, 1987). Since the net energy production from the conversion of glucose to ethanol is theoretically only two ATP, whereas the full respiratory oxidation of glucose to carbon dioxide and water yields a theoretical net of 36 ATP, aerobic growth is clearly the more energetically favorable physiology for the manufacturing of recombinant proteins. Fermentation data from a number of sources support the concept of a fed-batch mode of operation for most recombinant *E. coli* and *S. cerevisiae* fermentations of industrial interest (Fieschko and Ritch, 1986; Fieschko *et al.*, 1987), although pure continuous fermentation operations are possible. High yield fed-batch fermentations are achieved by feeding the glucose (or other carbon source) at a limiting rate, which slows the metabolism and therefore the respiration rate of the yeast. Fieschko *et al.* (1987) state that at growth rates greater than $0.2-0.25$ h^{-1}, ethanol accumulates in the medium; under these conditions, the cells convert from fully aerobic metabolism to oxidative fermentation, leading to further accumulation of ethanol. As cell density increases, constant growth rate is maintained by adjustment of the nutrient feed rate. Growth rates less than $0.2-0.25$ h^{-1} also enable the reproducible maintenance of high dissolved oxygen levels.

If fermenters are not available, yeast transformants can be grown in large (2.8-liter Fernbach) flasks to produce enough recombinant protein for small-scale purification. Obviously, factors that lead to maximal protein production in fermenters, such as limiting glucose feeds, pH control, and monitoring of dissolved oxygen, cannot be controlled in shake flasks, so cell densities and protein yields are generally lower. The standard yeast media (Sherman, 1991) contain 2% glucose, which for healthy strains is depleted approximately $12-14$ h after inoculation (1% inoculum); at this point, growth obviously slows, cellular proteases are de-repressed (Section II,B,4), and other aspects of metabolism change (e.g., de-repression of the *ADH2* promoter; Price *et al.*, 1990; see Section II,A,2). To maintain growth and recombinant protein production, either higher initial glucose concentrations can be used (for constitutive promoters) or another carbon source (galactose for the *GAL1* or *GAL10* promoter) must be added when the glucose is depleted. As noted above, the Crabtree effect (ethanol accumulation under aerobic conditions) is seen at glucose concentrations higher than 5%, yet Ichikawa *et al.* (1989) reported highest yields of secreted human lysozyme at 10% initial glucose concentration in shake flasks. Similarly, we often use 6% initial glucose in shake flask cultures with high yields of both cell mass and recombinant protein; apparently, the Crabtree effect is transient in these cultures.

Since high dissolved oxygen levels are important for high productivity, it is essential that shake flask cultures receive adequate aeration. In 2.8-liter flasks, culture volume should not exceed approximately 1 liter, agitated at 200–250 rpm on a rotary platform. As discussed above, baffled flasks provide better aeration but also lead to substantial foaming, which can denature exported recombinant proteins. Antifoam agents can be used, but should be tested first for their effects on the recombinant protein. Standard antifoam agents include many different surface active agents, but the most common are polypropylene glycols and silicone oils. Finally, pH can be maintained at 5–7 by the addition of 0.1 M citrate-phosphate, sodium succinate, potassium phthalate, etc., but the inclusion of such a high concentration of buffer can interfere with purification of exported recombinant proteins. The high ionic strength of the cell-free culture supernatant must be reduced by dilution, dialysis, protein precipitation, etc., before most ion-exchange chromatography methods.

E. Recovery and Processing

Protection of the recombinant protein's conformation and prevention of denaturation or modification is essential during harvest and recovery. Therefore, temperature, pH, and the extraction buffer are always critical parameters, which vary with the recombinant protein. Continuous flow centrifugation is a highly reliable method to separate yeast cells from the culture supernatant. (Membrane methods for cell or broth harvesting have proven to be more reliable in mammalian cell processes than in yeast or bacterial recoveries. After repeated use with microbial fermentations, these membranes frequently suffer shear fatigue, fouling, or medium related abrasion.) After centrifugation, the culture supernatant is filtered through 0.45-μm membrane filters, such as those available from AG Technology or Millipore or depth filters (e.g., from CUNO, Meriden, Connecticut), to remove residual cells and is then ready for protein purification.

Recovery of a cell-associated recombinant protein requires physical disruption of the yeast cells with glass bead mills (DynoMill), mixers (Braun), or pressure drop homogenizers (French press). The first two are equipped with cooling jackets to prevent thermal denaturation of the recombinant protein during disruption. High pressure homogenizers generally require multiple passes, with cooling in between, to achieve efficient breakage, but temperature effects remain a concern. Disruption may be enhanced by combining it with digestion with cell wall lytic enzymes (Baldwin and Robinson, 1992), although the economy of this approach has not been considered and disruption resistant yeast tend to remain resistant during subsequent homogenization passes. Key issues for industry in the choice of disruption device are the reproducibility of cell breakage and the ease of cleaning. After disruption, the cell extract is clarified by centrifugation (preceded by dilution of the extract for large-scale extraction).

F. Initial Purification

A purification strategy for the recombinant protein from either the cell-free culture supernatant or the clarified cell extract will depend on the unique properties of the heterologous protein. For exported proteins, the first step is usually high capacity ion-exchange chromatography. Ammonium sulfate fractionation of the supernatant (for small volumes) is usually unsuccessful due to the low protein concentration and the high solubility of the heavily glycosylated yeast proteins, although these proteins precipitate efficiently with 2 vol of 95% ethanol (MacKay *et al.*, 1988). For cell extracts, some initial fractionation may be achieved by pH adjustments, chilling out (of yeast proteins), or precipitation/crystallization cycles before chromatography. Initial chromatographic steps often involve the use of very short flow paths in columns that are designed to hold relatively fast flowing resins (Pharmacia) and in columns that operate in a radial flow configuration which also maximizes the throughput rate in the chromatographic step (Sepragen).

For therapeutic proteins expressed in yeast, it is essential to obtain high levels of purity with less than 5–10 ppm (parts per million) contaminating yeast protein. Obviously, with purity levels of 99.9999%, the validation and quality control challenges are significant. In addition, residual endotoxin specifications for many recombinant protein therapeutics must be less than 10 endotoxin units per applied dose, often a difficulty for recombinant proteins produced in *E. coli*. Finally, after purification, long-term storage of stable protein must be designed for each individual protein; lyophilization in appropriate buffers is frequently effective.

III. SUMMARY

The production of recombinant proteins in the yeast *S. cerevisiae* has proven to be an effective process for many proteins of both commercial and academic interest. As described in this chapter, at this time there is considerable information and many alternative strategies for yeast expression. The choice of the appropriate expression system (bacteria, yeast, fungi, insect cell lines, mammalian cells, transgenic animals) for a particular protein will depend primarily on the properties of the protein, levels required, and its intended purpose and secondarily on the availability of expertise, biological materials, equipment, etc. Perhaps the most exciting use of S. cerevisiae is in cell biology research, which takes advantage of the decades of classical genetics and more recent molecular genetic advances to understand complex problems in human biology. Thus, expression of recombinant proteins in yeast is likely to remain an important technology for many years.

ACKNOWLEDGMENTS

We thank Anne Bell, Bruce Carter, John Forstrom, Meher Irani, Mette Jars, Tanya Karwaki, Gary McKnight, Chris Raymond, and Andrzej Sledziewski for allowing use of their unpublished data; Molly Bernard for her cheerful and efficient literature searching; and Hal Blumberg, Diane Durnam, Don Foster, and Chris Raymond for valuable comments on the manuscript.

REFERENCES

Alber, T., and Kawasaki, G. (1982) *J. Mol. Appl. Genet.* **1**, 419–434.
Alberghina, L., Porro, D., Martegani, E., and Ranzi, B. M. (1991) *Biotechnol. Appl. Biochem.* **14**, 82–92.
Ammerer, G. (1983) *Methods Enzymol.* **101**, 192–201.
Ammerer, G., Hunter, C. P., Rothman, J. H., Saari, G. C., Valls, L. A., and Stevens, T. H. (1986) *Mol. Cell. Biol.* **6**, 2490–2499.
Antebi, A., and Fink, G. R. (1992) *Mol. Biol. Cell* **3**, 633–654.
Bailey, J. E., and Ollis, D. F. (1986) *Biochemical Engineering Fundamentals.* McGraw, New York.
Baldari, C., Murray, J. A., Ghiara, P., Cesareni, G., and Galeotti, C. L. (1987) *EMBO J.* **6**, 229–234.
Baldwin, C., and Robinson, C. W. (1992) *Lett. Appl. Microbiol.* **15**, 59–62.
Ballou, C. E. (1982) In *The Molecular Biology of the Yeast Saccharomyces: Metabolism and Gene Expression* (J. N. Strathern, E. W. Jones, and J. R. Broach, Eds.), pp. 335–360. Cold Spring Harbor Laboratory, Cold Spring Harbor, NY.
Ballou, C. E. (1990) *Methods Enzymol.* **185**, 440–470.
Bankaitis, V. A., Johnson, L. M., and Emr, S. D. (1986) *Proc. Natl. Acad. Sci. U.S.A.* **83**, 9075–9079.
Barr, P. J., Cousens, L. S., Lee-Ng, C. T., Medina-Selby, A., Masiarz, F. R., Hallewell, R. A., Chamberlain, S. H., Bradley, J. D., Lee, D., Steimer, K. S., Poulter, L., Burlingame, A. L., Esch, F., and Baird, A. (1988) *J. Biol. Chem.* **263**, 16471–16478.
Barr, P. J., Steimer, K. S., Sabin, E. A., Parkes, D., George-Nascimento, C., Stephans, J. C., Powers, M. A., Gyenes, A., Van Nest, G. A., Miller, E. T., Higgins, K. W., and Luciw, P. A. (1987) *Vaccine* **5**, 90–101.
Bathurst, I. C., Moen, L. K., Lujan, M. A., Gibson, H. L., Feucht, P. H., Pichuantes, S., Craik, C. S., Santi, D. V., and Barr, P. J. (1990) *Biochem. Biophys. Res. Commun.* **171**, 589–595.

Bauer, R., Paltauf, F., and Kohlwein, S. D. (1992) *Yeast* (Spec. Issue) 8, S564.

Bayne, M. L., Applebaum, J., Chicchi, G. G., Hayes, N. S., Green, B. G., and Cascieri, M. A. (1988) *Gene* 66, 235–244.

Becker, D. M., and Guarente, L. (1991) *Methods Enzymol.* 194, 182–187.

Beggs, J. D. (1978) *Nature (London)* 275, 104–109.

Beggs, J. D. (1981) *Proc. Alfred Benzon Symp.* 16, 383–396.

Benito, B., Portillo, F., and Lagunas, R. (1992) *FEBS Lett.* 300, 271–274.

Bennett, M. K., and Scheller, R. H. (1993) *Proc. Natl. Acad. Sci. U.S.A.* 90, 2559–2563.

Bennetzen, J. L., and Hall, B. D. (1982) *J. Biol. Chem.* 257, 3026–3031.

Better, M., and Horwitz, A. H. (1989) *Methods Enzymol.* 178, 476–496.

Bishop, P. D., Teller, D. C., Smith, R. A., Lasser, G. W., Gilbert, T., and Seale, R. L. (1990) *Biochemistry* 29, 1861–1869.

Bitter, G. A., Chen, K. K., Banks, A. R., and Lai, P.-H. (1984) *Proc. Natl. Acad. Sci. U.S.A.* 81, 5330–5334.

Bitter, G. A., and Egan, K. M. (1984) *Gene* 32, 263–274.

Bitter, G. A., and Egan, K. M. (1988) *Gene* 69, 193–207.

Bitter, G. A., Egan, K. M., Burnette, W. N., Samal, B., Fieschko, J. C., Peterson, D. L., Downing, M. R., Wypych, J., and Langley, K. E. (1988) *J. Med. Virol.* 25, 123–140.

Boeke, J. D., LaCroute, F., and Fink, G. R. (1984) *Mol. Gen. Genet.* 197, 345–346.

Boeke, J. D., and Sandmeyer, S. B. (1991) In *The Molecular and Cellular Biology of the Yeast Saccharomyces: Genome Dynamics, Protein Synthesis, and Energetics* (J. R. Broach, J. R. Pringle, and E. W. Jones, Eds.), pp. 193–261. Cold Spring Harbor Laboratory, Cold Spring Harbor, NY.

Bourbonnais, Y., Ash, J., Daigle, M., and Thomas, D. Y. (1993) *EMBO J.* 12, 285–294.

Bourbonnais, Y., Danoff, A., Thomas, D. Y., and Shields, D. (1991) *J. Biol. Chem* 266, 13203–13209.

Brake, A. J. (1990) *Methods Enzymol.* 185, 408–421.

Brake, A. J., Merryweather, J. P., Coit, D. G., Heberlein, U. A., Masiarz, F. R., Mullenbach, G. T., Urdea, M. S., Valenzuela, P., and Barr, P. J. (1984) *Proc. Natl. Acad. Sci. U.S.A.* 81, 4642–4646.

Broach, J. R. (1983) *Methods Enzymol.* 101, 307–325.

Broach, J. R., and Volkert, F. C. (1991) In *The Molecular and Cellular Biology of the Yeast Saccharomyces: Genome Dynamics, Protein Synthesis, and Energetics* (J. R. Broach, J. R. Pringle, and E. W. Jones, Eds.), pp. 297–331. Cold Spring Harbor Laboratory, Cold Spring Harbor, NY.

Bröker, M., Ragg, H., and Karges, H. E. (1987) *Biochim. Biophys. Acta* 908, 203–213.

Buckholz, R. G., and Gleeson, M. A. G. (1991) *Bio/Technol.* 9, 1067–1072.

Cabezón, T., De Wilde, M., Herion, P., Loriau, R., and Bollen, A. (1984) *Proc. Natl. Acad. Sci. U.S.A.* 81, 6594–6598.

Calderón-Cacia, M., Tekamp-Olson, P., Allen, J., and George-Nascimento, C. (1992). *Biochem. Biophys. Res. Commun.* 187, 1193–1199.

Caplan, S., Green, R., Rocco, J., and Kurjan, J. (1991) *J. Bacteriol.* 173, 627–635.

Casolaro, M. A., Fells, G., Wewers, M., Pierce, J. E., Ogushi, F., Hubbard, R., Sellers, S., Forstrom, J., Lyons, D., Kawasaki, G., and Crystal, R. G. (1987) *J. Appl. Physiol.* 63, 2015–2023.

Chang, C. N., Matteucci, M., Perry, L. J., Wulf, J. J., Chen, C. Y., and Hitzeman, R. A. (1986) *Mol. Cell. Biol.* 6, 1812–1819.

Chattoo, B. B., Sherman, F., Azubalis, D. A., Fjellstadt, T. A., Mehnert, D., and Ogur, M. (1979) *Genetics* 93, 51–66.

Chaudhuri, B., Steube, K., and Stephan, C. (1992) *Eur. J. Biochem.* 206, 793–800.

Chen, C. Y., Oppermann, H., and Hitzeman, R. A. (1984) *Nucleic Acids Res.* 12, 8951–8970.

Chisholm, V., Chen, C. Y., Simpson, N. J., and Hitzeman, R. A. (1990) *Methods Enzymol.* 185, 471–482.

Cigan, A. M., Pabich, E. K., and Donahue, T. F. (1988) *Mol. Cell. Biol.* **8**, 2964–2975.

Clare, J. J., Romanos, M. A., Rayment, F. B., Rowedder, J. E., Smith, M. A., Payne, M. M., Sreekrishna, K., and Henwood, C. A. (1991) *Gene* **105**, 205–212.

Cleves, A. E., and Bankaitis, V. A. (1992) *Adv. Microbial Physiol.* **33**, 73–144.

Coghlan, D., Jones, G., Denton, K. A., Wilson, M. T., Chan, B., Harris, R., Woodrow, J. R., and Ogden, J. E. (1992) *Eur. J. Biochem.* **207**, 931–936.

Coppella, S. J., and Dhurjati, P. (1987) *Biotechnol. Bioeng.* **29**, 679–689.

Coppella, S. J., and Dhurjati, P. (1990) *Biotechnol. Bioeng.* **35**, 356–374.

Cousens, L. S., Shuster, J. R., Gallegos, C., Ku, L. L., Stempien, M. M., Urdea, M. S., Sanchez-Pescador, R., Taylor, A., and Tekamp-Olson, P. (1987) *Gene* **61**, 265–275.

Darbre, A. (1986) In *Practical Protein Chemistry—A Handbook* (A. Darbre, Ed.), pp. 227–335. Wiley, New York.

Demolder, J., Fiers, W., and Contreras, R. (1992) *Gene* **111**, 207–213.

Derynck, R., Singh, A., and Goeddel, D. V. (1983) *Nucleic Acids Res.* **11**, 1819–1837.

Deshaies, R. J., Koch, B. D., Werner-Washburne, M., Craig, E. A., and Schekman, R. (1988) *Nature (London)* **332**, 800–805.

Dohmen, R. J., Strasser, A. W., Höner, C. B., and Hollenberg, C. P. (1991) *Yeast* **7**, 691–692.

Donahue, T. F., and Cigan, A. M. (1990) *Methods Enzymol.* **185**, 366–372.

Donovan, M. G., Veldman, S. A., and Bodley, J. W. (1992) *Yeast* **8**, 629–633.

Ecker, D. J., Stadel, J. M., Butt, T. R., Marsh, J. A., Monia, B. P., Powers, D. A., Gorman, J. A., Clark, P. E., Warren, F., Shatzman, A., and Crooke, S. T. (1989) *J. Biol. Chem.* **264**, 7715–7719.

Egel-Mitani, M., Hansen, M. T., Norris, K., Snel, L., and Fiil, N. P. (1988) *Gene* **73**, 113–120.

Elliott, S., Giffin, J., Suggs, S., Lau, E. P., and Banks, A. R. (1989) *Gene* **79**, 167–180.

Emr, S. D., Schekman, R., Flessel, M. C., and Thorner, J. (1983) *Proc. Natl. Acad. Sci. U.S.A.* **80**, 7080–7084.

Erhart, E., and Hollenberg, C. P. (1983) *J. Bacteriol.* **156**, 625–635.

Ernst, J. F. (1986) *DNA* **5**, 483–491.

Ernst, J. F. (1988) *DNA* **7**, 355–360.

Ernst, J. F., and Kawashima, E. (1988) *J. Biotechnol.* **7**, 1–10.

Esmon, B., Novick, P., and Schekman, R. (1981) *Cell* **25**, 451–460.

Etcheverry, T. (1990) *Methods Enzymol.* **185**, 319–329.

Etcheverry, T., Forrester, W., and Hitzeman, R. (1986) *Bio/Technol.* **4**, 726–730.

Eugster, H.-P., Sengstag, C., Hinnen, A., Meyer, U. A., and Würgler, F. E. (1991) *Biochem. Pharmacol.* **42**, 1367–1372.

Fieschko, J., and Ritch, T. (1986) *Chem. Eng. Commun.* **45**, 229–240.

Fieschko, J. C., Egan, K. M., Ritch, T., Koski, R. A., Jones, M., and Bitter, G. A. (1987) *Biotechnol. Bioeng.* **29**, 1113–1121.

Finnis, C., Goodey, A., Courtney, M., and Sleep, D. (1992) *Yeast* **8**, 57–60.

Fox, T. D., Folley, L. S., Mulero, J. J., McMullin, T. W., Thorsness, P. E., Hedin, L. O., and Costanzo, M. C. (1991) *Methods Enzymol.* **194**, 149–165.

Franzusoff, A., Rothblatt, J., and Schekman, R. (1991) *Methods Enzymol.* **194**, 662–674.

Fuller, R. S., Brake, A. J., Julius, D. J., and Thorner, J. (1985) In *Protein Transport and Secretion* (M.-J. Gething, Ed.), pp. 97–102. Cold Spring Harbor Laboratory, Cold Spring Harbor, NY.

Furukawa, K., Heinzle, E., and Dunn, I. J. (1983) *Biotechnol. Bioeng.* **25**, 2293–2318.

Gabriel, O., and Wang, S. F. (1969) *Anal. Biochem.* **27**, 545–554.

Gabrielsen, O. S., Reppe, S., Sæther, O., Blingsmo, O. R., Sletten, K., Gordeladze, J. O., Høgset, A., Gautvik, V. T., Alestrøm, P., Øyen, T. B., and Gautvik, K. M. (1990) *Gene* **90**, 255–262.

Garrido-Sanchez, L. E., Dantigny, P., and Pons, M.-N. (1988) *Biotechnol. Tech.* **2**, 17–22.

George-Nascimento, C., Gyenes, A., Halloran, S. M., Merryweather, J., Valenzuela, P., Steimer, K. S., Masiarz, F. R., and Randolph, A. (1988) *Biochemistry* **27**, 797–802.

Gibbons, W. R., and Westby, C. A. (1986) *Biotechnol. Lett.* **8**, 657–662.

Gietz, D., St. Jean, A., Woods, R. A., and Schiestl, R. H. (1992) *Nucleic Acids Res.* **20**, 1425.

Gill, G. S., Zaworski, P. G., Marotti, K. R., and Rehberg, E. F. (1990) *Bio/Technol.* **8**, 956–958.

Goeddel, D. V., Ed. (1990) *Methods Enzymol.*, **185**, 1–681.

Goff, C. G., Moir, D. T., Kohno, T., Gravius, T. C., Smith, R. A., Yamasaki, E., and Taunton-Rigby, A. (1984) *Gene* **27**, 35–46.

Goldstein, A., and Lampen, J. O. (1975) *Methods Enzymol.* **42**, 504–511.

Gooley, A. A., Classon, B. J., Marschalek, R., and Williams, K. L. (1991) *Biochem. Biophys. Res. Commun.* **178**, 1194–1201.

Gritz, L., and Davies, J. (1983) *Gene* **25**, 179–188.

Guarente, L. (1983) *Methods Enzymol.* **101**, 181–191.

Guisez, Y., Tison, B., Vandekerckhove, J., Demolder, J., Bauw, G., Haegeman, G., Fiers, W., and Contreras, R. (1991) *Eur. J. Biochem.* **198**, 217–222.

Guthrie, C., and Fink, G. R., Eds. (1991) *Methods Enzymol.* **194**, 1–933.

Hallewell, R. A., Imlay, K. C., Lee, P., Fong, N. M., Gallegos, C., Getzoff, E. D., Tainer, J. A., Cabelli, D. E., Tekamp-Olson, P., Mullenbach, G. T., and Cousens, L. S. (1991) *Biochem. Biophys. Res. Commun.* **181**, 474–480.

Hallewell, R. A., Mills, R., Tekamp-Olson, P., Blacher, R., Rosenberg, S., Ötting, F., Masiarz, F. R., and Scandella, C. J. (1987) *Bio/Technol.* **5**, 363–366.

Han, M., Kim, U. J., Kayne, P., and Grunstein, M. (1988) *EMBO J.* **7**, 2211–2218.

Haselbeck, A., and Tanner, W. (1983) *FEBS Lett.* **158**, 335–338.

Hasnain, S., Hirama, T., Tam, A., and Mort, J. S. (1992) *J. Biol. Chem.* **267**, 4713–4721.

Heinemann, J. A., and Sprague, G. F., Jr. (1991) *Methods Enzymol.* **194**, 187–195.

Henderson, R. C. A., Cox, B. S., and Tubb, R. (1985) *Curr. Genet.* **9**, 133–138.

Hill, J., Donald, K. A., and Griffiths, D. E. (1991) *Nucleic Acids Res.* **19**, 5791.

Hinnen, A., Hicks, J. B., and Fink, G. R. (1978) *Proc. Natl. Acad. Sci. U.S.A.* **75**, 1929–1933.

Hinnen, A., Meyhack, B., and Heim, J. (1989) *Biotechnology* **13**, 193–213.

Hiramatsu, R., Yamashita, T., Aikawa, J., Horinouchi, S., and Beppu, T. (1990). *Appl. Environ. Microbiol.* **56**, 2125–2132.

Hitzeman, R. A., Hagie, F. E., Hayflick, J. S., Chen, C. Y., Seeburg, P. H., and Derynck, R. (1982) *Nucleic Acids Res.* **10**, 7791–7808.

Hitzeman, R. A., Chen, C. Y., Hagie, F. E., Patzer, E. J., Liu, C.-C., Estell, D. A., Miller, J. V., Yaffe, A., Kleid, D. G., Levinson, A. D., and Oppermann, H. (1983a) *Nucleic Acids Res.* **11**, 2745–2763.

Hitzeman, R. A., Leung, D. W., Perry, L. J., Kohr, W. J., Levine, H. L., and Goeddel, D. V. (1983b) *Science* **219**, 620–625.

Hitzeman, R. A., Chen, C. Y., Hagie, F. E., Lugovoy, J. M., and Singh, A. (1984) In *Recombinant DNA Products: Insulin, Interferon, and Growth Hormone* pp. 47–65. CRC Press, Boca Raton, FL.

Hitzeman, R. A., Chen, C. Y., Dowbenko, D. J., Renz, M. E., Liu, C., Pai, R., Simpson, N. J., Kohr, W. J., Singh, A., Chisholm, V., Hamilton, R., and Chang, C. N. (1990) *Methods Enzymol.* **185**, 421–440.

Holland, J. P., and Holland, M. J. (1980) *J. Biol. Chem.* **255**, 2596–2605.

Horwitz, A. H., Chang, C. P., Better, M., Hellstrom, K. E., and Robinson, R. R. (1988) *Proc. Natl. Acad. Sci. U.S.A.* **85**, 8678–8682.

Ichikawa, K., Komiya, K., Suzuki, K., Nakahara, T., and Jigami, Y. (1989) *Agric. Biol. Chem.* **53**, 2687–2694.

Iijama, S., Yamashita, S., Matsunaga, K., Miura, H., Morikawa, M., Shimizu, K., Matsubara, M., and Kobayashi, T. (1987) *J. Chem. Technol. Biotechnol.* **40**, 203–213.

Ito, H., Fukuda, Y., Murata, K., and Kimura, A. (1983) *J. Bacteriol.* **153**, 163–168.

Jabbar, M. A., Sivasubramanian, N., and Nayak, D. P. (1985) *Proc. Natl. Acad. Sci. U.S.A.* **82**, 2019–2023.

Jacobs, E., Rutgers, R., Voet, P., Dewerchin, M., Cabezón, T., and de Wilde, M. (1989) *Gene* 80, 279–291.

Jigami, Y., Muraki, M., Harada, N., and Tanaka, H. (1986) *Gene* 43, 273–279.

Johnston, M. (1987). *Microbiol. Rev.* 51, 458–476.

Jones, E. W. (1990) *Methods Enzymol.* 185, 372–386.

Jones, E. W. (1991) *Methods Enzymol.* 194, 428–453.

Julius, D., Schekman, R., and Thorner, J. (1984) *Cell* 36, 309–318.

Kaster, K. R., Burgett, S. G., and Ingolia, T. D. (1984) *Curr. Genet.* 8, 353–358.

Kawase, Y., and Moo-Young, M. (1987) *Biotechnol. Bioeng.* 30, 345–347.

Kelly, J. D., Raines, E. W., Ross, R., and Murray, M. J. (1985) *EMBO J.* 4, 3399–3405.

Kendall, R. L., Yamada, R., and Bradshaw, R. A. (1990) *Methods Enzymol.* 185, 398–407.

Kim, J. Y., and Lee, Y. H. (1988) *Biotechnol. Bioeng.* 31, 755–758.

King, K., Dohlman, H. G., Thorner, J., Caron, M. G., and Lefkowitz, R. J. (1990) *Science* 250, 121–123.

Kingsman, S. M., Cousens, D., Stanway, C. A., Chambers, A., Wilson, M., and Kingsman, A. J. (1990) *Methods Enzymol.* 185, 329–341.

Kniskern, P. J., Hagopian, A., Montgomery, D. L., Burke, P., Dunn, N. R., Hofmann, K. J., Miller, W. J., and Ellis, R. W. (1986) *Gene* 46, 135–141.

Kohno, T., Carmichael, D. F., Sommer, A., and Thompson, R. C. (1990) *Methods Enzymol.* 185, 187–195.

Köhrer, K., and Domdey, H. (1991) *Methods Enzymol.* 194, 398–405.

Kotula, L., and Curtis, P. J. (1991) *Bio/Technol.* 9, 1386–1389.

Kramer, R. A., DeChiara, T. M., Schaber, M. D., and Hilliker, S. (1984) *Proc. Natl. Acad. Sci. U.S.A.* 81, 367–370.

Krezdorn, C. H., Watzele, G., Kleene, R. B., Ivanov, S. X., and Berger, E. G. (1993) *Eur. J. Biochem.* 212, 113–120.

Kronstad, J. W., Holly, J. A., and MacKay, V. L. (1987) *Cell* 50, 369–377.

Kukuruzinska, M. A., Bergh, M. L., and Jackson, B. J. (1987) *Annu. Rev. Biochem.* 56, 915–944.

Kurjan, J., and Herskowitz, I. (1982) *Cell* 30, 933–943.

Livi, G. P., Ferrara, A. A., Roskin, R., Simon, P. L., and Young, P. R. (1990) *Gene* 88, 297–301.

Livi, G. P., Lillquist, J. S., Miles, L. M., Ferrara, A., Sathe, G. M., Simon, P. L., Meyers, C. A., Gorman, J. A., and Young, P. R. (1991) *J. Biol. Chem.* 266, 15348–15355.

Lopes, T. S., Hakkaart, G.-J., Koerts, B. L., Raué, H. A., and Planta, R. J. (1991) *Gene* 105, 83–90.

Lopes, T. S., Klootwijk, J., Veenstra, A. E., Van der Aar, P. C., Van Heerikhuizen, H., Raué, H. A., and Planta, R. J. (1989) *Gene* 79, 199–206.

MacKay, V. L. (1983) *Methods Enzymol.* 101, 325–343.

MacKay, V. L. (1987) In *Biological Research on Industrial Yeasts* (G. G. Stewart, I. Russell, R. D. Klein, and R. R. Hiebsch, Eds.), Vol. 2, pp. 27–36. CRC Press, Boca Raton, FL.

MacKay, V. L., Armstrong, J., Yip, C., Welch, S., Walker, K., Osborn, S., Sheppard, P., and Forstrom, J. (1991) In *Structure and Function of the Aspartic Proteinases* (B. M. Dunn, Ed.), pp. 161–172. Plenum, New York.

MacKay, V. L., Welch, S. K., Insley, M. Y., Manney, T. R., Holly, J., Saari, G. C., and Parker, M. L. (1988) *Proc. Natl. Acad. Sci. U.S.A.* 85, 55–59.

MacKay, V. L., Yip, C., Welch, S., Gilbert, T., Seidel, P., Grant, F., and O'Hara, P. (1990) In *Recombinant Systems in Protein Expression* (K. K. Alitalo, M.-L. Huhtala, J. Knowles, and A. Vaheri, Eds.), pp. 25–36. Elsevier Science, Amsterdam.

Martens, F. B., Egberts, G. T. C., Kempers, J., Robles de Medina, M., and Welten, H. G. J. (1987) *Monogr. Eur. Brew. Conv.* 12, 95–107.

Maruyama, T., Gojobori, T., Aota, S., and Ikemura, T. (1986) *Nucleic Acids Res.* 14(Suppl.), r151–r197.

McAleer, W. J., Buynak, E. B., Maigetter, R. Z., Wampler, D. E., Miller, W. J., and Hilleman, M. R. (1984) *Nature (London)* 307, 178–180.

McHale, M. M., Cieslinski, L. B., Eng, W.-K., Johnson, R. K., Torphy, T. J., and Livi, G. P. (1991) *Mol. Pharmacol.* 39, 109–113.

Mellor, J., Dobson, M. J., Roberts, N. A., Kingsman, A. J., and Kingsman, S. M. (1985) *Gene* 33, 215–226.

Melnick, L. M., Turner, B. G., Puma, P., Price-Tillotson, B., Salvato, K. A., Dumais, D. R., Moir, D. T., Broeze, R. J., and Avgerinos, G. C. (1990) *J. Biol. Chem.* 265, 801–807.

Merril, C. R. (1990) *Methods Enzymol.* 182, 477–488.

Meyhack, B., Bajwa, W., Rudolph, H., and Hinnen, A. (1982) *EMBO J.* 1, 675–680.

Miller, J. (1972) *Experiments in Molecular Genetics.* Cold Spring Harbor Laboratory, Cold Spring Harbor, NY.

Miyajima, A., Otsu, K., Schreurs, J., Bond, M. W., Abrams, J. S., and Arai, K. (1986) *EMBO J.* 5, 1193–1197.

Miyanohara, A., Toh-e, A., Nozaki, C., Hamada, F., Ohtomo, N., and Matsubara, K. (1983) *Proc. Natl. Acad. Sci. U.S.A.* 80, 1–5.

Moir, D. T., and Davidow, L. S. (1991) *Methods Enzymol.* 194, 491–507.

Moir, D. T., and Dumais, D. R. (1987) *Gene* 56, 209–217.

Mylin, L. M., Hofmann, K. J., Schultz, L. D., and Hopper, J. E. (1990) *Methods Enzymol.* 185, 297–308.

Nyiri, L. K., Jefferis, R. P., III, and Humphrey, A. E. (1974) *Biotechnol. Bioeng. Symp.* 4, 613–628.

O'Connor, G. M., Sanchez-Riera, F., and Cooney, C. L. (1992) *Biotechnol. Bioeng.* 39, 293–304.

Ogden, J. E., Coghlan, D., Jones, G., Denton, K. A., Harris, R., Chan, B., Woodrow, J., and Wilson, M. T. (1992) *Biomat. Artif. Cells Immobilization Biotechnol.* 20, 473–475.

Ogden, J. E., Woodrow, J., Perks, K., Harris, R., Coghlan, D., and Wilson, M. T. (1991) *Biomat. Artif. Cells Immobilization Biotechnol.* 19, 457.

Orlean, P., Kuranda, M. J., and Albright, C. F. (1991) *Methods Enzymol.* 194, 682–697.

Orr-Weaver, T. L., Szostak, J. W., and Rothstein, R. J. (1983) *Methods Enzymol.* 101, 228–245.

Östman, A., Bäckström, G., Fong, N., Betsholtz, C., Wernstedt, C., Hellman, U., Westermark, B., Valenzuela, P., and Heldin, C.-H. (1989) *Growth Factors* 1, 271–281.

Oura, E. (1973) *Biotechnol. Bioeng. Symp.* 4, 117–127.

Picard, D., Schena, M., and Yamamoto, K. R. (1990) *Gene* 86, 257–261.

Phillips, D. H., and Johnson, M. J. (1961) *J. Biochem. Microbiol. Technol. Eng.* 3, 261–275.

Price, V., Mochizuki, D., March, C. J., Cosman, D., Deeley, M. C., Klinke, R., Clevenger, W., Gillis, S., Baker, P., and Urdal, D. (1987) *Gene* 55, 287–293.

Price, V. L., Taylor, W. E., Clevenger, W., Worthington, M., and Young, E. T. (1990) *Methods Enzymol.* 185, 308–318.

Raymond, C. K., Howald-Stevenson, I., Vater, C. A., and Stevens, T. H. (1992) *Mol. Biol. Cell* 3, 1389–1402.

Redding, K., Holcomb, C., and Fuller, R. S. (1991) *J. Cell Biol.* 113, 527–538.

Rine, J., Hansen, W., Hardeman, E., and Davis, R. W. (1983) *Proc. Natl. Acad. Sci. U.S.A.* 80, 6750–6754.

Rols, J. L., Condoret, J. S., Fonade, C., and Goma, G. (1990) *Biotechnol. Bioeng.* 35, 427–435.

Romanos, M. A., Scorer, C. A., and Clare, J. J. (1992) *Yeast* 8, 423–488.

Rose, A. B., and Broach, J. R. (1990) *Methods Enzymol.* 185, 234–279.

Rose, M., and Botstein, D. (1983) *Methods Enzymol.* 101, 167–180.

Rosenberg, S., Coit, D., and Tekamp-Olson, P. (1990) *Methods Enzymol.* 185, 341–351.

Rothman, J. H., and Stevens, T. H. (1986) *Cell* 47, 1041–1051.

Rothstein, R. (1991) *Methods Enzymol.* 194, 281–301.

Rothstein, R. J. (1983) *Methods Enzymol.* 101, 202–211.

Ruby, S. W., Szostak, J. W., and Murray, A. W. (1983) *Methods Enzymol.* **101**, 253–269.
Rudolph, H. K., Antebi, A., Fink, G. R., Buckley, C. M., Dorman, T. E., LeVitre, J., Davidow, L. S., Mao, J. I., and Moir, D. T. (1989) *Cell* **58**, 133–145.
Russell, D. W., Smith, M., Cox, D., Williamson, V. M., and Young, E. T. (1983) *Nature (London)* **304**, 652–654.
Sardonini, C. A., and DiBasio, D. (1987) *Biotechnol. Bioeng.* **29**, 469–475.
Schekman, R. (1985) *Annu. Rev. Cell Biol.* **1**, 115–143.
Schena, M., Picard, D., and Yamamoto, K. R. (1991) *Methods Enzymol.* **194**, 389–398.
Schneider, J. C., and Guarente, L. (1991) *Methods Enzymol.* **194**, 373–388.
Schultz, L. D., Tanner, J., Hofmann, K. J., Emini, E. A., Condra, J. H., Jones, R. E., Kieff, E., and Ellis, R. W. (1987) *Gene* **54**, 113–123.
Settineri, C. A., Medzihradsky, K. F., Masiarz, F. R., Burlingame, A. L., Chu, C., and George-Nascimento, C. (1990) *Biomed. Environ. Mass Spectrom.* **19**, 665–676.
Sharp, P. M., Tuohy, T. M., and Mosurski, K. R. (1986) *Nucleic Acids Res.* **14**, 5125–5143.
Sherman, F. (1991) *Methods Enzymol.* **194**, 3–21.
Sherman, F., and Hicks, J. (1991) *Methods Enzymol.* **194**, 21–37.
Shuster, J. R. (1991) *Curr. Opin. Biotechnol.* **2**, 685–690.
Shuster, J. R., Lee, H., and Moyer, D. L. (1990) *Yeast* **6**, S79.
Singh, A., Lugovoy, J. M., Kohr, W. J., and Perry, L. J. (1984) *Nucleic Acids Res.* **12**, 8927–8938.
Sledziewski, A. Z., Bell, A., Kelsay, K., and MacKay, V. L. (1988) *Bio/Technol.* **6**, 411–416.
Sledziewski, A. Z., Bell, A., Yip, C., Kelsay, K., Grant, F. J., and MacKay, V. L. (1990) *Methods Enzymol.* **185**, 351–366.
Sleep, D., Belfield, G. P., and Goodey, A. R. (1990) *Bio/Technol.* **8**, 42–46.
Smith, R. A., Duncan, M. J., and Moir, D. T. (1985) *Science* **229**, 1219–1224.
Stearns, T., Ma, H., and Botstein, D. (1990) *Methods Enzymol.* **185**, 280–297.
Steube, K., Chaudhuri, B., Märki, W., Merryweather, J. P., and Heim, J. (1991) *Eur. J. Biochem.* **198**, 651–657.
Strahl-Bolsinger, S., Immervoll, T., and Tanner, W. (1992) *Yeast* (Spec. Issue) **8**, S489.
Strahl-Bolsinger, S., and Tanner, W. (1991) *Eur. J. Biochem.* **196**, 185–190.
Suomalainen, H. (1969) *Antonie van Leeuwenhoek; J. Microbiol. Serol.* **35**(Suppl. Yeast Symp.), 83–111.
Takabayashi, K., Märki, W., Wolf, D. H., and Heim, J. (1990) *Yeast* (Spec. Issue) **6**, S490.
Taniguchi, M., Hoshino, K., Itoh, T., Kumakura, H., and Fujii, M. (1992) *Biotechnol. Bioeng.* **39**, 886–890.
Taniyama, Y., Yamamoto, Y., Nakao, M., Kikuchi, M., and Ikehara, M. (1988) *Biochem. Biophys. Res. Commun.* **152**, 962–967.
Thim, L., Hansen, M. T., Norris, K., Hoegh, I., Boel, E., Forstrom, J., Ammerer, G., and Fiil, N. P. (1986) *Proc. Natl. Acad. Sci. U.S.A.* **83**, 6766–6770.
Tokunaga, T., Iwai, S., Gomi, H., Kodama, K., Ohtsuka, E., Ikehara, M., Chisaka, O., and Matsubara, K. (1985) *Gene* **39**, 117–120.
Towbin, H., Staehelin, T., and Gordon, J. (1979) *Proc. Natl. Acad. Sci. U.S.A.* **76**, 4350–4354.
Trueheart, J., Boeke, J. D., and Fink, G. R. (1987) *Mol. Cell. Biol.* **7**, 2316–2328.
Tsai, P. K., Frevert, J., and Ballou, C. E. (1984) *J. Biol. Chem.* **259**, 3805–3811.
Tschopp, J. F., and Cregg, J. M. (1991) *Biotechnology* **18**, 305–322.
Van der Aar, P. C., Van den Heuvel, J. J., Roling, W. F., Raue, H. A., Stouthamer, A. H., and Van Verseveld, H. W. (1992) *Yeast* **8**, 47–55.
Vicentini, A. M., Kieffer, B., Matthies, R., Meyhack, B., Hemmings, B. A., Stone, S. R., and Hofsteenge, J. (1990) *Biochemistry* **29**, 8827–8834.
Wagenbach, M., O'Rourke, K., Vitez, L., Wieczorek, A., Hoffman, S., Durfee, S., Tedesco, J., and Stetler, G. (1991) *Bio/Technol.* **9**, 57–61.
Wahleithner, J. A., Li, L. M., and Lagarias, J. C. (1991) *Proc. Natl. Acad. Sci. U.S.A.* **88**, 10387–10391.

Wang, H. Y., Lawless, R. J., Jr., and Lin, J.-E. (1988) *Process Biochem.* 23, 23–27.

Webster, T. D., and Dickson, R. C. (1983) *Gene* 26, 243–252.

Wilkinson, K. D. (1990) *Methods Enzymol.* 185, 387–397.

Winston, F., Chumley, F., and Fink, G. R. (1983) *Methods Enzymol.* 101, 211–228.

Wittrup, K. D., Bailey, J. E., Ratzkin, B., and Patel, A. (1990) *Biotechnol. Bioeng.* 35, 565–577.

Woolford, C. A., Daniels, L. B., Park, F. J., Jones, E. W., Van Arsdell, J. N., and Innis, M. A. (1986) *Mol. Cell. Biol.* 6, 2500–2510.

Wu, R., Grossman, L., and Moldave, K., Eds. (1983) *Methods Enzymol.* 101, 1–746.

Zaworski, P. G., Marotti, K. R., MacKay, V., Yip, C., and Gill, G. S. (1989) *Gene* 85, 545–551.

Ziegler, F. D., Maley, F., and Trimble, R. B. (1988) *J. Biol. Chem.* 263, 6986–6992.

Zsebo, K. M., Lu, H.-S., Fieschko, J. C., Goldstein, L., Davis, J., Duker, K., Suggs, S. V., Lai, P.-H., and Bitter, G. A. (1986) *J. Biol. Chem.* 261, 5858–5865.

5

In Vitro Mutagenesis

THIERRY VERNET*

ROLAND BROUSSEAU†

*Institut de biologie structurale, CEA/CNRS, 41 avenue des martyrs, 38027 Grenoble Cedex 1, France; †Biotechnology Research Institute, National Research Council, 6100 Royalmount Avenue, Montreal, Quebec, Canada H4P 2R2

I. INTRODUCTION

Advances in X-ray crystallography and molecular modeling have led to the detailed description of numerous protein structures. Unfortunately, seeing the intricate shape of a protein does little to advance our knowledge of how this unique shape is built (the folding problem) nor how the protein functions. At best, careful observation of the structure and comparison with closely related structures provide a series of hypotheses to be tested. These tests can be carried out using the perturbation analysis approach that is familiar to scientists. The change of one or a few amino acid side chains within a protein is followed by an assessment of the consequences of these alterations. The fact that this can now be done routinely, with the help of commercial kits, testifies to the profound impact of the techniques of *in vitro* site-directed mutagenesis.

Over the past 10 years *in vitro* site-directed modification of DNA sequences has evolved into a central method of molecular biology and is having a major impact in both academic research and commercial biotechnology. Figure 1 gives examples of the areas of application of this technique. Noncoding DNA sequences can be manipulated to probe regulatory regions of DNA (Oliphant and Struhl, 1987; Gaal *et al.*, 1989) or to alter

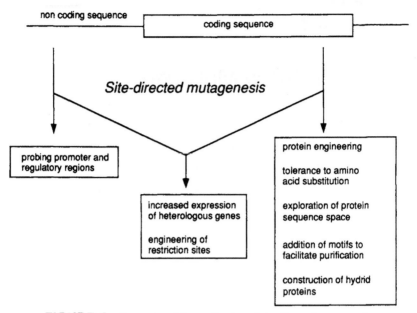

FIGURE 1   Examples of the application of site-directed mutagenesis.

the level of gene expression (Min *et al.*, 1988). Coding sequences are now routinely modified in a variety of ways to get insights on fundamental properties of proteins including stability (Albert, 1989; Shortle, 1989), protein folding (Goldenberg *et al.*, 1989), and enzyme specificity (Carter *et al.*, 1989; Evnin *et al.*, 1990; Khouri *et al.*, 1991) and mechanism (Hermes *et al.*, 1990; Wagner and Benkovic, 1990; Ménard *et al.*, 1991). Variants of proteins have also been designed or selected to facilitate purification and identification (Ong *et al.*, 1989).

Efficient *in vitro* mutagenesis procedures are also essential for the identification of neutral mutations. Determination of the spectrum of functional amino acid substitutions in proteins had classically relied upon comparative studies of protein variants isolated from natural sources. This limitation has now been removed and it is now possible to introduce any desired amino acid substitution at any given position within a sequence. More recently, efficient *in vitro* saturation mutagenesis has been instrumental in probing the informational content of noncoding sequences (Zaruchi-Schultz *et al.*, 1982; Hill *et al.*, 1986) and protein sequences (Reidhaar-Olson and Sauer, 1988; Lim and Sauer, 1989).

Many methods have been developed for modification of the structure of DNA *in vitro*. They can be classified in two groups: those methods using base-specific chemical mutagens such as sodium bisulfite and those where a piece of DNA (synthetic or PCR-amplified) is used as a mutagen to modify another piece of DNA. Recent advances include reliable synthesis of large oligonucleotides (up to 120 nucleotides), the control of the synthesis of mixture of oligonucleotides, the use of PCR for *in vitro* mutagenesis, and the development of selection systems for mutated DNA.

This chapter does not aim at an exhaustive review of the literature; indeed, even the outstanding authoritative review of Smith (1985) had to renounce comprehensiveness in favor of emphasizing recent developments and trends. Other recent reviews such as the one by Smith (1990) will no doubt benefit the reader requiring more extensive coverage of the proliferating literature on this topic. We have focused on recent improvements and uses of the *in vitro* site-directed mutagenesis. We mainly cover methods using synthetic DNA or DNA amplified by the polymerase chain reaction (PCR) as mutagens. Chemical modifications of the DNA will only be briefly mentioned. Some applications of the techniques will be cited to illustrate recent advances. A summary of key methodological references and indications about some useful commercial products will be given for those who may be entering the field (Appendix).

II. *IN VITRO* MUTAGENESIS: PRINCIPLE AND VARIATIONS

Before the advent of genetic engineering the only practical way available to mutate genes required the treatment of whole cells with various mutagenic agents (chemicals, UV or gamma radiation) followed by phenotypic characterization and genetic analysis. This still powerful means to identify new cellular functions is now used together with *in vitro* mutagenesis to generate high resolution information on the structure/function relationships of genetic loci.

The basic technique underlying most of the recent work in *in vitro* mutagenesis can be traced back to the work of Hutchison and Edgell (1971), who annealed DNA fragments from the double-stranded form of phage ϕX174 to intact single-stranded phage to introduce mutations. Once synthetic oligonucleotides and rapid DNA sequencing became widely available in the late seventies, the method was modified accordingly (Hutchison *et al.*, 1978) and rapidly blossomed into the powerful tool we know today.

The modern methods of *in vitro* DNA manipulations have considerably broadened the scope of mutagenesis by allowing the construction of specific alterations. The DNA fragment of interest is first isolated from the cell and introduced into a replicon using recombinant DNA procedures. Targeted modification of the cloned DNA is then performed *in vitro* using either chemical or synthetic DNA mutagens. The modified gene is then introduced back into the cell genome or as an episome and the consequences of the genetic change are evaluated. A comparison of the classical and modern procedures is schematized in Fig. 2.

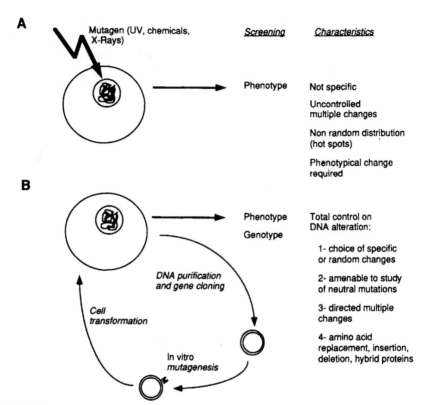

FIGURE 2   Comparison of cellular and *in vitro* mutagenesis. (A) Mutagenesis of live cells. (B) Site-directed *in vitro* mutagenesis. (M) Site of mutagenesis.

The availability of synthetic DNA has enhanced dramatically the potential of *in vitro* mutagenesis. The desired change is built in a fragment of DNA, usually a synthetic oligonucleotide (often called oligo for short), prior to its incorporation into the DNA to be mutated. Alternatively, oligonucleotides can be used as primers in PCR and the amplified DNA used in the mutagenesis reaction. The type of mutations may be restricted to a single base, or equally well the researcher may create deletions or insertions of almost any size, or generate a collection of random mutants. The incorporation of the mutagenic DNA in the template is achieved *in vitro* either by ligation (Fig. 3B; Section II,C examines this in detail) or by a single (Zoller

FIGURE 3   Comparison of cassette and heteroduplex mutagenesis. (A) Heteroduplex mutagenesis. The possible origin of the ssDNA template is indicated. (B) Cassette mutagenesis. Synthetic dsDNA containing the desired mutation and protruding flanking restriction sites (S1 and S2) is ligated into a plasmid vector as replacement of the wild-type sequence.

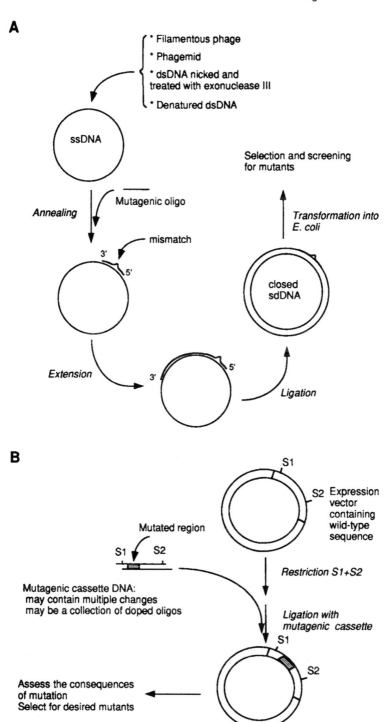

A

* Filamentous phage
* Phagemid
* dsDNA nicked and treated with exonuclease III
* Denatured dsDNA

ssDNA

Annealing

Mutagenic oligo

mismatch

3'
5'

Extension

3'   5'

Ligation

closed sdDNA

Transformation into E. coli

Selection and screening for mutants

B

S1

S2 Expression vector containing wild-type sequence

Mutated region

S1   S2

Mutagenic cassette DNA:
may contain multiple changes
may be a collection of doped oligos

Restriction S1+S2

Ligation with mutagenic cassette

S1

S2

Assess the consequences of mutation
Select for desired mutants

and Smith, 1982; Section II,B) or multiple polymerase reaction cycles (Fig. 4; Section II,D; see also Mullis *et al.*, 1986; Saiki *et al.*, 1988). The efficiency of mutagenesis, that is, the ratio of mutant to wild-type obtained, is comparable for both methods and in some cases reaches close to 100%. This level of efficiency alleviates the need for phenotypical or physical screening for the isolation of mutants.

A. *In Vitro* Chemical Mutagenesis

Chemical modification of DNA is efficient for the introduction of random mutations. For instance, the transition of cytosine to thymine has been obtained by treatment of the single-stranded DNA (ssDNA) target by sodium bisulfite (Shortle and Botstein, 1983). Targeting of the mutation has been achieved through various elegant procedures based on localized exposition of ssDNA within a double-stranded DNA (dsDNA) (reviewed by Pine and Huang, 1987). However, the procedure is usually cumbersome and control of the distribution and efficiency of mutation is difficult. Identification of mutants relies in most cases upon phenotypical or physical analysis. Consequently, the chemical method is poorly adapted for the identification of neutral mutations. An elegant if somewhat labor-intensive procedure has been devised by Myers *et al.* (1985) to circumvent these difficulties and allow for mutant screening in the absence of phenotypic selection. In one instance (Hermes *et al.*, 1990), a direct comparison between the chemical method of Myers *et al.* (1985) and the oligonucleotide method was performed; better results were obtained for the oligonucleotide method, especially in terms of obtaining an even distribution of mutants throughout the target region.

B. Oligonucleotide-Based Mutagenesis

In their landmark paper Zoller and Smith (1982) described the basic steps for *in vitro* oligonucleotide site-directed mutagenesis. First, a mismatched oligonucleotide is hybridized to the target region of a ssDNA template. The heteroduplex is then extended by a DNA polymerase and the double-stranded DNA closed by the T4 DNA ligase of *Escherichia coli*. Transfection of *E. coli* is followed by screening for phage plaques containing the mutation. These various steps are depicted in Fig. 3A. Screening is customarily performed through oligonucleotide hybridization. Several references are available to predict the melting temperature of oligonucleotide duplexes; the reader is referred to the review of Smith (1985) for a discussion of this topic.

The large fragment of the *E. coli* polymerase (Klenow fragment) has been widely used since it produces complete duplex, even in the presence of DNA secondary structures, due to an intrinsic helicase activity. Moreover,

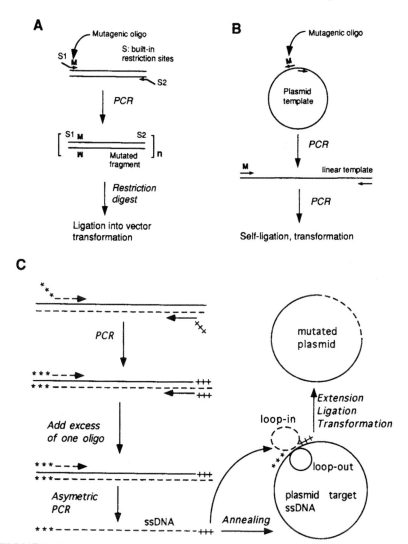

FIGURE 4   Some examples of the use of PCR in site-directed mutagenesis. (A) Incorporation of the mutagenic oligonucleotide by PCR is followed by cloning as in the cassette mutagenesis procedure. (B) Incorporation of mutagenic oligonucleotide by PCR using full plasmid as
template. The circular template DNA is converted into linear dsDNA in the first few cycles of
the reaction and thereafter amplified. The linear amplified DNA is recircularized before transformation into *E. coli.* (C) Application of the PCR for simultaneous large insertion and deletion. A source mutagenic ssDNA fragment is first amplified by PCR using oligonucleotide
complementary to the flanking region to be deleted in the target DNA. A second round of
asymmetric PCR provides the mutagenic ssDNA. This ssDNA is used in the heteroduplex
mutagenesis reaction as in Fig. 3A. The two strands of DNA are differentiated into plain and
interrupted lines.

the engineered enzyme lacks the 5' exonuclease editing activity, a positive feature for mutagenesis inasmuch as it prevents nucleolysis of the oligonucleotide mutagenic primers. Displacement of the mutagenic oligonucleotide has been reduced by controlling the ratio of Klenow fragment to ligase in the reaction (Gillam and Smith, 1979), by using a second primer located 5' to the mutagenic primer (Schold *et al.*, 1984; Norris *et al.*, 1983) or by using partial ssDNA exposed region for template (Kramer *et al.*, 1982). The phage T4 and T7 DNA polymerases have also been used for this application since in addition to being devoid of 5' exonuclease activity they do not have the strand displacement capabilities. Detailed protocols for the reaction have been published (Carter, 1987; Kramer and Fritz, 1987; Kunkel *et al.*, 1987).

The principle of the method remains unchanged but over the years the efficiency has been improved by the introduction of various schemes allowing direct selection for the mutant (see Section II,G). Also, the method has been extended to double-stranded DNA templates (for a review see Smith, 1985), eliminating the time-consuming step of recloning the mutated DNA into proper vectors. Many multipurpose plasmids incorporating a filamentous phage origin of replication have been built, again to avoid a cloning step (see Section II,E,2).

The types of mutations that are now introduced into DNA have been extended from single amino acid replacement to multiple substitutions, deletions, or insertions of almost any size, and localized random mutagenesis. Protocols derived from the primer extension method are still predominantly used. Indeed, the method is simple, reasonably inexpensive, and does not require conveniently positioned restriction sites as does cassette mutagenesis (Section II,C) nor it is as susceptible to unwanted mutations as PCR-based procedures (see Section II,D,5). Computer programs are now available to assist the design of mutagenesis experiment (Zheng *et al.*, 1988) and the management of mutants generated by saturation mutagenesis (Waye and Gupta, 1989; Dubnick *et al.*, 1991).

C. Cassette Mutagenesis

In its simplest form two complementary DNA strands containing the desired genetic alteration are chemically synthesized and incorporated by ligation in place of the wild-type corresponding DNA region (Fig. 3B). The method is very efficient but requires conveniently positioned restriction sites. Its main application has been for localized random mutagenesis where one or a few amino acids are exchanged by all possible replacements (Wells *et al.*, 1985; Estell *et al.*, 1985; Lim and Sauer, 1989; Karlsson *et al.*, 1991). A few researchers have also used this method to explore more complex problems; these include the study of structure–function relationships in proteins of unknown structure (Clarke *et al.*, 1988), the exploration of the structural requirements of signal peptides using homopolymeric substitutions (Chou and Kendall, 1990), and the discovery of distinct functional roles for the two phosphatase-like domains of protein tyrosine phosphatase (PTPase) (Streuli *et al.*, 1990).

A procedure requiring the chemical synthesis of DNA of only one strand has been developed (Hill *et al.*, 1987). This method has been adapted for the cloning of sequences with any degree of degeneracy and has been used to define regulatory DNA consensus sequences (Hill *et al.*, 1986).

D. PCR-Based Mutagenesis

The PCR methodology has had a truly revolutionary impact on molecular genetics, and this impact has naturally been felt as well in the specialized area of site-specific mutagenesis. Primer extension and cassette mutagenesis procedures have recently been greatly simplified by the PCR procedure. The PCR technique, initially developed for the specific amplification of DNA, is based upon the use of two oligonucleotides that hybridize each to one strand of the target DNA. Exponential amplification of the DNA region that encompasses the oligonucleotides is achieved by repeated cycles of denaturation → annealing → DNA polymerization. The isolation of thermostable DNA polymerases (Saiki *et al.*, 1988) and automation of the cycles on commercially available equipment have greatly simplified the procedure that has rapidly been adapted for *in vitro* mutagenesis (Higuchi *et al.*, 1988).

PCR-based procedures are usually rapid, simple, and efficient in the generation of mutants. Base substitutions, insertions, or deletions have been incorporated into the primer oligonucleotides. This is possible as mismatched primers are still capable of priming the PCR reaction. Indeed, this property is at the basis of cloning (Lee *et al.*, 1988) and detection of DNA based on degenerate oligonucleotides sequences.

1. Use of PCR for Point Mutations, Small Insertions, and Deletions

These have become so ubiquitous as to defy enumeration, therefore only a few representative cases are mentioned here. Several of the early protocols reported in the literature required cloning of the PCR-mutated fragment as a cassette between unique restriction sites included in the design of the oligonucleotide primers (Kadowaki *et al.*, 1989; Vallette *et al.*, 1989; Dulau *et al.*, 1989). These methods resulted in creating a PCR variant of the cassette mutagenesis approach mentioned earlier, but with the important difference that the cassette can now easily encompass several hundred bases, instead of the 90- to 120-base limitation inherent in synthetic oligonucleotides (Fig. 4A). The requirement for conveniently located restriction sites has been eliminated in more recent protocols that include a two-step PCR protocol (Kammann *et al.*, 1989), a procedure based upon overlap extension (Higuchi *et al.*, 1988, Ho *et al.*, 1989), the use of asymmetric PCR (Perrin and Gilliland, 1990), and DNA extension of ssDNA template using PCR amplified DNA (Near, 1992). In fact, PCR amplification of entire plasmids eliminates the need for cloning of DNA fragments and requires a simple self-ligation step (Hemsley *et al.*, 1989; Imai *et al.*, 1991; Street *et al.*, 1991; Jones and Howard, 1991) (Fig. 4B).

Herlitze and Koenen (1990) introduced all three types of mutations, i.e., substitutions, deletions, and insertions, with an efficiency of between 50 and 90% using PCR and conventional cloning techniques. Landt *et al.* (1990) offer a method suitable for a short gene cloned between two regions where universal primers are available. Nelson and Long (1989) use four primers but are not limited to any specialized cloning vector. Sarkar and Sommer (1990) use three primers, one of which contains an SP6 phage promoter allowing for *in vitro* expression without the need for cloning. Near (1992) used PCR-amplified fragments as mutagenic DNA in Kunkel-type mutagenesis to assemble rapidly a library of variable region immunoglobulin genes. PCR-mediated site-directed mutagenesis has also been used to diagnose human genetic mutations (Schwartz *et al.*, 1991; Groppi *et al.*, 1990) through the creation of distinctive restriction sites in the amplified DNA from the mutant phenotypes.

2. Introduction of Large Deletions and Insertions

The use of single-stranded DNA amplified by asymmetric PCR (Kreitman and Landweber, 1989) has been extended to create large insertions, large deletions, or both simultaneously. We have used this strategy to exchange, in a single step, DNA regulatory regions or protein domains. The protocol, schematized in Fig. 4C, was used to exchange baculovirus promoters in the transfer plasmid IpDC127 (Vernet *et al.*, 1990). The 144-bp polyhedrin promoter was replaced by the 356-bp immediate early 1 (IE-1) promoter. Similarly we have substituted the 321-bp gene fragment encoding the Pro region of papain for the 294-bp fragment encoding the Pro region of cathepsin S (D. Tessier, T. Vernet, and D. Y. Thomas, unpublished results). Sequencing of the loop-in region and flanking sequences revealed a single unexpected point mutation for one of the constructions that was later corrected by site-directed mutagenesis.

3. Construction of Hybrid Proteins: Overlap Extension Approach to Protein Domain Switching

Following the publication of Higuchi *et al.* (1988) on the use of PCR in mutagenesis, Ho *et al.* (1989) developed the so-called splicing overlap extension approach to build first a 600-bp hybrid gene, then a 900-bp hybrid gene (Horton *et al.*, 1989) via PCR. Villarreal and Long (1991) applied the overlap extension PCR approach to the construction of a hybrid gene of moderate length (approximately 600 bp). Their construction was made more difficult by the presence of regions of high G + C content in their target genes, which are known to impair proper amplification. The technical solution used was to perform the amplifications in the presence of 10% dimethyl sulfoxide (DMSO), although this may have caused additional errors (see below, Section II,D,6).

4. Introduction of Random Mutations through Absence of Proofreading or through Chemical Means

The PCR reaction performed with nonproofreading polymerases such as the one isolated from *Thermus aquaticus* (*Taq* polymerase) can be used as a method for introducing randomly spaced point mutations. The *Taq* polymerase lacks a 3' to 5' exonucleolytic editing activity (Tindall and Kunkel, 1988) and this property has been exploited for random mutagenesis (Leung *et al.*, 1989; Zhou *et al.*, 1991). Random mutagenesis has also been achieved using a hybrid protocol combining chemical saturation mutagenesis and PCR (Diaz *et al.*, 1992).

5. Considerations on the Error Rate of Thermostable DNA Polymerases

The actual error rates of thermostable DNA polymerases have been the subject of several reports, with a considerable range of estimates being given by the various authors, as can be seen from the results summarized in Table I. This range can be explained with the understanding that researchers used different methods to assess fidelity, the principal ones being sequencing of several isolates of the same amplified target fragment versus the mutational assay of Tindall and Kunkel (1988). One must also consider that the error rate for a given polymerase is influenced by several factors such as the nature of the substrate, concentrations of dNTPs, concentration of Mg^{2+}, temperature, pH, and the duration of extension cycle. What are the take-home lessons from these various error rates?

a. The error rates for the *Taq* polymerase will in fact vary from experiment to experiment, dependent upon the factors mentioned above.

b. If the numbers available so far from the literature are looked upon for an average number, one is led to expect one error every 500 to 600 nucleotides after a typical 25-cycle amplification, which is high enough to make sequencing imperative. Indeed, we strongly urge full sequencing of the target gene or DNA region even when performing classical site-specific mutagenesis experiments, where the more faithful Klenow polymerase is used with a single cycle of elongation.

c. There are not yet enough data on the newer thermostable polymerases to make a definite recommendation in their favor on the basis of their claimed higher fidelity.

A more detailed review on the fidelity of DNA polymerases used in PCR is to be found in the publication of Eckert and Kunkel (1991).

TABLE I   Reported Error Rates per Incorporated Nucleotide for Various DNA Polymerases

Enzyme	Error rate (10^{-6})	Reference
Klenow fragment of *Escherichia coli* DNA polymerase I	40[a]	Tindall and Kunkel, 1988
Thermus aquaticus (*Taq*) Polymerase	200[b]	Saiki *et al.*, 1988
	183[b]	Dunning *et al.*, 1988
	110[a]	Tindall and Kunkel, 1988
	109[b]	Villareal and Long, 1991
	76[b]	Ciccarelli *et al.*, 1991
	37[b]	Zhou *et al.*, 1991
	16[b]	Imai *et al.*, 1991
	15[b]	Kwiatowski *et al.*, 1991
	13[b]	Williams, 1989
	13[b]	Innis *et al.*, 1988
	<8.2[b]	Herlitze and Koenen, 1990
	<4.4[b]	Fucharoen *et al.*, 1989
Thermococcus litoralis	30[a]	Mattila *et al.*, 1991
Pyrococcus furiosus	1.6[a]	Lundberg *et al.*, 1991
Thermus flavus	103[a]	Mattila *et al.*, 1991

Note. The values in this table must be interpreted with caution inasmuch as the assay methods and conditions vary significantly between the various publications. For those publications that only reported the overall error frequency found by DNA sequencing, the formula given by Kwiatowski *et al.* (1991) was used to calculate the error rate per nucleotide polymerized. This formula is (2 × observed error frequency/number of cycles).
[a] Mutational reversion frequency assay as per Tindall and Kunkel (1988).
[b] Determined by DNA sequencing.

6. Difficulties and Limitations of the PCR Approach

Many of the so-called "general methods of mutagenesis by PCR" published until now require in fact the presence of specific conditions such as conveniently located unique restriction sites, proximity to universal primer sites, and the like. Users will soon learn from experience which protocol applies to their particular requirements and are best advised to keep an open mind with regard to literature claims.

The *Taq* DNA polymerase was shown to add an extra adenine at the end of most PCR-amplified DNA fragments (Clark, 1988). Undesired mutations can be avoided either by adding an exonuclease step in the protocol (Hemsley *et al.*, 1989) or by careful design of the mutagenic primer (Kuipers *et al.*, 1991).

As mentioned earlier, the major disadvantage in using PCR for mutagenesis still resides in the occurrence of unexpected mutations. Consequently, the sequence of the entire amplified region has to be verified.

E. Type of Templates

1. Double-Stranded DNA

One of the advantages of cassette mutagenesis, that is, the absence of specialized vector requirements for ssDNA production, has been extended to the oligonucleotide-mediated site-directed methods. Annealing of the mutagenic DNA is carried out following denaturation of the linearized dsDNA; repair is performed with the two-primer method. Recently, closing of the linear denatured DNA by a mutagenic oligonucleotide has been reported (Slilaty *et al.*, 1990). Alternatively, the dsDNA can be completely or partially converted into ssDNA by selective degradation of one strand using exonuclease III. For all these methods the extension of the primer is carried out using similar conditions (for a review, see Smith, 1985).

Overall, methods using dsDNA are usually not as efficient as methods based on ssDNA. However, their versatility and simplicity compensate for this disadvantage for some applications. The performance of the method has been markedly improved through the use of selection procedures for the mutated plasmids (see below, Section II,G).

2. Single-Stranded DNA

In the initial protocols the fragment of DNA to be mutated had to be cloned into vectors derived from filamentous phages to generate ssDNA. Following mutagenesis the mutated DNA often had to be excised from the replicating form of the phage and recloned into the proper vector. Detailed knowledge of the biology of filamentous phages has led to the development of hybrid plasmid vectors that can be converted into circular ssDNA. These vectors, often referred to as phagemids, include a filamentous phage origin of replication. They are propagated in *E. coli* either as closed circular dsDNA or as ssDNA upon infection with a filamentous phage. Preferential packaging of the phagemid DNA has been achieved by using genetically modified helper phage (Dotto and Horiuchi, 1981; Dente *et al.*, 1983).

Phagemids for a variety of host cells, including *E. coli* (Dente *et al.*, 1983; Dente and Cortese, 1987; Zagurski and Baumeister, 1987), yeast (Baldari and Cesareni, 1985; Vernet *et al.*, 1987), or insect cells (Vialard *et al.*, 1990; Tessier *et al.*, 1991), have been constructed. The availability of ssDNA for many plasmids and simple direct selection methods based on uracil-containing template (see Section II,G) have been combined in a powerful, efficient, reliable yet simple protocol for site-directed mutagenesis.

3. Direct Mutagenesis at the Chromosome Level

In vivo site-directed mutagenesis of yeast chromosome by transformation of yeast with synthetic oligonucleotides has been reported (Moerschell *et al.*, 1991). The method bypasses the usual *in vitro* steps for building the mutation. The method can be used to isolate functional revertants or defective mutants when selection is possible. Alternatively, identification of the mutants can be carried out by hybridization following cotransformation of the cells with a selectable plasmid-based marker. Multiple mismatches have been constructed using oligonucleotides with an optimized size of 40 bases. The main advantage of the procedure resides in its speed but the requirement for specific selection systems adapted to each case probably limits the utilization of the method.

F. Saturation Mutagenesis

The original oligonucleotide mutagenesis method of Zoller and Smith (1982) used oligonucleotides of defined sequence designed to introduce a specific mutation, insertion, or deletion within the target DNA. Some basic guidelines toward the design of oligonucleotides to be used in such experiments can be found in the publications of Carter (1987) and of Kramer and Fritz (1987). This approach, although very useful when applied to elucidate the role of previously defined critical amino acids in a protein structure (or bases in a noncoding sequence), only allows for screening of a limited number of mutants, as it is impractical to perform separate mutation experiments with a large number of oligonucleotides.

A very useful approach called saturation mutagenesis was introduced by several groups around 1986 (Smith, 1985; Derbyshire *et al.*, 1986; Oliphant *et al.*, 1986; Hutchison *et al.*, 1986; Hill *et al.*, 1987), all of whom used oligonucleotides modified with small amounts of contaminating nucleotides at some or all of the positions. This modification has also been referred to as "doped," "poisoned," "spiked," "random," or just plain "mixed" oligonucleotide synthesis. The principle of the method is illustrated in Fig. 5.

This approach broadens enormously the scope of mutagenesis: positions can be mutagenized in any given ratio; critical regions can be synthesized without any modification while surrounding regions are thoroughly scanned. Inherent to this approach are some interesting questions in terms of oligonucleotide design.

Stop codons are often undesirable if the mutagenesis is targeted within a protein coding sequence; therefore researchers have suggested various ways to avoid them when performing mixed oligonucleotide synthesis. Schultz and Richards (1986) pointed out that by using A,C,G, T at the first (also designated as N) and second positions but only G and C at the third position one eliminates two of the three stop codons (TAA and TGA) while retaining all 20 amino acids as choices. Mandecki (1990) used RNN (for nomenclature of ambiguous nucleotide positions see Table II; see also *Eur. J. Biochem.* 150, 1–5, 1985) as codon together with its complement NNY to obtain DNA guaranteeing open reading frames in both directions, while only losing two amino acids (tryptophan TGG and glutamine CAR) of the possible 20. Another method was suggested by Little (1990), who eliminated some of the

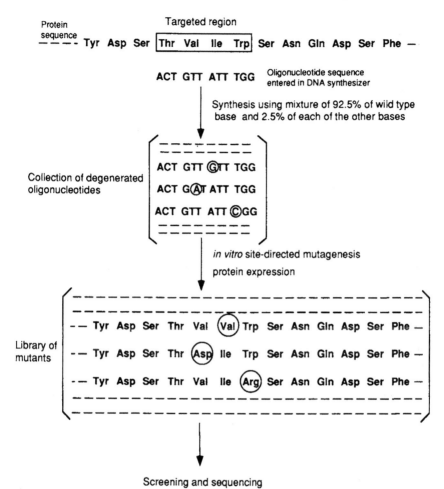

FIGURE 5   Site-directed mutagenesis using degenerated oligonucleotides. A sequence of four amino acids is targeted for random mutagenesis. The ratio of nucleotides used for the addition of each nucleotide base during synthesis can be calculated to favor statistical occurrence of a single amino acid change per clone screened at the end of the procedure. The percentage of "doping" shown in this example is arbitrary; other ratios would be selected to obtain a higher or lower frequency of mutations. In this example manually prepared premixed reagents are used. In the future advanced DNA synthesizers will achieve accurate and variable mixing of nucleotides at any given position in an oligonucleotide sequence.

stop codons through treatment with restriction enzymes that recognize TAG as part of their cleavage site.

Researchers have attempted to create formulas that would govern the choice and level of nucleotide mixing to achieve specific results in saturation mutagenesis. Derbyshire *et al.* (1986) gave a statistical formula from which

TABLE II Nomenclature for Incompletely
Specified Bases in Nucleic Acid Sequences
According to the IUB Nomenclature Committee[a]

R	G or A	puRine
Y	T or C	pYrimidine
M	A or C	aMino
K	G or T	Keto
S	G or C	Strong interaction (3 H bonds)
W	A or T	Weak interaction (2 H bonds)
H	A or C or T	not G; H follows G in alphabet
B	G or T or C	not A; B follows A in alphabet
V	G or C or A	not T; (not U); V follows U in alphabet
D	G or A or T	not C; D follows C in alphabet
N	G or A or T or C	aNy

[a] (1985) *Eur. J. Biochem.* 150, 1–5.

the average frequency of mutation per nucleotide can be calculated from a given level of contamination of the starting nucleotide reagents. A more comprehensive computational analysis has been accomplished by Arkin and Youvan (1992). This analysis gave predictions based on the performance and limitations of the most currently available commercial DNA synthesizers, which can be readily set to achieve 25, 33, or 50% mixing but are unable to accurately measure ratios below 25%. Newer synthesizers do offer the capability of mixing nucleotides down to the 10% range and these machines are likely to become widely used for saturation mutagenesis work. It is however possible, given some cooperation from the DNA chemists, to perform synthesis with manually mixed reagents to obtain any desired ratio down to perhaps 0.1%, at the expense of rather a great deal of bottle-switching and line rinsing on the DNA synthesizers.

The literature differs somewhat in terms of assessing whether the actual experimental results in terms of base incorporation differ from the computed probabilities. Horwitz and DiMaio (1990), in their reference study, noted an overrepresentation of G → A transitions as well as an underrepresentation of T → C. Schultz and Richards (1986), working with equimolar ratios of A,C,G, and T in the first two positions and of G and C in the third position of the codon to be mutagenized, observed an imbalance in favor of purines (68 purines of 101 clones sequenced) in the first, but not in the second, position (50 purines of 101). The ratio at the third position was close to predicted, with 54 Gs versus 47 Cs. Although the sample size is small, these results raise the possibility of bias arising within the organism as a result of proximity effects related to DNA replication. It must be noted that in this experiment there was no selection pressure toward functional protein and therefore no bias toward certain codons based on the corresponding amino acid regenerating an active protein.

Other researchers such as York *et al.* (1991) report base frequencies very close to the expected values in the absence of phenotypic selection. Hermes *et al.* (1989) have also addressed this question and found that an approximately statistical distribution did take place in their experiment, although A was somewhat overrepresented. Small deletions and insertions have also been noted by various workers, presumably as consequences of *in vivo* replication errors caused by mismatches in the original duplex formed (Horwitz and DiMaio, 1990).

Several computational tools are available to help the researcher design his/her mutagenesis strategy. Commercial software such as PC/GENE (Intelligenetics Co., Palo Alto, CA) or the University of Wisconsin Genetics Computer Group package (UWGCG) offers programs to find potential sites capable of being transformed into restriction sites through silent mutations. Most packages also allow a putative mutagenic oligonucleotide to be checked against the sequence of commonly used vectors to avoid undesired matches. Predictions of secondary structures in messenger RNAs, especially important near the translation initiation site, can also be performed routinely with these packages. These together with the usual common sense guidelines for designing oligonucleotides (avoiding self-complementary structures, for instance) allow an unusually high degree of confidence in the outcome of this type of experiment.

G. Favoring the Mutants: Current Selection Strategies

This technical aspect of site-specific mutagenesis can be considered relatively mature, with the uracil selection procedure (Kunkel, 1985) being the most commonly used. Detailed procedures have been published for several of the fundamental strategies such as the uracil procedure (Kunkel *et al.*, 1987), the α-phosphorothioate approach (Taylor *et al.*, 1985; Sugimoto *et al.*, 1989; Olsen and Eckstein, 1990), the gapped duplex approach (Kramer and Fritz, 1987), and the two primer procedure (Norris *et al.*, 1983; Zoller and Smith, 1987). The reader who wishes to enter the field will find useful, practical information in several review articles that are conveniently regrouped in Volume 154 of *Methods in Enzymology*. In all these methods the goal of achieving a mutation frequency high enough (at least above 1%, to close to 100%) to obviate the need for phenotypical selection is readily achieved, with frequencies approaching the theoretical maximum in many cases.

Recent work in this area has focused on removing the need for cloning the DNA to be mutagenized in special vectors; an example is the procedure of Slilaty *et al.* (1990), which allows the use of double-stranded plasmids of the pUC8 type to achieve the same high mutation frequencies (40–60%) typical of the single-stranded vectors. Another aspect of this work has seen the creation of many dual-mode vectors that can be readily produced in either the single or the double strand form required (see above, Section II,E,2).

The ubiquitous PCR approach has found applications in mutant selection as well. Lee *et al.* (1991) have applied the PCR approach to the selection procedure and claim a 100% yield of mutant using a two-step PCR protocol, although this requires unique restriction sites that may not always be conveniently available. Shen *et al.* (1991) devised a coupled marker method that similarly claims 100% efficacy.

III. EXAMPLES OF APPLICATIONS

A. Tolerance of Protein to Amino Acid Replacement: Exploring the Sequence Universe

Recent developments in the field of mutagenesis tend less toward refining the technique; instead, efforts are directed toward designing strategies to explore rationally the extremely vast diversity of the "sequence space" made of all possible combinations of amino acids even for a small protein. In other words, the emphasis has shifted to the protein design and engineering aspect (Leatherbarrow and Fersht, 1986), with the understanding that the technical side will deliver whichever mutants are deemed of interest by the design team.

One such attempt to rationalize protein structure has been put forward by DuBose and Hartl (1989), who called their approach "directed replacement." In this case two mutually compensating frameshift mutations are introduced into a structural element such as an α-helix segment via Kunkel-type mutagenesis (DuBose et Hartl, 1989).

Already an important finding that emerges from saturation mutagenesis experiments has been highlighted by Mandecki (1990) in that many proteins can tolerate a high degree of amino acid substitution. A similar conclusion was reached by Climie *et al.* (1990) based on their results on thymidylate synthase and by Reidhaar-Olson and Sauer (1988) and Lim and Sauer (1989) following their studies of the phage λ repressor. The redundancy of information in amino acid sequences has been illustrated by multiple replacements of side-chain residues by alanine in T4 lysozyme (Heinz *et al.*, 1992). These conclusions agree with observations on families of natural proteins such as the subtilisin-like serine proteases. In these enzymes the size of the catalytic domain can vary from 268 to 511 residues, and of these only 18 positions are highly conserved (11 are absolutely conserved and 7 allow only one or two substitutions; Siezen *et al.*, 1991).

Yet there is no gainsaying the fact that protein structures can also be exceedingly sensitive to seemingly minor changes; Hermes *et al.* (1990) demonstrate that the E165D mutation in triose phosphate isomerase displaces the corresponding carboxylate group by less than 1 Å yet causes a 1000-fold drop in the k_{cat}. The view that proteins are composed of a few very critical residues held together by a highly variable scaffolding emerges from these and other studies.

B. Tailoring and Delivering Protein to Industry

The commercially important enzyme subtilisin has been the object of much structural study, with upward of 400 site-specific mutants already made and for which kinetic data are available (Wells and Estell, 1990). Perhaps the most spectacular success to date has been the engineering of an oxidation-resistant subtilisin through substitution of Met-222 by Ala or Ser (Estell *et al.*, 1985). Interestingly, the conservative substitutions based on the tables of Dayhoff *et al.* (1978) were only weakly active, much less so than the more radical M222A or M222S mutants. This raises the question whether long-term evolutionary pressures can be adequately modeled by short-term mutagenesis experiments. However, it also offers the possibility that many commercially important enzymes have not in fact been optimized for maximum *in vitro* activity by evolution, thereby offering the would-be genetic engineering scope for substantial gains through judicious use of protein design.

Experiments aimed at changing substrate specificity, pH response, and thermal stability of subtilisin have also shown progress (reviewed in Wells and Estell, 1990). Another proteolytic enzyme, trypsin, has been the subject of a large effort aiming at detailed understanding of its structure–function relationships (Evnin *et al.*, 1990; Craik *et al.*, 1987).

C. Mutagenesis of Noncoding Sequences of DNA

An example of the power of mixed oligonucleotide mutagenesis in the study of promoter regions is found in the work of Gaal *et al.* (1989), who created as library of 290 mutant *rrnB* P1 promoters from two cloning experiments using oligonucleotides mixed at the 1.7% level. The synthetic approach combined regions of correct sequence with mutated regions, allowing for precise dissection of the various regions within the wild-type promoter.

An interesting combination of classical chemical mutagenesis and saturation mutagenesis was used by Dobson and Berg (1989) in their study of the I segment of transposon Tn5 (Is50). This very short fragment (19 bases) was mutagenized at every position using three different methods. The hydrazine chemical mutagenesis of Myers *et al.* (1985) gave 5 mutants, the bisulfite method of Shortle and Botstein (1983) produced 2 mutants, and the mixed oligonucleotide approach provided the remaining 21 mutants used in the study.

Milton *et al.* (1990) provided an unusually thorough study of a 60-bp fragment of SV40 virus, making all possible single base pair substitutions in addition to three double mutations and two triple mutations. Each mutation was made in a defined manner by site-specific mutagenesis (Kunkel, 1985) using an oligonucleotide with a single base substitution in each case. The results were interpreted in terms of the influence of the mutations on the degree of bending of this region and detailed conclusions could be drawn.

D. Mapping Functional Domains of Protein

A systematic strategy for amino acid replacement, dubbed alanine scanning, has been introduced by Cunningham and Wells (1989) and used by several other groups (Ashkenazi *et al.*, 1990; Gloss *et al.*, 1992). This approach chose alanine as the amino acid least likely to perturb secondary structure; positions where substitution of the wild-type amino acid with alanine leads to an altered phenotype are taken as evidence of being critical.

As mentioned earlier (Section III,A), site-specific mutagenesis has been widely used to explore in a general sense the sequence universe available to proteins. However, it is also widely used to elucidate more specific problems in terms of functional domains of proteins. Of particular interest are efforts to define the structural requirements of export sequences, either in bacteria (Chou and Kendall, 1990) or in eukaryotic cells (import sequences of yeast mitochondria; Bedwell *et al.* 1989). Saturation mutagenesis using mixed oligonucleotides is commonly used for these tasks. The work of Streuli *et al.* (1990) on the functional domains of protein tyrosine phosphatase is another example of defining distinct structural domains through the use of mutagenesis.

IV. CONCLUSIONS

If *in vitro* mutagenesis allows scientists to "play evolution," one has to remember that it is not possible to evaluate the true fitness of the mutated proteins. Indeed, it is impossible to recreate in a lab the complex selection mechanisms that are operating in nature over long periods of time. Within this limitation, however, truly important knowledge can be gained in terms of the process of protein folding, of thermal, pH, and proteolytic activity, and of binding domains, enzymatic activity, and many other important features of proteins.

The power of the method should not lead one to forget its limitations. One of the most significant limitations resides in the fact that the mutated piece of DNA must be sequenced in its entirety not only to ensure the presence of the desired mutation but also to ensure that the procedure has not introduced unexpected other changes. The requirements for DNA sequencing become increasingly severe as researchers attempt to explore larger subsets of the "sequence space" (Hermes *et al.*, 1990) represented by the amino acid choices at any given position within a protein.

Another limitation is that in some instances the mutated protein may become unstable and may not be expressible in recombinant organism use in a particular study, although it might be stable in its native host. It must also be remembered that a protein that evolved over perhaps 2 billion years may well represent an optimum compromise in terms of solubility, stability, and activity; most of the possible mutations to such a structure, even if planned with all the available help from molecular modeling, are likely to be less active and less stable than the original.

APPENDIX: SOME KEY LITERATURE REFERENCES AND COMMERCIAL PRODUCTS FOR THOSE WISHING TO USE *IN VITRO* SITE-DIRECTED MUTAGENESIS

[Note: The mention of a commercial product here relies solely upon the authors's experience and does not imply superiority of these over competing products.]

A. Some Key "How-to" References

1. Site-specific mutagenesis
    i. Two-primer method: Zoller, M. J., and Smith, M. (1987) *Methods in Enzymology* 154, 329–351. Very detailed, reliable protocols are given.
    ii. Uracil selection method: Kunkel, T. A., Roberts, J. D., and Zakour, R. A. (1987) *Methods in Enzymology* 154, 367–383. Here also, in the best tradition of the *Methods in Enzymology* series, the protocols are thoroughly described and potential problems and pitfalls clearly identified.
2. PCR methodology

    The following manual is virtually indispensable to those entering the field of PCR, whether or not they apply PCR to mutagenesis: Innis, M. A., Gelfand, D. H., Sninsky, J. J., and White, T. J. (Eds.) (1990) *PCR Protocols: A Guide to Methods and Applications*. Academic Press, San Diego.

3. Saturation mutagenesis using "spiked" oligonucleotides: Horwitz, B. H., and DiMaio, D. (1990) *Methods in Enzymology* 185, 599–611.

B. Some Useful Commercial Products

1. Muta-Gene *in vitro* mutagenesis kits from Bio-Rad (Catalog Nos. 170-3571 and 170-3576). Based on the Kunkel method. Provides M13-based vectors or phagemids. Easy to use and reliable, high yield of mutants.
2. Mutagenesis kit from Amersham Life Sciences (Catalog No. RPN 1523) based on the thionucleotide incorporation and selection of mutated strand by restriction endonuclease followed by exonuclease III. Very reliable, often gives 100% yield of mutants.

ACKNOWLEDGMENTS

We acknowledge the excellent technical assistance of Daniel Tessier. Issued as National Research Council of Canada Publication No. 39909.

REFERENCES

Albert, T. (1989) *Annu. Rev. Biochem.* **58**, 765–798.
Arkin, A. P., and Youvan, D. C. (1992) *Bio/Technol.* **10**, 297–300.
Ashkenazi, A., Presta, L. G., Marsters, S. A., Camerato, T. R., Rosenthal, K. A., Fendly, B. M., and Capon, D. J. (1990) *Proc. Natl. Acad. Sci. U.S.A.* **87**, 7150–7154.
Baldari, C., and Cesareni, G. (1985) *Gene (Amst.)* **35**, 27–32.
Bedwell, D. M., Strobell, S. S., Yun, K., Jongeward, G. D., and Emr, S. D. (1989) *Mol Cell. Biol.* **9**, 1014–1025.
Carter, P. (1987) *Methods Enzymol.* **154**, 382–403.
Carter, P., Nilsson, B., Burnier, J. P., Burdick, D., and Wells, J. (1989) *Proteins Struct. Funct. Genet.* **6**, 240–248.
Chou, M. M., and Kendall, D. A. (1990) *J. Biol. Chem.* **265**, 2873–2880.
Ciccarelli, R. B., Gunyuzlu, P., Huang, J., Scott, C., and Oakes, F. T. (1991) *Nucleic Acids Res.* **19**, 6007–6013.
Clark, J. M. (1988) *Nucleic Acids Res.* **16**, 9677–9686.
Clarke, N. D., Lien, D. C., and Schimmel, P. (1988) *Science (Washington)* **240**, 521–523.
Climie, S., Ruiz-Perez, L., Gonzalez-Pacanowska, D., Prapunwattana, P., Cho, S. W., Stroud, R., and Santi, D. V. (1990) *J. Biol. Chem.* **265**, 18776–18779.
Craik, C. S., Roczniak, S., Sprang, S., Fletterick, R., and Rutter, W. (1987) *J. Cell. Biochem.* **33**, 199–211.
Cunningham, B. C., and Wells, J. A. (1989) *Science (Washington)* **244**, 1081–1085.
Dayhoff, M. O., Schwartz, R. M., and Orcutt, B. C. (1978) In *Atlas of Protein Sequence and Structure* (M. O. Dayhoff, Ed.), Vol. 5, Suppl. 3, pp. 345–352. National Biomedical Research Foundation, Georgetown University Medical Center, Washington, DC.
Derbyshire, K. M., Salvo, J. J., and Grindley, N. D. F. (1986) *Gene (Amst.)* **46**, 145–152.
Dente, L. Cesareni, G., and Cortese, R. (1983) *Nucleic Acids Res.* **11**, 1645–1655.
Dente, L., and Cortese, R. (1987) *Methods Enzymol.* **155**, 111–119.
Diaz, J-J., Rhoads, D. D., and Roufa, D. J. (1992) *BioTechniques* **11**, 204–211.
Dobson, K. W., and Berg, D. E. (1989) *Gene (Amst.)* **85**, 75–82.
Dotto, G. P., and Horiuchi, K. (1981) *J. Mol. Biol.* **153**, 169–176.

DuBose, R. F., and Hartl, D. L. (1989) *Proc. Natl. Acad. Sci. U.S.A.* 86, 9966–9970.

Dubnick, M., Thliveris, A. T., and Mount, D. W. (1991) *Gene (Amst.)* 105, 1–7.

Dulau, L., Cheyrou, A., and Aigle, M. (1989) *Nucleic Acids Res.* 17, 2873.

Dunning, A. M., Talmud, P., and Humphries, S. E. (1988) *Nucleic Acids Res.* 16, 10393.

Eckert, K. A., and Kunkel, T. A. (1991) *PCR Methods Appl.* 1, 17–24.

Estell, D. A., Graycar, T. P., and Wells, J. A. (1985) *J. Biol. Chem* 260, 6518–6521.

Evnin, L. B., Vaskez, J. R., and Craik, C. S. (1990) *Proc. Nato. Acad. Sci. U.S.A.* 97, 6659–6663.

Fucharoen, S., Fucharoen, G., Fucharoen, P., and Fukumaki, Y. (1989) *J. Biol. Chem.* 264, 7780–7783.

Gaal, T., Barkei, J., Dickson, R. R., deBoer, H., deHaseth,k P. L., Alavi, H., and Gourse, R. L. (1989) *J. Bacteriol.* 171, 4852–4861.

Gillam, S., and Smith, M. (1979) *Gene (Amst.)* 8, 81–97.

Gloss, L. M., Planas, A., and Kirsch, J. F. (1992) *Biochemistry* 31, 32–39.

Goldenberg, D. P., Frieden, R. W., Haak, J. A., and Morrison, T. B. (1989) *Nature (London)* 338, 127–132.

Groppi, A., Beguerret, J., and Iron, A. (1990) *Clin. Chem.* 36, 1765–1768.

Heinz, D. W., Baase, W. A., and Matthews, S. W. (1992) *Proc. Natl. Acad. Sci. U.S.A.* 89, 3751–3755.

Herlitze, S., and Koenen, M. (1990) *Gene (Amst.)* 91, 143–147.

Hermes, J. D., Blacklow, S. C., and Knowles, J. (1990) *Proc. Natl. Acad. Sci. U.S.A.* 87, 696–700.

Hermes, J. D., Parekh, S. M., Blacklow, S. C., Koster, H., and Knowles, J. R. (1989) *Gene (Amst.)* 84, 143–151.

Hemsley, A., Arnheim, N., Toney, M. D., Cortopassi, G., and Galas, D. J. (1989) *Nucleic Acids Res.* 17, 6545–6551.

Higuchi, R., Krummel, B., and Saiki, R. K. (1988) *Nucleic Acids Res.* 16, 7351–7357.

Hill, D. E., Hope, J. P., Macke, J. P., and Struhl, K. (1986) *Science (Washington)* 234, 451–457.

Hill, D. E., Oliphant, A. R., and Struhl, K. (1987) *Methods Enzymol.* 155, 559–568.

Ho, S. N., Hunt, H. D., Horton, R. M., Pullen, J. K., and Pease, L. R. (1989) *Gene (Amst.)* 77, 51–59.

Horton, R. M., Hunt, H. D., Ho, S. N., Pullen, J. K., and Pease, L. R. (1989) *Gene (Amst.)* 77, 61–68.

Horwitz, B. H., and DiMaio, D. (1990) *Methods Enzymol.* 185, 599–611.

Hutchison, C. A., III, and Edgell, M. H. (1971) *J. Virol.* 8, 181–189.

Hutchison, C. A., III, Nordeen, S. K., Vogt, K., and Edgell, M. H. (1986) *Proc. Natl. Acad. Sci. U.S.A.* 83, 710–714.

Hutchison, C. A., III, Philips, S., Edgell, M. H., Gillam, S., Jahnke, P., and Smith, M. (1978) *J. Biol. Chem.* 253, 6551–6560.

Imai, Y., Matsushima, Y., Sugimura, T., and Terada, M. (1991) *Nucleic Acids Res.* 19, 2785.

Innis, M. A., Myambo, K. B., Gelfand, D. H., and Brow, M. A. D. (1988) *Proc. Natl. Acad. Sci. U.S.A.* 85, 9436–9440.

Jones, D. H., and Howard, B. H. (1991) *BioTechniques* 10, 62–66.

Kadowaki, H., Kadowaki, T., Wondisford, F. E., and Taylor, S. I. (1989) *Gene (Amst.)* 76, 161–166.

Kammann, M., Laufs, J., Schell, J., and Gronenborn, B. (1989) *Nucleic Acids Res.* 17, 5404.

Karlsson, R. G., Nordling, M., Pascher, T., Tsai, L. C., Sjölin, L., and Lundberg, L. G. (1991) *Protein Eng.* 4, 343–349.

Khouri, H. E., Vernet, T., Ménard, R., Parlati, F., Laflamme, P., Tessier, D. C., Thomas, D. Y., and Storer, A. C. (1991) *Biochemistry* 30, 8929–8936.

Kramer, W., and Fritz, H. J. (1987) *Methods Enzymol.* 154, 351–367.

Kramer, W., Schughart, K., and Fritz, H.-J. (1982) *Nucleic Acids Res.* 10, 6475–6485.

Kreitman, M., and Landweber, L. F. (1989) *Gene Anal. Techn.* **6**, 84–88.

Kuipers, O., Boot, H. J., and de Vos, W. (1991) *Nucleic Acids Res.* **19**, 4558.

Kunkel, T. A. (1985) *Proc. Natl. Acad. Sci. U.S.A.* **82**, 488–492.

Kunkel, T. A., Roberts, J. D., and Zakour, R. A. (1987) *Methods Enzymol.* **154**, 367–383.

Kwiatowski, J., Skarecky, D., Hernandez, S., Pham, D., Quijas, F., and Ayala, F. J. (1991) *Mol. Biol. Evol.* **8**, 884–887.

Landt, O., Grunert, H. P., and Hahn, U. (1990) *Gene (Amst.)* **96**, 125–128.

Leatherbarrow, R. J., and Fersht, A. R. (1986) *Protein Eng.* **1**, 7–16.

Lee, C. C., Wu, X. W., Gibbs, R. A., Cook, R. G., Muzny, D. M., and Caskey, T. M. (1988) *Science (Washington)* **239**, 1288–1291.

Lee, N., Liu, J., He, C., and Testa, D. (1991) *Appl. Env. Microbiol.* **57**, 2888–2890.

Leung, D., Chen, E., and Goeddel, D. (1989) *Technique* **1**, 11–15.

Lim, W. A., and Sauer, R. T. (1989) *Nature (London)* **339**, 31–36.

Little, J. M. (1990) *Gene (Amst.)* **88**, 113–116.

Lundberg, K. S., Shoemaker, D. D., Adams, M. W. W., Short, J. M., Sorge, J. A., and Mathur, E. J. (1991) *Gene (Amst.)* **108**, 1–6.

Mandecki, W. (1990) *Protein Eng.* **3**, 221–226.

Mattila, P., Korpela, J., Tenkane, T., and Pitk nen, K. (1991) *Nucleic Acids Res.* **19**, 4967–4973.

Ménard, R., Khouri, H. E., Plouffe, C., Laflamme, P., Dupras, D., Vernet, T., Tessier, D. C., Thomas, D. Y., and Storer, A. C. (1991) *Biochemistry* **30**, 5531–5538.

Milton, D. L., Casper, M. L., and Gesteland, R. F. (1990) *J. Mol. Biol.* **213**, 135–140.

Min, K. T., Kim, M. H., and Lee, D. S. (1988) *Nucleic Acids. Res.* **16**, 5075–5088.

Moerschell, R. P., Goutam, D., and Sherman, F. (1991) *Methods Enzymol.* **194**, 362–369.

Mullis, K. F., Faloona, S., Scharf, R., Saiki, R., Horn, G., and Erlich, H. (1986) In *Cold Spring Harbor Symposia on Quantitative Biology,* Vol. 51. Cold Spring Harbor Laboratory, Cold Spring Harbor, NY.

Myers, R. M., Lerman, L. S., and Maniatis, R. (1985) *Science (Washington)* **229**, 242–247.

Near, R. I. (1992) *BioTechniques* **12**, 88–97.

Nelson, R. M., and Long, G. L. (1989) *Anal. Biochem.* **180**, 147–151.

Norris, K., Norris, F., Christiansen, L., and Fill, N. (1983) *Nucleic Acids Res.* **11**, 5103–5112.

Oliphant, A. R., Nussbaum, A. L., and Struhl, K. (1986) *Gene (Amst.)* **44**, 177–183.

Oliphant, A. R., and Struhl, K. (1987) *Methods Enzymol.* **155**, 568–586.

Olsen, D. B., and Eckstein, F. (1990) *Proc. Natl. Acad. Sci. U.S.A.* **87**, 1451–1455.

Ong, E., Greenwood, J. M., Gilkes, N. R., Kilburn, D. G., Miller, R. C., Jr., and Warren, R. A. J. (1989) *Trends Biotech.* **7**, 239–243.

Perrin, S., and Gilliland, G. (1990) *Nucleic Acids Res.* **18**, 7433–7438.

Pine, R., and Huang, P. C. (1987) *Methods Enzymol.* **154**, 415–430.

Reidhaar-Olson, J. F., and Sauer, R. T. (1988) *Science (Washington)* **241**, 53–57.

Saiki, R. K., Gelfand, D. H., Stoffel, S., Scharf, S., Higuchi, R., Horn, G. T., Mullis, K. B., and Erlich, H. A. (1988) *Science (Washington)* **239**, 487–491.

Sarkar, G., and Sommer, S. S. (1990) *BioTechniques* **8**, 404–407.

Schold, M., Colombero, A., Reyes, A. A., and Wallace, R. B. (1984) *DNA* **3**, 469–477.

Schultz, S. C., and Richards, J. H. (1986) *Proc. Natl. Acad. Sci. U.S.A.* **83**, 1588–1592.

Schwartz, E. I., Shevtsov, S. P., Kuchinski, A. P., Kovalev, Y. P., Putalov, O. V., and Berlin, Y. A. (1991) *Nucleic Acids Res.* **13**, 3752.

Shen, T.-J., Zhu, L.-Q., and Sun, X. (1991) *Gene (Amst.)* **103**, 73–77.

Shortle, D. (1989) *J. Biol. Chem.* **264**, 5315–5318.

Shortle, D., and Botstein, D. (1983) *Methods Enzymol.* **100**, 457–468.

Siezen, R. J., de Vos, W. M., Leunissen, J. A. M., and Dijkstra, B. W. (1991) *Protein Eng.* **4**, 719–737.

Slilaty, S., Fung, M., Shen, S., and Lebel, S. (1990) *Anal. Biochem.* **185**, 184–190.

Smith, M. (1990) In *Protein Form and Function* (R. A. Bradshaw and M. Purton, Eds.), pp. 21–30. Elsevier, Cambridge, England.

Smith, M. (1985) *Annu. Rev. Genet.* 19, 423–463.

Street, I. P., Coffman, H. R., and Poulter, C. D. (1991) *Tetrahedron* 47, 5919–5924.

Streuli, M., Krueger, N. X., Thai, T., Tang, M., and Saito, H. (1990) *EMBO J.* 9, 2399–2407.

Sugimoto, M., Esaki, N., Tanaka, H., and Soda, K. (1989) *Anal. Biochem.* 179, 309–311.

Taylor, J. W., Schmidt, W., Cosstick, R., Okruszek, A., and Eckstein, F. (1985) *Nucleic Acids Res.* 13, 8749–8764.

Tessier, D. C., Thomas, D. Y., Laliberté F., Khouri, H. E., and Vernet, T. (1991) *Gene (Amst.)* 98, 177–183.

Tindall, K. R., and Kunkel, T. A. (1988) *Biochemistry* 27, 6008–6013.

Vallette, F., Mege, E., Reiss, A., and Adesnik, M. (1989) *Nucleic Acids. Res.* 17, 723–733.

Vernet, T., Dignard, D., and Thomas, D. Y. (1987) *Gene (Amst.)* 52, 225–233.

Vernet, T., Tessier, D. C., Richardson, C., Laliberté, F., Khouri, H. E., Bell, A. W., Storer, A. C., and Thomas, D. Y. (1990) *J. Biol. Chem.* 265, 16661–16666.

Vialard, J., Lalumière, M., Vernet, T., Briedis, D., Alkhalib, G., Henning, D., Levin, D., and Richardson, C. (1990) *J. Virol.* 69, 37–50.

Villareal, X. C., and Long, G. L. (1991) *Anal. Biochem.* 197, 362–367.

Wagner, C. R., and Benkovic, S. J. (1990) *TIBS* 8, 263–270.

Waye, M. M. Y., and Gupta, A. K. (1989) *BioTechniques* 7, 604–606.

Wells, J. A., and Estell, D. A. (1990) In *Proteins: Form and Function* (R. A. Bradshaw and M. Purton, Eds.), pp. 21–29. Elsevier, Cambridge, England.

Wells, J. A., Vasser, M., and Powers, D. B. (1985) *Gene (Amst.)* 34, 315–323.

Williams, J. F. (1989) *BioTechniques* 7, 762–768.

York, J. D., Li, P., and Gardell, S. J. (1991) *J. Biol. Chem.* 266, 8495–8500.

Zagurski, R., and Baumeister, K. (1987) *Methods Enzymol.* 155, 139–155.

Zaruchi-Schultz, T., Tsai, S. Y., Itakura, I., Soberon, X., Wallace, R. B., Tsai, M. J., Woo, S. L. C., and O'Malley, B. W. (1982) *J. Biol. Chem.* 257, 11070–11077.

Zheng, J., Jiang, K., Watterson, D. M., Craig, T. A., and Higgins, S. B. (1988) *Comput. Biol. Med.* 18, 409–418.

Zhou, Y., Zhang, X., and Ebright, R. H. (1991) *Nucleic Acids Res.* 19, 6052.

Zoller, M. J., and Smith, M. (1982) *Nucleic Acids Res.* 10, 6487–6500.

Zoller, M. J., and Smith, M. (1987) *Methods Enzymol.* 154, 329–351.

Part III

CHARACTERIZATION

6

Determination of Structures of Larger Proteins in Solution by Three- and Four-Dimensional Heteronuclear Magnetic Resonance Spectroscopy[1]

G. MARIUS CLORE

ANGELA M. GRONENBORN

Laboratory of Chemical Physics, Building 5, National Institute of Diabetes and Digestive and Kidney Diseases, National Institutes of Health, Bethesda, Maryland 20892-0520

I. INTRODUCTION

A complete understanding of protein function and mechanism of action can only be accomplished with a knowledge of its three-dimensional structure at atomic resolution. At present there are two methods available for determining such structures. The first method, which has been established for many years, is X-ray diffraction of protein single crystals. The second method has only blossomed in the past 5 years and is based on the application of nuclear magnetic resonance (NMR[2]) spectroscopy of proteins in

---

[1]An earlier version of this chapter appeared in *Science* **252**, 1390–1399 (1991).

[2]Abbreviations used: NMR, nuclear magnetic resonance; NOE, nuclear Overhauser enhancement; NOESY, nuclear Overhauser enhancement spectroscopy; HOHAHA, homonuclear Hartmann–Hahn spectroscopy; TOCSY, total correlation spectroscopy; HMQC; heteronuclear multiple quantum coherence.

solution. The driving force for the development of an alternative to X-ray crystallography was threefold. First, many proteins do not crystallize, and even when they do the crystals may diffract poorly or difficulties in solving the phase problem (e.g., finding suitable heavy atom derivatives) may be encountered. Second, there may be significant and possibly important functional differences between structures in the crystal state and in solution. Third, dynamic processes ranging from the picosecond to second time scales are amenable to study by NMR. Despite these attractive features it should be borne in mind that, just like crystallography, NMR also has a number of limitations. In particular, the protein under investigation must be soluble and should not aggregate up to a concentration of at least 1 mM. Further, the dependence of linewidth on rotational correlation time probably sets an intrinsic upper molecular mass limit of about 50–60 kDa for the applicability of current solution NMR technology.

The advent of two-dimensional (2D) NMR (Jeener, 1971; Aue et al., 1976; Jeener et al., 1979; Bax and Lerner, 1986; Ernst et al., 1987) set the stage for the determination of the first low-resolution structures of small proteins in the mid 1980s (Williamson et al., 1985; Clore et al., 1985, 1986a; Kaptein et al., 1985). Subsequent improvements over the next few years led to a tremendous increase in the precision and accuracy of such protein structure determinations such that it is now possible, using 2D NMR methods, to determine structures of proteins up to about 100 residues, which are comparable in quality to 2- to 2.5 Å resolution X-ray structures (Clore and Gronenborn, 1991b). This progress has been summarized in a number of reviews (Wüthrich, 1986, 1990a,b; Clore and Gronenborn, 1987, 1989, 1991a). As proteins get larger than about 100 residues, however, conventional 2D methods can no longer be applied successfully. This chapter focuses on the novel methodological developments of heteronuclear 3D and 4D NMR that have been designed to overcome the limitations imposed by the increased molecular weight and spectral complexity inherent to these proteins, and describes their application, which has recently culminated in the determination of the first high-resolution NMR structure of a protein greater than 150 residues (Clore et al., 1991a).

The concept of increasing spectral dimensionality to extract information can perhaps most easily be understood by analogy. Consider, for example, the encyclopedia *Britannica*. In a one-dimensional representation, all the information (i.e., words and sentences arranged in a particular set order) present in the encyclopedia would be condensed into a single line. If this line were expanded to two-dimensions in the form of a page, the odd word may be resolved but the vast majority would still be superimposed on each other. When this page is expanded into a book (i.e., three dimensions) comprising a set number of lines and words per page, as well as a fixed number of pages, some pages may become intelligible, but many words will still lie on top of each other. The final expansion to the multivolume book (i.e., four dimensions) then makes it possible to extract in full all the information present in the individual entries of the encyclopedia.

II. THE NATURE OF THE STRUCTURAL DATA DERIVED FROM NMR MEASUREMENTS

The principal source of geometric information used in NMR protein structure determination lies in short approximate interproton distance restraints derived from nuclear Overhauser enhancement (NOE) measurements (Noggle and Schirmer, 1971). The physical basis for the NOE effect is relatively simple and is based on the fact that each proton spin possesses a property known as magnetization. Magnetization is exchanged between the spins by a process termed cross-relaxation, and the rate constant for this process is directly related to r^6. The chemical analogy to such a system is one with a large number of interconverting species in equilibrium with each other. The phenomenon of cross-relaxation is observed by perturbing the magnetization of a particular spin and observing the resulting change in magnetization, known as the NOE, of the other spins as the equilibrium is reestablished. Thus, the time dependence of the NOE is governed by a set of coupled first-order differential equations. The experiment is therefore similar in spirit to chemical relaxation kinetics (e.g., a temperature jump experiment) except that the initial perturbation in the chemical system involves rate constants rather than species concentration. If the NOE is observed only a short time after the perturbation, the size of the NOE is proportional to the cross-relaxation rate and hence to r^6. Because of the r^6 dependence, the magnitude of these effects decreases rapidly as the interproton distances increase, so that effects are generally not observable beyond 5 Å. The interproton distance restraints derived from the NOE measurements may also be supplemented by backbone and side-chain torsion angle restraints derived from three-bond coupling constants and appropriate NOEs (Wagner *et al.*, 1987; Kraulis *et al.*, 1989; Güntert *et al.*, 1989; Nilges *et al.*, 1990).

With the approximate interproton distance and torsion angle restraints in hand, a number of computational strategies can be applied to locate the minimum of a target function comprising terms for the experimental restraints, covalent geometry (i.e., bonds, angles, planes, and chirality), and nonbonded contacts (e.g., a van der Waals repulsion term to prevent atoms from coming too close together). The types of algorithms employed operate either in *n*-dimensional distance space followed by projection into real space [e.g., metric matrix distance geometry (Havel *et al.*, 1983)] or directly in real space [e.g., minimization in torsion angle space with a variable target function (Braun and Go, 1985), dynamical simulated annealing (Nilges *et al.*, 1988a,b,c), and restrained molecular dynamics (Clore *et al.*, 1985, 1986b; Brünger *et al.*, 1986; Kaptein *et al.*, 1985). All real space methods require initial structures that can be random structures with correct covalent geometry, structures very far from the final structure (e.g., an extended strand), structures made up of a completely random array of atoms, or structures generated by distance space methods. The key requirements of all these methods is that they have large radii of convergence and that they fully sample in an unbiased fashion the conformational space consistent with the experimental, geometrical, and van der Waals restraints (Clore and Gronenborn, 1989). The various methods have been described in detail in a number of reviews (Braun, 1987; Crippen and Havel, 1988; Clore and Gronenborn, 1989) and have all been successfully applied to NMR structure determinations.

To assess the accuracy and precision of an NMR structure determination, it is essential to calculate a large number of structures independently with the same experimental data set using different starting structures or conditions. The spread of structures consistent with the experimental data can be assessed qualitatively from a visual inspection of a best-fit superposition of a series of computed conformers, and quantitatively by calculating the average atomic rms distribution of the individual structures about the mean coordinate positions. The representation of NMR solution structures as an ensemble of conformers in which each individual member is compatible with the experimental data may still be regarded as unusual, insofar as one has long been accustomed to the traditional single-chain trace representation of X-ray structures. In the latter case, a single polypeptide chain is usually fitted to the electron-density map, although one has to bear in mind that the density arises from a linear superposition of all the different conformers present within the crystal.

The power of the NMR method and an illustration of the progressive improvement in NMR protein structures as a function of increasing *only* the number, rather than the precision, of experimental restraints are shown in Fig. 1. Thus, in all the examples shown in Fig. 1, the NOE-derived interproton distance restraints are only approximate and classified into three broad ranges, 1.8–2.7 Å, 1.8–3.3 Å, and 1.8–5.0 Å, corresponding to strong, medium, and weak NOEs, respectively. First-generation structures, such as α1-purothionin (45 residues) (Clore *et al.*, 1986a), were based on 5–7 experimental restraints per residue. The atomic rms distribution of the backbone atoms is rather large ($\sim$1.5 Å) so that although a cartoon-like representation of the polypeptide fold is obtained, details of both the local backbone and side-chain conformations are essentially obscured by the large variations between the structures. As the experimental restraints are increased to $\sim$10 per residue, including some backbone ϕ torsion angle restraints derived from a qualitative interpretation of the $^3J_{HN\alpha}$ couplings, a significant improvement is observed and the resulting second-generation structures have a backbone atomic rms distribution of $\sim$0.9 Å (Driscoll *et al.*, 1989a). The side-chain conformations, however, are still rather blurred. Additional improvement requires stereospecific assignments of β-methylene protons and χ_1 side-chain torsion angle restraints. In a simplified approach

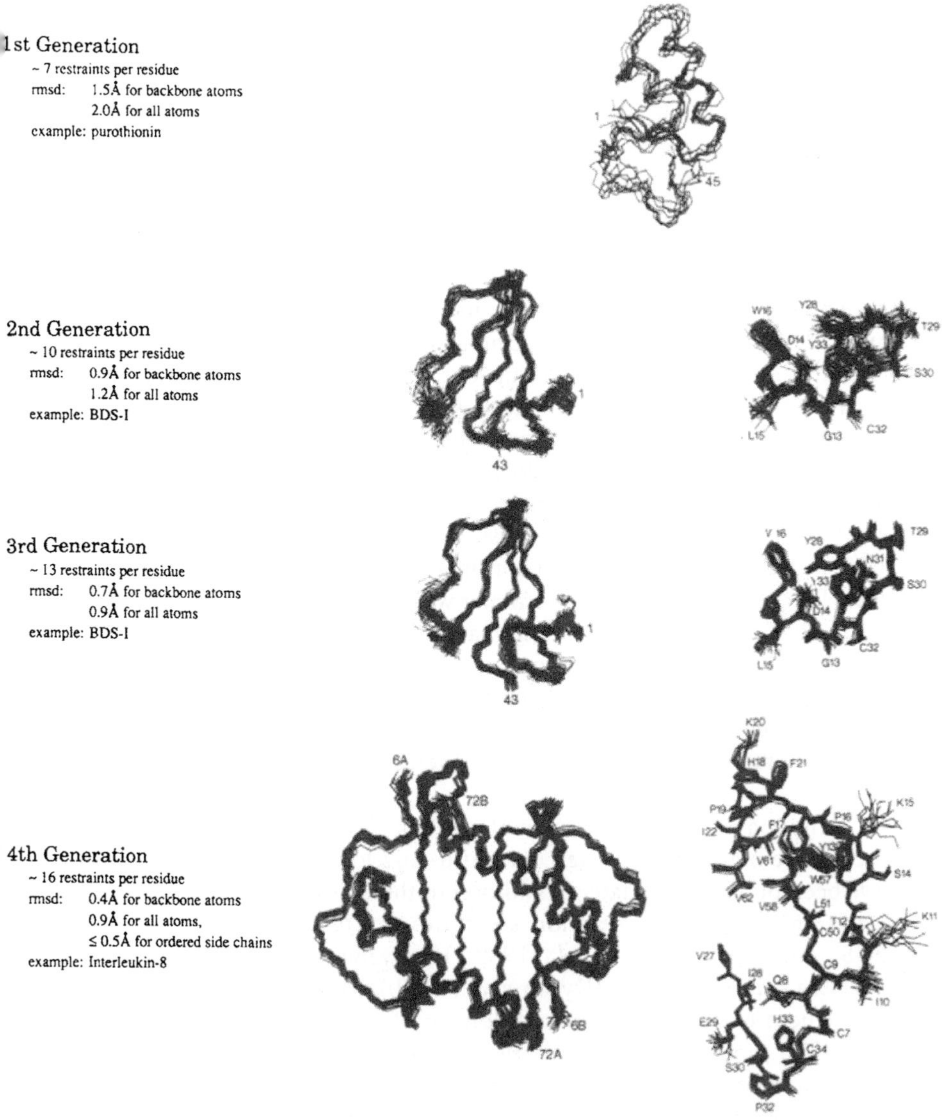

1st Generation
~ 7 restraints per residue
rmsd: 1.5Å for backbone atoms
        2.0Å for all atoms
example: purothionin

2nd Generation
~ 10 restraints per residue
rmsd: 0.9Å for backbone atoms
        1.2Å for all atoms
example: BDS-1

3rd Generation
~ 13 restraints per residue
rmsd: 0.7Å for backbone atoms
        0.9Å for all atoms
example: BDS-1

4th Generation
~ 16 restraints per residue
rmsd: 0.4Å for backbone atoms
        0.9Å for all atoms,
        ≤ 0.5Å for ordered side chains
example: Interleukin-8

FIGURE 1   Illustration of the progressive improvement in the precision and accuracy of NMR structure determinations with increasing number of experimental restraints. The fourth-generation structures are equivalent in quality to ~ 2.5 Å resolution X-ray structures. Each protein is represented by a best-fit superposition of a number of independently computed conformers: there are 9 for α1-purothionin, 31 for the second generation version of BDS-1, 42 for the third generation of BDS-1, 30 for the backbone atoms of interleukin-8, and 15 for the side chains of interleukin-8. In all the examples shown, the NOE-derived interproton distance restraints are only classified into three broad ranges, 1.8–2.7 Å, 1.8–3.3 Å, and 1.8–5.0 Å, corresponding to strong, medium, and weak NOEs. The structure of α1-purothionin (Clore *et al.*, 1986a) was calculated by metric matrix distance geometry (Havel *et al.*, 1983) followed by restrained molecular dynamics (Clore *et al.*, 1985), while the structures of BDS-I (Driscoll *et al.*, 1989a,b) and interleukin-8 (Clore *et al.*, 1989) were calculated using the hybrid distance geometry-simulated annealing method (Nilges *et al.*, 1988a). rmsd is the average atomic rms distribution of the individual calculated structures about the mean coordinate positions.

these are obtained from a qualitative interpretation of $^3J_{\alpha\beta}$ coupling constants and intraresidue NOEs (Wagner et al., 1987). The resulting third-generation structures are based on ~13 experimental restraints per residue and display a marked improvement not only in the definition of the backbone atoms, but more importantly in the side chains as well (Driscoll et al., 1989b). This is readily apparent by comparing the second- (Driscoll et al., 1989a) and third-generation (Driscoll et al., 1989b) structures of the same protein BDS-I (43 residues). Finally, in the fourth-generation structures, a much larger number of stereospecific assignments and loose torsion angle restraints are obtained by carrying out conformational grid searches of ϕ,ψ,χ_1 torsion angle space on the basis of the $^3J_{HN\alpha}$ and $^3J_{\alpha\beta}$ coupling constants and intraresidue and sequential interresidue NOEs involving the NH, C$^\alpha$H, and C$^\beta$H protons (Nilges et al., 1990). This leads to 16–20 experimental restraints per residue, a backbone atomic rms distribution of ≤0.4 Å, and an atomic rms distribution of ≤0.5 Å for ordered side chains. The errors in the atomic coordinates of such fourth-generation structures, like that of interleukin-8 (a dimer with 72 residues per subunit) (Clore et al., 1989) shown in Fig. 1, are similar to those of ~2.5-Å resolution X-ray structures (Clore and Gronenborn, 1991b). Indeed, the solution structure of interleukin-8 was used by necessity to solve the X-ray structure by molecular replacement, as traditional methods based on heavy atom derivatives had proved unsuccessful, despite several years of effort (Baldwin et al., (1991).

Although there are many examples of first- and second-generation structures in the literature, relatively few third- and fourth-generation structures have been published to date. In addition to the two structures shown in Fig. 1, examples of third-generation structures are tendamistat (Kline et al., 1988), hirudin (Folkers et al., 1989; Haruyama et al., 1989), and Escherichia coli thioredoxin (Dyson et al., 1990). Examples of fourth-generation structures are the C-terminal domain of cellobiohydrolase (Kraulis et al., 1989), the homeodomain of the Antennapedia protein (Billeter et al., 1990), a zinc finger domain from a human enhancer binding protein (Omichinski et al., 1990), human thioredoxin (Forman-Kay et al., 1991), the IgG binding domain of protein G (Gronenborn et al., 1991), and interleukin-1β (Clore et al., 1991a).

III. GENERAL PRINCIPLES INVOLVED IN DERIVING EXPERIMENTAL RESTRAINTS FROM NMR DATA

The experimental restraints derived from the NMR data require the identification of specific interactions between proton pairs, which may be either through-space (i.e., via the NOE) or through-bonds (i.e., via coupling constants). The power of the NMR method, compared to other spectroscopic techniques, lies in the fact that each proton gives rise to a specific resonance in the spectrum. Thus, a key aspect of any NMR structure determination is the requirement to assign each resonance to an individual proton and then to identify uniquely each pairwise through-space NOE interaction. In principle this can be accomplished in a relatively straightforward manner using correlation experiments to identify resonances belonging to different amino acid types via through-bond connectivities, and NOE experiments subsequently to link these residues in a sequential manner along the polypeptide chain on the basis of sequential and short-range interresidue NOEs involving the NH, $C^\alpha H$, and $C^\beta H$ protons (Wüthrich, 1986; Clore and Gronenborn, 1987, 1989). In particular, the types of NOE interactions that are most instructive for this purpose involve $NH(i)-NH(i+1,2)$, $C^\alpha H(i)-NH(i+1,2,3,4)$, $C^\beta H(i)-NH(i+1)$, and $C^\alpha H(i)-C^\beta H(i+3)$ connectivities, and the pattern of observed NOEs provides a very good indication of the different secondary structure elements along the polypeptide chain. With the resonance assignments in hand, one can then proceed to identify long-range NOE interactions between protons belonging to residues far apart in the sequence but close together in space, which yield the crucial information for determining the tertiary structure of the protein.

Whereas the principles of sequential resonance assignment are simple, the practice is difficult. Even for a small protein of say 50 residues, there are likely to be some 300–400 protons for which resonances have to be uniquely assigned. This number goes up linearly with the number of residues, so that for a 150-residue protein there will be 900–1200 protons. Because of the large number of protons, there is an extensive degree of resonance overlap and chemical-shift degeneracy. As a result, one-dimensional NOE and decoupling experiments can only be applied with any degree of confidence for peptides up to about 10 residues, and even then there may be serious difficulties. The major conceptual advance in the application of NMR as a method of protein structure determination was the introduction of 2D NMR (reviewed by Ernst et al., 1987). By spreading out the correlations in two ^1H frequency dimensions, each interaction is labeled by two chemical shifts, namely the frequencies of the originating and destination protons. This not only results in a tremendous increase in spectral resolution but equally importantly enables one to detect and interpret effects that would not have been possible in one dimension.

All 2D experiments can be reduced to the same basic conceptual scheme shown in Fig. 2. This comprises a preparation pulse, an evolution period (t_1) during which the spins are labeled according to their chemical shifts, a mixing period (M_1) during which the spins are correlated with one another, and finally a detection period (t_2). The experiment is repeated several times with successively linearly incremented values of the evolution period t_1 to yield a data matrix $s(t_1,t_2)$. Fourier transformation in the t_2 dimension yields a set of n 1D spectra in which the intensities of the resonances are si-

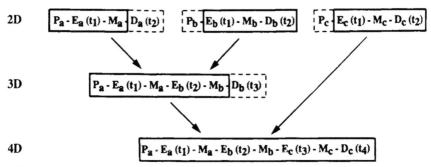

FIGURE 2   General representation of pulse sequences used in multidimensional NMR illustrating the relationship between the basic schemes used to record 2D, 3D, and 4D NMR spectra. Note how 3D and 4D experiments are constructed by the appropriate linear combination of 2D ones. Abbreviations: P, preparation; E, evolution; M, mixing; D, detection. In 3D and 4D NMR, the evolution periods are incremented independently.

nusoidally modulated as a function of the t_1 duration. Subsequent Fourier transformation in the t_1 dimension yields the desired 2D spectrum $S(\omega_1,\omega_2)$.

Two-dimensional methods have proved extremely powerful for the structure determination of small proteins. The largest proteins (in terms of number of residues) where this approach has been successfully applied to achieve a complete 3D structure determination are *E. coli* (Dyson *et al.,* 1990) and human thioredoxin (Forman-Kay *et al.,* 1991), which have 108 and 105 residues, respectively. Beyond this limit of about 100 residues, 2D methods soon break down owing to two fundamental problems. The first is associated with extensive spectral overlap due to the larger number of resonances. Consequently, the contour plot of a 2D spectrum of a 150-residue protein assumes the appearance of an intangible network of intersecting circles as an ever larger number of cross-peaks merge into one another. The result is that the 2D spectra of such a protein can no longer be interpreted. The second is a sharp decrease in the efficiency with which magnetization can be transferred through the small three-bond ^{1}H–^{1}H J couplings (3–12 Hz) as the linewidths become larger than the couplings due to increasing rotational correlation time. This leads to incomplete delineation of spin systems (i.e., amino acid types) in through-bond correlation experiments.

Solutions to both these problems are obtained by extending the dimensionality of the NMR spectra to remove resonance overlap and degeneracy, and by making use of through-bond correlations via heteronuclear couplings that are larger than the linewidths. This necessitates the use of uniformly ^{15}N- and/or ^{13}C-labeled protein. In proteins that can be overexpressed in bacterial systems, such labeling can be readily achieved by growing the organism in minimal medium supplemented by ^{15}NH$_4$Cl and/or ^{13}C$_6$-glucose as the sole nitrogen and carbon sources, respectively.

IV. BASICS OF 3D AND 4D NMR

The design and implementation of higher dimensionality NMR experiments can be carried out by the appropriate combination of 2D NMR experiments, as illustrated schematically in Fig. 2 (Oschkinat *et al.*, 1988). A 3D experiment is constructed from two 2D pulse schemes by leaving out the detection period of the first experiment and the preparation pulse of the second. This results in a pulse train comprising two independently incremented evolution periods t_1 and t_2, two corresponding mixing periods M_1 and M_2, a detection period t_3. Similarly, a 4D experiment is obtained by combining three 2D experiments in an analogous fashion. Thus, conceptually *n*-dimensional NMR can be conceived as a straightforward extension of 2D NMR. The real challenge, however, of 3D and 4D NMR is twofold: first, to ascertain which 2D experiments should be combined to best advantage, and second, to design the pulse sequences in such a way that undesired artifacts, which may severely interfere with the interpretation of the spectra, are removed. This task is far from trivial.

The first application of 3D NMR to a small protein, namely α1-purothionin (45 residues), was presented in 1988 (Oschkinat *et al.*, 1988). The experiment was of the proton homonuclear variety in which a through-bond correlation experiment (homonuclear Hartmann–Hahn Spectroscopy, HOHAHA) was combined with a through-space one (nuclear Overhauser enhancement Spectroscopy,, NOESY). Although this experiment demonstrated the potential of the methodology, it suffered from a number of drawbacks that severely limited its application to larger proteins. First, the correlation portion of the experiment relied on small ^1H–^1H couplings. Second, all homonuclear 3D spectra are substantially more difficult to interpret than the equivalent 2D versions, as the number of cross-peaks present in the former far exceeds that in the latter.

Fortunately, heteronuclear 3D and 4D NMR experiments do not suffer from any of these disadvantages and yield important additional information in the form of ^{15}N and ^{13}C chemical shifts. They exploit a series of large one-bond heteronuclear couplings for magnetization transfer through-bonds, which are summarized in Fig. 3. This, together with the fact that the ^1H nucleus is always detected, renders these experiments very sensitive. Indeed, high-quality 3D and 4D heteronuclear-edited spectra can easily be obtained on samples of 1–2 mM uniformly labeled protein in a time frame that is limited solely by the number of increments that have to be collected for appropriate digitization and the number of phase cycling steps that have to be used to reduce artifacts to an acceptably low level. Typical measurement times are 1.5 to 3 days for 3D experiments and 2.5 to 5 days for 4D ones. A detailed technical review of heteronuclear multidimensional NMR has been provided by Clore and Gronenborn (1991c).

Many of the 3D and 4D experiments are based on heteronuclear editing

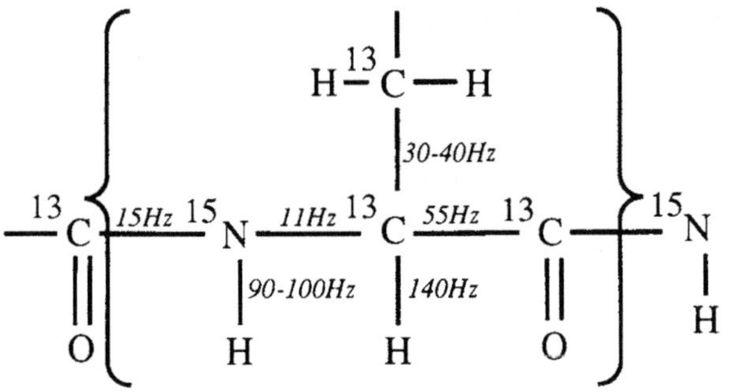

FIGURE 3 Summary of the one-bond heteronuclear couplings along the polypeptide chain utilized in 3D and 4D NMR experiments. The backbone torsion angles ϕ and ψ involve rotations about the $N_i-C^{\alpha}_i$ and $C^{\alpha}_i-C_i$ bonds, respectively, while the side chain torsion angle χ_1 involves a rotation about the $C^{\alpha}_i-C^{\beta}_i$ bond.

of $^1H-^1H$ experiments so that the general appearance of conventional 2D experiments is preserved and the total number of cross-peaks present is the same as that in the 2D equivalents. The progression from a 2D spectrum to 3D and 4D heteronuclear-edited spectra is depicted schematically in Fig. 4. Consider, for example, the cross-peaks involving a particular ^1H frequency in a 2D NOESY spectrum, a 3D ^{15}N- or ^{13}C-edited NOESY spectrum, and finally a 4D $^{15}N/^{13}C$- or $^{13}C/^{13}C$-edited NOESY spectrum. In the 2D spectrum a series of cross-peaks will be seen from the originating proton frequencies in the F_1 dimension to the single destination ^1H frequency along the F_2 dimension. From the 2D experiment it is impossible to ascertain whether these NOEs involve only a single destination proton or several destination protons with identical chemical shifts. By spreading the spectrum into a third dimension according to the chemical shift of the heteronucleus attached to the destination proton(s), NOEs involving different destination protons will appear in distinct $^1H-^1H$ planes of the 3D spectrum. Thus each interaction is simultaneously labeled by three chemical-shift coordinates along three orthogonal axes of the spectrum. The projection of all these planes onto a single plane yields the corresponding 2D spectrum. For the purposes of sequential assignment, heteronuclear-edited 3D spectra are often sufficient for analysis. However, when the goal of the analysis is to assign NOEs between protons far apart in the sequence, a 3D ^{15}N- or ^{13}C-edited NOESY spectrum will often prove inadequate. This is because the originating protons are only specified by their ^1H chemical shifts, and more often than not, there are several protons that resonate at the same frequencies. For example, in the case of the 153-residue protein interleukin-1β, there are about 60 protons that resonate in a 0.4 ppm interval between 0.8 and 1.2 ppm (Clore *et al.*, 1990a). Such ambiguities can then be resolved by spreading out the 3D spectrum still further into a fourth dimension accord-

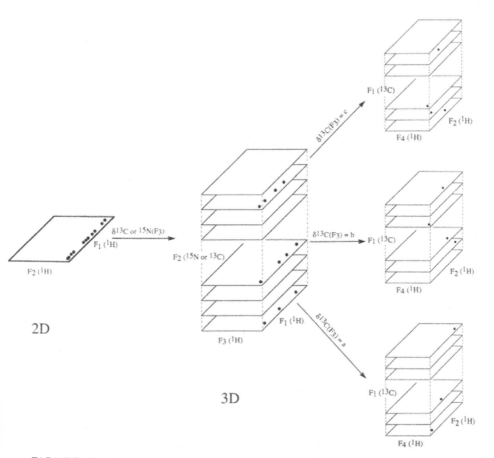

FIGURE 4 Schematic illustration of the progression and relationship among 2D, 3D, and 4D heteronuclear NMR experiments. The closed circles represent NOE cross peaks. In the example shown there are 11 NOEs originating from 11 different protons in the F_1 dimension to a single-frequency position in the F_2 dimension. In the 2D spectrum, it is impossible to ascertain whether there is only one destination proton or several in the F_2 dimension. By spreading the spectrum into a third dimension (labeled F_2), according to the chemical shift of the hetero-nucleus attached to the destination proton, it can be seen that the NOEs now lie in three distinct $^1H(F_1)$–$^1H(F_3)$ planes, indicating that three different destination protons are involved. How-ever, the ^1H chemical shifts still provide the only means of identifying the originating protons. Hence the problem of spectral overlap still prevents the unambiguous assignment of these NOEs. By extending the dimensionality of the spectrum to four, each NOE interaction is labeled by four chemical shifts along four orthogonal axes. Thus, the NOEs in each plane of the 3D spectrum are now spread over a cube in the 4D spectrum according to the chemical shift of the heteronucleus directly attached to the originating protons. Adapted from Clore *et al.*, (1991b).

ing to the chemical shift of the heteronucleus attached to the originating protons, so that each NOE interaction is simultaneously labeled by four chemical-shift coordinates along four orthogonal axes, namely those of the originating and destination protons and those of the corresponding heteronuclei directly bonded to these protons (Kay *et al.*, 1990; Clore *et al.*, 1991b; Zuiderweg *et al.*, 1991). The result is a 4D spectrum in which each plane of the 3D spectrum constitutes a cube in the 4D spectrum.

For illustration purposes it is also useful to compare the type of information that can be extracted from a very simple system using 2D, 3D, and 4D NMR. Consider a molecule with only two NH and two aliphatic protons in which only one NH proton is close to an aliphatic proton. In addition, the chemical shifts of the NH protons are degenerate, as are those of the aliphatic protons, so that only two resonances are seen in the one-dimensional spectrum. In the 2D NOESY spectrum, an NOE will be observed between the resonance position of the NH protons and the resonance position of the aliphatic protons, but it will be impossible to ascertain which one of the four possible NH-aliphatic proton combinations gives rise to the NOE. By spreading the spectrum into a third dimension, for example, by the chemical shift of the ^{15}N atoms attached to the NH protons, the number of possibilities will be reduced to two, provided, of course, that the chemical shifts of the two nitrogen atoms are different. Finally, when the fourth dimension corresponding to the chemical shift of the ^{13}C atoms attached to the aliphatic protons is introduced, a unique assignment of the NH-aliphatic proton pair giving rise to the NOE can be made.

V. HETERONUCLEAR 3D AND 4D NMR IN PRACTICE

Figure 5A presents a portion of the 2D ^{15}N-edited NOESY spectrum of interleukin-1β (153 residues) illustrating NOE interactions between the NH protons along the F_2 axis and the $C^{\alpha}H$ protons along the F_1 dimension. Despite the fact that a large number of cross-peaks can be resolved, it can be seen that many of the cross-peaks have identical chemical shifts in one or other dimensions. For example, there are 15 cross-peaks involving NH protons at a $F_2(^1H)$ chemical shift of ~ 9.2 ppm. A single $^1H(F_1)-^1H(F_3)$ plane of the 3D ^{15}N-edited NOESY spectrum of interleukin-1β at $\delta^{15}N(F_2) = 123.7$ ppm is shown in Fig. 5B. Not only is the number of cross-peaks in this slice small, but at $\delta^1H(F_3) \sim 9.2$ ppm there is only a single-cross peak involving one NH proton. The correlations observed in the ^{15}N-edited NOESY spectrum (Fesik and Zuiderweg, 1988; Marion *et al.*, 1989) are through-space ones. Intraresidue correlations from the NH protons to the $C^{\alpha}H$ and $C^{\beta}H$ protons can similarly be resolved using a 3D ^{15}N-edited HOHAHA spectrum (Marion *et al.*, 1989; Clore *et al.*, 1991c) in which

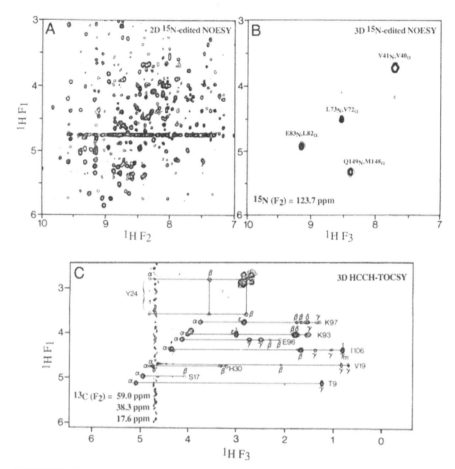

FIGURE 5 Example of 2D and 3D spectra of interleukin-1β recorded at 600 MHz (Driscoll *et al.*, 1990a; Clore *et al.*, 1990a). The 2D spectrum in (A) shows the $NH(F_2$ axis)–$C^\alpha H(F_1$ axis) region of a 2D ^{15}N-edited NOESY spectrum. The same region of a single $NH(F_3)$–$^1H(F_1)$ plane of the 3D ^{15}N-edited NOESY at $\delta^{15}N(F_2) = 123.7$ ppm is shown in (B). The actual 3D spectrum comprises 64 such planes and projection of these on a single plane would yield the same spectrum as in (A). (C) A single $^1H(F_3)$–$^1H(F_1)$ plane of the 3D HCCH–TOCSY spectrum at $\delta^{13}C(F_2) = 38.3 \pm nSW$ (where SW is the spectral width of 20.71 ppm in the ^{13}C dimension) illustrating both direct and relayed connectivities originating from the $C^\alpha H$ protons. Note how easy it is to delineate complete spin systems of long side chains such as Lys (i.e., cross peaks to the $C^\beta H$, $C^\gamma H$, $C^\delta H$, and $C^\epsilon H$ protons are observed) owing to the fact that magnetization along the side chain is transferred via large $^1J_{CC}$ couplings. Several features of the HCCH–TOCSY spectrum should be pointed out. First, extensive folding is employed, which does not obscure analysis, as ^{13}C chemical shifts for different carbon types are located in characteristic regions of the ^{13}C spectrum with little overlap. Second, the spectrum is edited according to the chemical shift of the heteronucleus attached to the originating proton rather than the distination one. Third, multiple cross-checks on the assignments are readily made by looking for the symmetry-related peaks in the planes corresponding to the ^{13}C chemical shifts of the destination protons in the original slice.

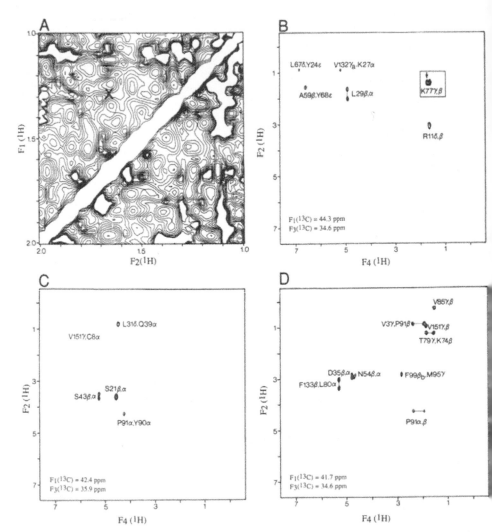

FIGURE 6 Comparison of 2D and 4D NMR spectra of interleukin-1β recorded at 600 MHz (Clore *et al.*, 1991b). The region between 1 and 2 ppm of the 2D NOESY spectrum is shown in (A). ¹H(F₂)–¹H(F₄) planes at several ¹³C(F₁) and ¹³C(F₃) frequencies of the 4D ¹³C/¹³C NOESY spectrum are shown in (B)–(D). No individual cross-peaks can be observed in the 2D spectrum and the letter X has ¹H coordinates of 1.39 and 1.67 ppm. In contrast, only two cross-peaks are observed in the boxed region in (B) between 1 and 2 ppm, one of which (indicated by an arrow) has the same ¹H coordinates as the letter X. Further analysis of the complete 4D spectrum reveals the presence of 7 NOE cross peaks superimposed at the ¹H coordinates of the letter X. This can be ascertained by looking at the ¹³C(F₁)–¹³C(F₃) plane taken at the ¹H coordinates of X. True diagonal peaks corresponding to magnetization that has not been transferred from one proton to another, as well as intense NOE peaks involving protons attached to the same carbon atom (i.e., methylene protons), appear in only a single ¹H(F₂)–¹H(F₄) plane of each ¹³C(F₁),¹H(F₂),¹H(F₄) cube at the carbon frequency where the originating and destination carbon atoms coincide (i.e., at F₁ = F₃). Thus, these intense reso-

efficient isotropic mixing sequences are used to transfer magnetization between protons via three-bond ^1H–^1H couplings.

The 3D ^{15}N-edited NOESY and HOHAHA spectra constitute only one
of several versions of a 3D heteronuclear-edited spectrum. Many alternative
through-bond pathways can be utilized to great effect. Consider, for example, the delineation of amino acid spin systems that involves grouping those
resonances that belong to the same residue. In 2D NMR, correlation experiments are used to delineate either direct or relayed connectivities via small
three-bond ^1H–^1H couplings. Even for proteins of 50–60 residues, it can be
difficult to delineate long-chain amino acids such as Lys and Arg in this
manner. In heteronuclear 3D NMR an alternative pathway can be employed
that involves transferring magnetization first from a proton to its directly
attached carbon atom via the large $^1J_{CH}$ coupling (~130 Hz), followed by
either direct or relayed transfer of magnetization along the carbon chain via
the $^1J_{CC}$ couplings (~30–40 Hz), before transferring the magnetization back
to protons (Bax *et al.*, 1990a,b; Fesik *et al.*, 1990). An example of such a
spectrum is the so called HCCH–TOCSY (Bax *et al.*, 1990b; Clore *et al.*,
1990a) shown in Fig. 5C. The $^1H(F_1)$–$^1H(F_3)$ plane at $\delta^{13}C(F_2) = 59$ ppm
illustrates both direct and relayed connectivities along various side chains
originating from $C^\alpha H$ protons. As expected, the resolution of the spectrum is
excellent and there is no spectral overlap. Just as importantly, however, the

---

nances no longer obscure NOEs between proton with similar or degenerate chemical shifts.
Two examples of such NOEs can be seen in (C)(between the $C^\alpha H$ protons of Pro-91 and Tyr-90)
and (D)(between one of the $C^\beta H$ protons of Phe-77 and the methyl protons of Met-95). These
various planes of the 4D spectrum also illustrate another key aspect of 3D and 4D NMR,
namely the importance of designing the pulse scheme to remove optimally undesired artifacts
that may severely interfere with the interpretation of the spectra. Thus, whereas the 4D
$^{13}C/^{13}C$-edited NOESY experiment is conceptually analogous to that of a 4D $^{13}C/^{15}N$-edited
one, the design of a suitable pulse scheme is actually much more complex in the $^{13}C/^{13}C$ case.
This is due to the fact that there are a large number of spurious magnetization transfer pathways
that can lead to observable signals in the homonuclear $^{13}C/^{13}C$ case. For example, in the 4D
$^{15}N/^{13}C$-edited case there are no "diagonal peaks" that would correspond to magnetization
that has not been transferred from one hydrogen to another, as the double heteronuclear
filtering (i.e., ^{13}C and ^{15}N) is extremely efficient at completely removing these normally very
intense and uninformative resonances. Such a double filter is not available in the $^{13}C/^{13}C$ case,
so that both additional pulses and phase cycling are required to suppress magnetization transfer
through these pathways. This task is far from trivial, as the number of phase cycling steps in 4D
experiments is severely limited by the need to keep the measurement time down to practical
levels (i.e., less than 1 week). The results of such care in pulse design can be clearly appreciated
from the artifact-free planes shown in (B)–(D). However, when a 4D $^{13}C/^{13}C$-edited NOESY
spectrum is recorded with the same pulse scheme as that used in the 4D $^{15}N/^{13}C$ experiment
(with the obvious replacement of ^{15}N pulses by ^{13}C pulses), a large number of spurious peaks
are observed along a pseudo-diagonal at $\delta^1H(F_2) = \delta^1H(F_4)$ in planes where the carbon
frequencies of the originating and destination protons do *not* coincide. As a result, it becomes
virtually impossible under these conditions to distinguish artifacts from NOEs between protons
with the same ^1H chemical shifts, as was possible with complete confidence in (C) and (D).

sensitivity of the experiment is extremely high and complete spin systems are readily identified in interleukin-1β even for long side chains, such as those of two lysine residues shown in the figure. Indeed, analyzing spectra of this king, it was possible to obtain complete ^{1}H and ^{13}C assignments for the side chains of interleukin-1β (Clore *et al.*, 1990a).

Three-dimensional NMR also permits one to devise experiments for sequential assignment that are based solely on through-bond connectivities via heteronuclear couplings (Ikura *et al.*, 1990) and thus do not rely on the NOESY experiment. This ability becomes increasingly important for larger proteins, as the types of connectivities observed in these correlation experiments are entirely predictable, whereas in the NOESY spectrum, which relies solely on close proximity of protons, it may be possible to confuse sequential connectivities with long-range ones. These 3D heteronuclear correlation experiments are of the triple resonance variety and make use of one-bond ^{13}CO(i-1)–^{15}N(i), ^{15}N(i)–^{13}C$^\alpha$(i), and ^{13}C$^\alpha$(i)–^{13}CO(i) couplings, as well as two-bond ^{13}C$^\alpha$(i-1)–^{15}N(i) couplings. In this manner multiple independent pathways for linking the resonances of one residue with those of its adjacent neighbor are available, thereby avoiding ambiguities in the sequential assignment.

The power of 4D heteronuclear NMR spectroscopy for unraveling interactions that would not have been possible in lower dimensional spectra is illustrated in Fig. 6 by the ^{13}C/^{13}C-edited NOESY spectrum of interleukin-1β (Clore *et al.*, 1991b). Figure 6A shows a small portion of the aliphatic region between 1 and 2 ppm of a conventional 2D NOESY spectrum of interleukin-1β. The overlap is so great that no single individual cross-peak can be resolved. One might therefore wonder just how many NOE interactions are actually superimposed, for example, at the ^{1}H chemical-shift coordinates of the letter X at 1.39 (F_1) and 1.67 (F_2) ppm. A ^{1}H(F_2)–^{1}H(F_4) plane of the 4D spectrum at δ^{13}C(F_1), δ^{13}C(F_3) = 44.3, 34.6 ppm is shown in Fig. 6B and the square box at the top right hand side of this panel encloses the region between 1 and 2 ppm. Only two cross-peaks are present in this region, and the arrow points to a single NOE between the C$^\gamma$H and the C$^\beta$H protons of Lys-77 with the same ^{1}H chemical-shift coordinates as the letter X in Fig. 6A. All the other NOE interactions at the same ^{1}H chemical-shift coordinates can be determined by inspection of a single ^{13}C(F_1)–^{13}C(F_3) plane taken at δ^1H(F_2),δ^1H(F_4) = 1.39,1.67 ppm. This reveals a total of seven NOE interactions superimposed at the ^{1}H chemical-shift coordinates of the letter X. Another feature of the 4D spectrum is illustrated by the two ^{1}H(F_2)–^{1}H(F_4) planes at different F_1 and F_3 ^{13}C frequencies shown in Figs. 6C and 6B. In both cases, there are cross-peaks involving protons with identical or near chemical shifts, namely that between Pro-91(C$^\alpha$H) and Tyr-90(C$^\alpha$H), diagnostic of a *cis*-proline, in Fig. 6C, and between Phe-99(C$^{\beta b}$H) and Met-95(C$^\gamma$H) in Fig. 6D. These interactions could not be resolved in either a 2D spectrum or a 3D ^{13}C-edited spectrum, as they would lie on the spectral diagonal (i.e., the region of the spectrum corresponding to magnetization that has not been transferred from one proton to another). In the 4D spectrum, however, they are easy to observe, provided, of course, that the ^{13}C chemical shifts of the directly bonded ^{13}C nuclei are different.

Because the number of NOE interactions present in each $^1H(F_4)-^1H(F_2)$ plane of 4D $^{13}C/^{15}N$ or $^{13}C/^{13}C$-edited NOESY spectra is so small, the inherent resolution in a 4D spectrum is extremely high, despite the low level of digitization. Indeed, spectra with equivalent resolution can be recorded at magnetic field strengths considerably lower than 600 MHz, although this would obviously lead to a reduction in sensitivity. Further, it can be calculated that 4D spectra with virtual lack of resonance overlap and good sensitivity can be obtained on proteins with as many as 400 residues. Thus, once complete ^1H, ^{15}N, and ^{13}C assignments are obtained, analysis of 4D spectra should permit the automated assignment of almost all NOE interactions.

VI. APPLICATION OF 3D AND 4D NMR TO THE DETERMINATION OF LARGER PROTEIN STRUCTURES

Although the potential of heteronuclear 3D and 4D NMR methods in resolving problems associated with both extensive resonance overlap and large linewidths is obvious, how does this new approach fare in practice? In this regard it should be borne in mind that resonance assignments are only a means to an end, and the true test of multidimensional NMR lies in examining its success in solving the problem that it was originally designed to tackle, namely the determination of high-resolution three-dimensional structures of larger proteins in solution. This goal has now been attained in the case of interleukin-1β, a protein of 153 residues and 17.4 kDa, which plays a central role in immune and inflammatory responses (Clore et al., 1991a). This protein is 50% larger than any other protein whose three-dimensional structure had been previously determined by NMR. Subsequently, structures of two other larger proteins have been determined, namely the complex of calmodulin with a target peptide from light-chain mysosin kinase (Ikura et al., 1992) and the cytokine interleukin-4 (Powers et al., 1992).

Despite extensive analysis of 2D spectra obtained at different pH values and temperatures, as well as examination of 2D spectra of mutant proteins, it did not prove feasible to obtain unambiguous ^1H assignment for more than about 30% of the residues of interleukin-4 (Driscoll et al., 1990a). Thus, any further progress could only be made by resorting to higher dimensionality heteronuclear NMR. A summary of the strategy we employed for determining its structure is shown in Fig. 7. The initial step involved the complete assignment of the ^1H, ^{15}N, and ^{13}C resonances of the backbone

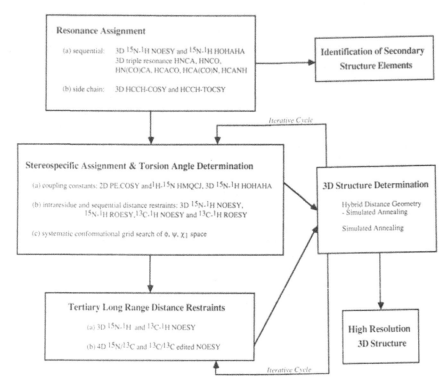

FIGURE 7   Outline of the general strategy employed in our laboratory to determine the three-dimensional structure of larger proteins such as interleukin-1β by 3D and 4D NMR. The various NMR experiments are as follows. Through-space interactions are detected in the heteronuclear-edited 3D and 4D NOESY and ROESY experiments. Direct and relayed scalar correlations from the $C^\alpha H$ and $C^\beta H$ protons to the NH protons via three-bond $^1H-^1H$ couplings are detected in the 3D ^{15}N-edited HOHAHA spectrum; direct and relayed scalar correlations between aliphatic protons via the large $^1J_{CC}$ couplings are observed in the 3D HCCH–COSY and HCCH–TOCSY experiments, respectively. $^{13}C^\alpha(i-1,i)-^{15}N(i)-^{15}N(i)$, $^{13}CO(i-1)-$ $^{15}N(i)-NH(i)$, and $C^\alpha H(i)-^{13}C^\alpha(i)-^{13}CO(i)$ correlations via direct through-bond connectivities are observed in the 3D HNCA, HNCO, and HCACO experiments, respectively. Finally, $^{13}C^\alpha(i-1)-^{15}N(i)-NH(i)$ and $C^\alpha H(i-1)-^{13}C^\alpha(i-1)-^{15}N(i)$ correlations via two successive one-bond couplings between the $^{13}C^\alpha(i-1)-^{13}CO(i-1)$ and $^{13}CO(i-1)-^{15}N(i)$ atoms are observed in the HN(CO)CA and HCA(CO)N relayed experiments, respectively, while $C^\alpha H(i-1,i)-^{15}N(i)-$ NH(i) correlations can be observed in the H(CA)NH experiment, which relays magnetization via the $^{13}C^\alpha$ atom. $^3J_{HN\alpha}$ and $^3J_{\alpha\beta}$ $^1H-^1H$ coupling constants, which are related to the torsion angles φ and χ_1 via empirical Karplus relationships (Pardi et al., 1984), are measured from 2D heteronuclear $^1H-^{15}N$ HMQC-J (Kay and Bax, 1990; Forman-Kay et al., 1990) and homonuclear $^1H-^1H$ PE.COSY (Mueller, 1987) correlation spectra, respectively. A semiquantitative measure of the $^3J_{\alpha\beta}$ couplings, which is sufficient for securing stereospecific assignments and χ_1 torsion angle restraints, can also be obtained from the relative magnitude of the NH–$C^\beta H$ correlations observed in the 3D ^{15}N-edited HOHAHA spectrum (Clore et al., 1991c). Stereospecific assignments and torsion angle restraints are obtained by comparing the relevant experimental data (i.e., coupling constants and intraresidue and sequential distance restraints involving the NH, $C^\alpha H$, and $C^\beta H$ protons) to the calculated values of these parameters present in two databases. The first is a systematic one covering the complete φ, ψ, and χ_1 conformational space

and side chains using the entire gammut of double- and triple-resonance 3D experiments listed in the top left-hand panel of the figure (Driscoll et al., 1990a,b; Clore et al., 1990a). In the second step backbone and side-chain torsion angle restraints, as well as stereospecific assignments for β-methylene protons, were obtained by means of a three-dimensional systematic grid search of ϕ,ψ,χ_1 space (Nilges et al., 1990). In the third step, approximate interproton distance restraints between nonadjacent residues were derived from analysis of 3D and 4D heteronuclear-edited NOESY spectra. Analysis of the 3D heteronuclear-edited NOESY spectra alone was sufficient to derive a low-resolution structure on the basis of a small number of NOEs involving solely NH, $C^{\alpha}H$, and $C^{\beta}H$ protons (Clore et al., 1990b). However, further progress using 3D NMR was severely hindered by the numerous ambiguities still present in these spectra, in particular for NOEs arising from the large number of aliphatic protons. Thus, the 4D heteronuclear-edited NOESY spectra proved to be absolutely essential for the successful completion of this task. In addition, the proximity of backbone NH protons to bound structural water molecules was ascertained from a 3D ^{15}N-separated ROESY spectrum that permits one to distinguish specific protein–water NOE interactions from chemical exchange with bulk solvent (Clore et al., 1990c). In this regard we should emphasize again that in our laboratory, all the NOE data are interpreted in as conservative a manner as possible, and are simply classified into three distance ranges, 1.8–2.7 Å, 1.8–3.3 Å, and 1.8–5.0 Å, corresponding to strong-, medium-, and weak-intensity NOEs.

With an initial set of experimental restraints in hand, 3D structure calculations were initiated. Typically we use the hybrid distance geometry-dynamical simulated annealing method in which an approximate polypeptide fold is obtained by projection of a subset of atoms from n-dimensional distance space into cartesian coordinate space followed by simulated annealing including all atoms (Nilges et al., 1988a). Alternatively we employ simulated annealing starting either from random structures with intact covalent geometry (Nilges et al., 1988b) or from a completely random array of atoms (Nilges et al., 1988c). All these simulated annealing protocols involve solving Newton's equations of motion subject to a simplified target function comprising terms for the experimental restraints, covalent geometry, and nonbonded contacts. The underlying principle lies in raising the temperature of the system followed by slow cooling in order to overcome false local

---

(in a three-dimensional grid spaced at 10° intervals) of a tripeptide fragment with idealized geometry, while the second comprises a library of tripeptide segments from high resolution X-ray structures (Nilges et al., 1990). This procedure is carried out for both possible stereospecific assignments, and when the experimental data are only consistent with one of the two possibilities, the correct stereospecific assignment, as well as allowed ranges for ϕ, ψ, and χ_1, is obtained. The minimum ranges that we employ for the torsion angle restraints are ±30°, ±50°, and ±20°, respectively.

minima and large potential energy barriers along the path toward the global minimum region of the target function, and to sample efficiently and comprehensively the conformation space consistent with the experimental restraints. A key aspect of the overall strategy lies in the use of an iterative approach, whereby the experimental data are reexamined in the light of the initial set of calculated structures in order to resolve ambiguities in NOE assignments, to obtain more stereospecific assignments (e.g., the α-methylene protons of glycine and the methyl groups of valine and leucine) and torsion angle restraints, and to assign backbone hydrogen bonds associated with slowly exchanging NH protons as well as with bound water molecules. The iterative cycle comes to an end when all the experimental data have been interpreted.

The final experimental data set for interleukin-1β comprised a total of 3146 approximate and loose experimental restraints made up of 2780 distance and 366 torsion angle restraints (Clore *et al.*, 1991a). This represents an average of ~21 experimental restraints per residue. If one takes into account that interresidue NOEs affect two residues, whereas intraresidue NOE and torsion angle restraints only affect individual residues, the average number of restraints influencing the conformation of each residue is ~33. A superposition of 32 independently calculated structures is shown in Fig. 8. All 32 structures satisfy the experimental restraints within their specified errors, display very small deviations from idealized covalent geometry, and have good nonbonded contacts. It can be seen that both the backbone and the ordered side chains are exceptionally well defined. Indeed, the atomic rms distribution about the mean coordinate positions is 0.4 Å for the backbone atoms, 0.8 Å for all atoms, and 0.5 Å for side chains with ≤40% of their surface (relative to that in a tripeptide Gly-X-Gly) accessible to solvent (Clore *et al.*, 1991a).

The structure of interleukin-1β itself resembles a tetrahedron and displays threefold internal pseudo-symmetry. There are 12 β-strands arranged in an exclusively antiparallel β-structure, and six of the strands form a β-barrel (seen in the front of Fig. 8A) that is closed off at the back of the molecule by the other six strands. Each repeating topological unit is composed of 5 strands arranged in an antiparallel manner with respect to each other, and one of these units is shown in Fig. 8B. Water molecules occupy very similar positions in all three topological units, as well as at the interface of the three units, and are involved in bridging backbone hydrogen bonds. Thus, in the case of the topological unit shown in Fig. 8B, the water molecule labeled W5 accepts a hydrogen bond from the NH of Phe-112 in stand IX and donates two hydrogen bonds to the backbone carbonyls of Ile-122 in strand X and Thr-144 in strand XII. The packing of some internal residues with respect to one another, as well as the excellent definition of internal side chains is illustrated in Fig. 8C. Because of the high resolution of the interleukin-1β structure, it was possible to analyze in detail side-chain–side-

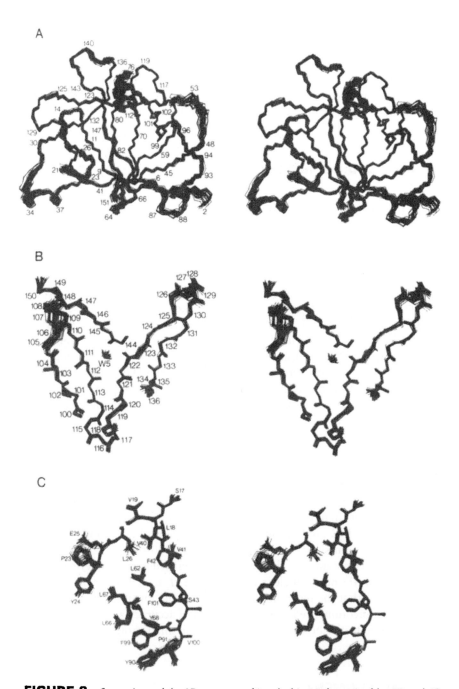

FIGURE 8 Stereoviews of the 3D structure of interluekin-1β determined by 3D and 4D heteronuclear NMR spectroscopy on the basis of a total of 3146 approximate and loose experimental NMR restraints (Clore *et al.*, 1991a). Best-fit superpositions of the backbone (N, Cα, C) atoms of residues 2–151, the backbone (N, Cα, C, O) atoms of one of the three repeating topological units including a water molecule (W5), and selected side chains of 32 simulated annealing structures are shown in (A), (B), and (C), respectively. The N-terminal residue and the two C-terminal ones (residues 152–153) are ill-defined.

chain interactions involved in stabilizing the structure. In addition, examination of the structure in the light of mutational data permitted us to propose the presence of three distinct sites involved in the binding of interleukin-1β to its cell surface receptor (Clore et al., 1991a).

VII. CONCLUDING REMARKS

In this chapter we have summarized the recent developments in heteronuclear 3D and 4D NMR that have been designed to extend the NMR methodology to medium-sized proteins in the 15- to 30-kDa range. The underlying principle of this approach consists of extending the dimensionality of the spectra to obtain dramatic improvements in spectral resolution while simultaneously exploiting large heteronuclear couplings to circumvent problems associated with larger linewidths. A key feature of all these experiments is that they do not result in any increase in the number of observed cross-peaks relative to their 2D counterparts. Hence, the improvement in resolution is achieved without raising the spectral complexity, rendering data interpretation straightforward. Thus, for example, in 4D heteronuclear-edited NOESY spectra, the NOE interactions between proton pairs are labeled not only by the ^{1}H chemical shifts but also by the corresponding chemical shifts of their directly bonded heteronuclei in four orthogonal axes of the spectrum. Also important in terms of practical applications is the high sensitivity of these experiments, which makes it feasible to obtain high-quality spectra in a relatively short time frame on 1–2 mM protein samples uniformly labeled with ^{15}N and/or ^{13}C.

Just as 2D NMR opened the application of NMR to the structure determination of small proteins of less than about 100 residues, 3D and 4D heteronuclear NMR applications provide the means of extending the methodology to medium-sized proteins in the 150- to 300-residue range. Indeed, the recent determination of the high-resolution structure of interleukin-1β using 3D and 4D heteronuclear NMR (Clore et al., 1991a) demonstrates beyond doubt that the technology is now available for obtaining the structures of such medium-sized proteins at a level of accuracy and precision that is comparable to the best results attainable for small proteins.

ACKNOWLEDGMENTS

We thank Ad Bax for many stimulating discussions. This work was supported in part by the AIDS Targeted Anti-Viral Program of the Office of the Director of the National Institutes of Health.

REFERENCES

Aue, W. P., Bartholdi, E., and Ernmst, R. R. (1976) *J. Chem. Phys.* **64**, 2229–2246.

Baldwin, E. T., Weber, I. T., St. Charles, R., Zuan, J. C., Appella, E., Matsushima, K., Edwards, B. F. P., Clore, G. M., Gronenborn, A. M., and Wldoawer (1991) *Proc. Natl. Acad. Sci. U.S.A.* **88**, 502–506.

Bax, A., Clore, G. M., Driscoll, P. C., Gronenborn, A. M., Ikura, M., and Kay, L. E. (1990a) *J. Magn. Reson.* **87**, 620–627.

Bax, A., Clore, G. M., and Gronenborn, A. M. (1990b) *J. Magn. Reson.* **88**, 425–431.

Bax, A., and Lerner, E. L. (1986) *Science* **232**, 960–970.

Billeter, M., Qian, Y., Otting, G., Müller, M., Gehring, W. J., and Wüthrich, K. (1990) *J. Mol. Biol.* **214**, 183–197.

Braun, W. (1987) *Q. Rev. Biophys.* **19**, 115–157.

Braun, W., and Go, N. (1985) *J. Mol. Biol.* **186**, 611–626.

Brünger, A. T., Clore, G. M., Gronenborn, A. M., and Karplus, M. (1986) *Proc. Natl. Acad. Sci. U.S.A.* **83**, 3801–3805.

Clore, G. M., and Gronenborn, A. M. (1987) *Protein Eng.* **1**, 275–288.

Clore, G. M., and Gronenborn, A. M. (1989) *CRC Crit. Rev. Biochem. Mol. Biol.* **24**, 479–564.

Clore, G. M., and Gronenborn, A. M. (1991a) *Annu. Rev. Biophys. Biophys. Chem.* **21**, 29–63.

Clore, G. M., and Gronenborn, A. M. (1991b) *J. Mol. Biol.* **221**, 47–53.

Clore, G. M., and Gronenborn, A. M. (1991c) *Prog. NMR Spectr.* **23**, 43–92.

Clore, G. M., Gronenborn, A. M., Brünger, A. T., and Karplus, M. (1985) *J. Mol. Biol.* **186**, 435–455.

Clore, G. M., Nilges, M., Sukuraman, D. K., Brünger, A. T., Karplus, M., and Gronenborn, A. M. (1986a) *EMBO J.* **5**, 2728–2735.

Clore, G. M., Brünger, A. T., Karplus, M., and Gronenborn, A. M. (1986b) *J. Mol. Biol.* **191**, 523–551.

Clore, G. M., Appella, E., Yamada, M., Matsushima, K., and Gronenborn, A. M. (1989) *Biochemistry* **29**, 1689–1696.

Clore, G. M., Bax, A., Driscoll, P. C., Wingfield, P. T., and Gronenborn, A. M. (1990a) *Biochemistry* **29**, 8172–8184.

Clore, G. M., Driscoll, P. C., Wingfield, P. T., and Gronenborn, A. M. (1990b) *J. Mol. Biol.* **214**, 811–817.

Clore, G. M., Bax, A., Wingfield, P. T., and Gronenborn, A. M. (1990c) *Biochemistry* **29**, 5671–5676.

Clore, G. M., Wingfield, P. T., and Gronenborn, A. M. (1991a) *Biochemistry* **30**, 2315–2323.

Clore, G. M., Kay, L. E., Bax, A., and Gronenborn, A. M. (1991b) *Biochemistry* **30**, 12–18.

Clore, G. M., Bax, A., and Gronenborn, A. M. (1991c) *J. Biomol. NMR* **1**, 13–22.

Crippen, G. M., and Havel, T. F. (1988) *Distance Geometry and Molecular Conformation.* Wiley, New York.

Driscoll, P. C. Gronenborn, A. M., and Clore, G. M. (1989a) *FEBS Lett.* **243**, 223–233.

Driscoll, P. C., Gronenborn, A. M., Beress, L., and Clore, G. M. (1989b) *Biochemistry* **28**, 2188–2198.

Driscoll, P. C., Clore, G. M., Marion, D., Wingfield, P. T., and Gronenborn, A. M. (1990a) *Biochemistry* **29**, 3542–3556.

Driscoll, P. C., Gronenborn, A. M., Wingfield, P. T., and Clore, G. M. (1990b) *Biochemistry* **29**, 4468–4682.

Dyson, H. J., Gippert, G. P., Case, D. A., Holmgren, A., and Wright, P. E. (1990) *Biochemistry* **29**, 4129–4136.

Ernst, R. R., Bodenhausen, G., and Wokaun, A. (1987) *Principles of Nuclear Magnetic Resonance in One and Two Dimensions.* Clarendon, Oxford.

Fesik, S. W., Eaton, H. L., Olejniczak, E. T., Zuiderweg, E. R. P., McIntosh, L. P., and Dahlquist, F. W. (1990) *J. Amer. Chem. Soc.* 112, 886–887.

Fesik, S. W., and Zuiderweg, E. R. P. (1988) *J. Magn. Reson.* 78, 588–593.

Folkers, P. J. M., Clore, G. M., Driscoll, P. C., Dodt, J., Køohler, S., and Gronenborn, A. M. (1989) *Biochemistry* 28, 2601–2617.

Forman-Kay, J. D., Clore, G. M., Wingfield, P. T., and Gronenborn, A. M. (1991) 30, 2685–2698.

Foreman-Kay, J. D., Gronenborn, A. M., Kay, L. E., Wingfield, P. T., and Clore, G. M. (1990) *Biochemistry* 29, 1566.

Gronenborn, A. M., Filpula, D. R., Essig, N. Z., Achari, A., Whitlow, M., Wingfield, P. T., and Clore, G. M. (1991) *Science* 253, 657–661.

Güntert, P., Braun, W., Wider, W., and Wüthrich, K. (1989) *J. Amer. Chem. Soc.* 111, 3997–4004.

Haruyama, H., and Wüthrich, K. (1989) *Biochemistry* 28, 4301–4312.

Havel, T. F., Kuntz, I. D., and Crippen, G. M. (1983) *Bull. Math Biol.* 45, 665–720.

Ikura, M., Kay, L. E., and Bax, A. (1990) *Biochemistry* 29, 4659–4667.

Ikura, M., Clore, G. M., Gronenborn, A. M., Zhu, G., Klee, B., and Bax, A. (1992) *Science* 256, 632–638.

Jeener, J. (1971) *Ampere International Summer School, BaskoPolj, Yugoslavia.* Unpublished lecture.

Jeener, J., Meier, B. H., Bachmann, P., and Ernst, R. R. (1979) *J. Chem. Phys.* 71, 4546–4553.

Kaptein, R., Zuiderweg, E. R. P., Scheek, R. M., Boelens, R., and van Gunsteren, W. F. (1985) *J. Mol. Biol.* 182, 179–182.

Kay, L. E., and Bax, A. (1990) *J. Magn. Reson.* 86, 110–126.

Kay, L. E., Clore, G. M., Bax, A., and Gronenborn, A. M. (1990) *Science* 249, 411–414.

Kline, A. D., Braun, W., and Wüthrich, K. (1988) *J. Mol. Biol.* 204, 675–724.

Kraulis, P. J., Clore, G. M., Nilges, M., Jones, A. T., Petterson, G., Knowles, J., and Gronenborn, A. M. (1989) *Biochemistry* 28, 7241–7257.

Marion, D., Driscoll, P. C., Kay, L. E., Wingfield, P. T., Bax, A., Gronenborn, A. M., and Clore, G. M. (1989) *Biochemistry* 29, 6150–6156.

Mueller, L. (1987) *J. Magn. Reson.* 72, 191–196.

Nilges, M., Gronenborn, A. M., and Clore, G. M. (1988a) *FEBS Lett.* 229, 317–324.

Nilges, M., Gronenborn, A. M., Brünger, A. T., and Clore, G. M. (1988b) *Protein Eng.* 2, 27–38.

Nilges, M., Clore, G. M., and Gronenborn, A. M. (1988c) *FEBS Lett.* 239, 129–136.

Nilges, M., Clore, G. M., and Gronenborn, A. M. (1990) *Biopolymers* 29, 813–822.

Noggle, J. H., and Schirmer, R. E. (1971) *The Nuclear Overhauser Effect—Chemical Applications.* Academic Press, New York.

Omichinski, J. G., Clore, G. M., Appella, E., Sakaguchi, K., and Gronenborn, A. M. (1990) *Biochemistry* 29, 9324–9334.

Oschkinat, H., Griesinger, C., Kraulis, P. J., Sørensen, O. W., Ernst, R. R., Gronenborn, A. M., and Clore, G. M. (1988) *Nature (London)* 332, 374–376.

Pardi, A., Billetter, M., and Wüthrich, K. (1984) *J. Mol. Biol.* 180, 741–751.

Powers, R., Garrett, D. S., March, C. J., Frieden, E. A., Gronenborn, A. M., and Clore, G. M. (1992) *Science* 256, 1673–1677.

Wagner, G., Braun, W., Havel, T. F., Schaumann, T., Go, N., and Wüthrich, K. (1987) *J. Mol. Biol.* 196, 611–639.

Williamson, M. P., Havel, T. F., and Wüthrich, K. (1985) *J. Mol. Biol.* 182, 295–315.

Wüthrich, K. (1986) *NMR of Proteins.* Wiley, New York.

Wüthrich, K. (1990a) *J. Biol. Chem.* 265, 22059–22062.

Wüthrich, K. (1990b) *Science* 243, 45–50.

Zuiderweg, E. R. P., Petros, A. M., Fesik, S. W., and Olejniczak, E. T. (1991) *J. Amer. Chem. Soc.* 113, 370–371.

7

A Consumer's Guide to Protein Crystallography

DAGMAR RINGE

GREGORY A. PETSKO

Departments of Biochemistry and Chemistry and
Rosenstiel Basic Medical Sciences Research Center
Brandeis University
Waltham, Massachusetts 02254-9110

The atomic structures of large and small molecules can be determined by a technique called *X-ray diffraction* provided the molecule in question can be crystallized. Most of the three-dimensional structures of proteins, nucleic acids, and viruses that are known at present have been determined by X-ray crystallography. In favorable cases, the resolution of a crystallographic structure determination is such that the relative positions of all nonhydrogen atoms are known to a precision of a few tenths of an Angstrom unit.

Why should a nonspecialist care about the details of this technique? Why cannot the information be used without worrying about how it was obtained? There are two reasons:

First, the demand for structural information is so intense, and the competition to publish structures first has become so keen, that premature structures, incorrect in whole or in part, have appeared in the literature. Some of these have even found their way (briefly) into the Protein Data Bank, the repository for protein three-dimensional structural information maintained at the Brookhaven National Laboratory. If even experienced protein crystallographers can make mistakes, how can pharmaceutical chemists or cell biologists who seek to make use of their results be sure that the structures they need are accurate, without knowing something about the technique? If a protein crystallographer draws conclusions about the function of the mol-

Protein Engineering and Design

ecule from a crystal structure, how can noncrystallographers trust those conclusions unless they can read the paper from an informed, critical perspective?

Second, structural information has become so central to enzymology, cell biology, and immunology that many people in those fields are likely to be involved in a collaboration with a protein crystallographer at some point in their careers. At the start of such a collaboration, they will be faced with many important questions. How much protein must they supply to a protein crystallographer? How pure does it have to be? How long will it take before they see the structure? What conclusions can safely be drawn from it?

We think that the user of protein crystallographic information can easily learn enough about what underlies the method to become an intelligent and critical user, and an understanding and productive collaborator. In this chapter, we attempt to teach this basic information. Each section of the chapter has a concept heading summarizing the salient point that the user needs to remember. We begin with a discussion of the most important quantity in any structure determination, the resolution.

Resolution Determines the Information Content of a Structure: The Higher the Resolution, the More Precise and More Accurate the Structure Is Likely To Be

There are two end-products of a crystallographic structure determination. The objective end-product is something called an *electron density map*, which is nothing more than a contour plot indicating those regions in the crystal where the electrons in the molecule are to be found. Subjective human beings must interpret this electron density map in terms of an *atomic model*. The model allows the measurement of the atomic coordinates—the x, y, and z values relative to some defined origin—of the nonhydrogen atoms in the structure. Users rarely see the electron density map. They usually are given the atomic coordinate set and must infer from its characteristics and from information in the published papers how reliable the structure is. Much of our discussion will therefore deal with the atomic coordinates, but it is essential to realize that they are only as good as the electron density map from which they were derived.

It is also essential to appreciate the distinction between the *accuracy* of an experimentally determined structure and its *precision*. The latter is much easier to assess than the former. We define the precision of a structure determination in terms of the reproducibility of its atomic coordinates. If the crystallographic data allow the equilibrium position of each atom to be determined precisely, then the same structure done independently elsewhere should yield atomic coordinates that agree very closely with the first set. Precise coordinates are usually reproducible to within a few tenths of an angstrom: this is the sort of root-mean-square deviation found between the equivalent coordinates in two independent, high-quality determinations of the same protein structure. Accuracy refers to whether the structure is correct, but there are two ways to define correctness. A structure may be right but irrelevant: the protein may have crystallized in a conformation that is not the biologically active one. Such cases are very rare, but they have occurred. Those structures are "right" in that they represent accurate interpretation of the crystallographic data, but they do not represent the form of the molecule that functions in its biological context. In such cases, no human error has been made. However, it is also possible for a structure to be inaccurate due to human error: the crystallographer may interpret the data incorrectly. Various levels of inaccuracy are possible. The complete structure may be wrong (very rare but it has happened), some errors in chain connectivity or in sequence registration may be made (less rare but usually corrected soon after they occur), or certain side chains may be positioned incorrectly (quite common, especially in the early stages of a structure determination; usually corrected in the final stages of an analysis but not always detectable). For users, the most important thing to understand is that precision and accuracy are somewhat related, in the sense that the more precise a structure determination is, the less likely it is to have gross errors. What links precision with accuracy is the concept of resolution.

Crystallographers express the *resolution* of a structure in terms of a distance: if a structure has been determined at 2 Å resolution then any atoms separated by more than this distance will appear as separate maxima in the electron density contour plot, and their positions can be obtained directly to high precision. Atoms closer together than the resolution limit will appear as a fused electron density feature, and their exact positions must be inferred from the shape of the electron density and knowledge of the stereochemistry of amino acids or nucleotides. Since the average C—C single bond distance is 1.54 Å, the precision with which the individual atoms of, say, an inhibitor bound to an enzyme can be located will be much greater at 1.5 Å resolution than in a structure determined at, say, 2.5 Å resolution. Operationally, crystallographers categorize structures as being of *low resolution* (i.e., about 5–6 Å resolution) if all that can be determined is the overall shape of the molecule; as being of *medium resolution* (around 3 Å resolution) if the folding of the polypeptide chain and the relative positions of the side chains can be determined to about 0.5 Å precision, and as being of *high resolution* (2 Å resolution or better) if most of the atoms in the molecule can be located with a precision better than 0.3 Å (Fig. 1). Most structure determinations today, even of rather large structures like viruses, are carried out at medium or high resolution, but sometimes very important biological information can be obtained from even a low resolution picture (for example, the tunnel in the ribosome revealed by low-resolution structural analysis (Yonath *et al.*, 1993).

There is no substitute for high resolution. It makes the structure deter-

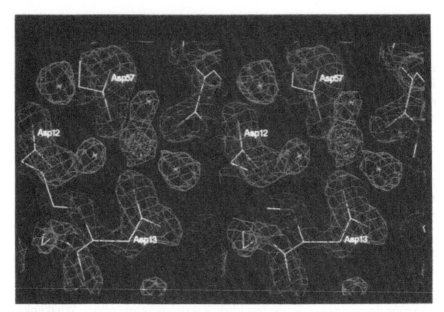

FIGURE 1   Stereo view of the electron density of the metal binding site of the bacterial chemoctactic response regulatory protein CheY at 1.8 Å resolution. In this excellent electron density map, the positions of a number of bound water molecules (shown as stars) are apparent, as is the location of the magnesium ion (stippled sphere). Electron density is typically displayed on computer graphics screens as "chicken wire" mesh at a particular level above background. The atomic model is rendered in skeletal form, with every covalent bond being represented as a line. Atomic positions occur at the ends of line segments, and the atoms are color-coded: dark gray for nitrogen, white for carbon, light gray for oxygen. The magnesium is octahedrally coordinated, with three protein ligands (two carboxylate oxygens and a carbonyl group) and three water ligands (one held in place by another carboxylate, Asp 57). At lower resolution these details would not have been apparent. Data from Stock *et al.*, (1993).

mination easier and more reliable: the closer to true atomic resolution (better than 1.5 Å), the less ambiguity in positioning every atom. Atomic resolution allows the detection of mistakes in the amino acid sequence, correction of preliminary incorrect chain connectivity, and identification of unexpected chemical features in the molecule (see Tsernoglou *et al.*, 1977, and Bjorkman *et al.*, 1987, for striking examples). Incorrect structures have been obtained at high resolution, but only very rarely. Most of the mistakes that have been made in protein crystallography have been made because a medium resolution structure has been misinterpreted or overinterpreted. Users should assume that the higher the resolution, the more accurate and precise the information is likely to be.

The steps in solving a protein crystal structure at high resolution are diagrammed in Fig. 2. First, the protein must be crystallized. Then, the X-ray diffraction pattern from the crystal must be recorded. Phase values

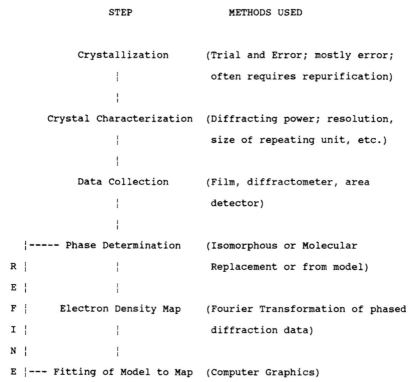

STEP   METHODS USED

Crystallization   (Trial and Error; mostly error;
                              often requires repurification)

Crystal Characterization   (Diffracting power; resolution,
                              size of repeating unit, etc.)

Data Collection   (Film, diffractometer, area
                             detector)

Phase Determination   (Isomorphous or Molecular
                             Replacement or from model)

Electron Density Map   (Fourier Transformation of phased
                             diffraction data)

Fitting of Model to Map   (Computer Graphics)

FIGURE 2   Schematic of the steps in the solution of a protein crystal structure. Refinement methods: Least-squares and/or simulated annealing.

must then be assigned to all of the recorded data; this can be done experimentally or, in some cases, computationally. The phased data are then used to generate an image of the electron density distribution of the molecule. Fitting an atomic model to the electron density map provides the first picture of the structure of the protein, which is improved by an iterative process called refinement until the model is free from gross errors and the precision theoretically attainable at the resolution of the data is achieved. We will use this flow-chart as an outline of our discussion.

But first, we need to discuss a principle from physics. It is the only bit of physics users need to know to understand X-ray crystallography, but it is essential. It is, in fact, the fundamental theorem of X-ray diffraction.

The X-Ray Diffraction Pattern of an Object Is the Fourier Transform of the Electron Density of the Object

X-rays are used to probe structure at the atomic level because their wavelengths are about the same dimension as the 1-Å radius of most atoms.

X-rays are electromagnetic waves and when they impinge upon the electrons in an object they cause those electrons to oscillate. The oscillations in turn make the electrons act as point sources to emit X-rays of the same wavelength; this phenomenon is called *scattering*. The scattered X-rays form a characteristic pattern that contains information about the relative positions of the atoms in the structure. If the distribution of electrons in the molecule can be expressed in terms of a density function, then *the scattering pattern is just the Fourier transformation of the electron density in the molecule.*

Fourier transformation is a mathematical procedure for decomposing a function into a sum of sine and cosine functions. It is very useful because it is reversible: if the scattering pattern from a molecule can be measured and expressed as a set of sine and cosine terms, then its Fourier transform will generate the electron density distribution in the molecule. Since electrons tend to be concentrated around the atomic nuclei, the resulting electron density image will give the relative positions of every atom. That is the basis of all X-ray structure determination: the diffraction pattern of a molecule is just the Fourier transform of its electron density, and vice versa.

But if single molecules scatter X-rays, why do we need a crystal? Unfortunately, the scattering from a single molecule is too weak to measure. In fact, it is too weak by about 17 powers of ten! But that does not seem too difficult a problem to solve: just take a powder or solution containing 10^{17} molecules and measure its scattering. That will not work unless the 10^{17} molecules are all in a defined orientation relative to one another, because if they are not, then their individual scattering patterns will superimpose in random orientations and will average out to a featureless smear. Scattering studies from solutions of protein molecules only provide information about the overall size of molecules, since that is the one parameter that is independent of orientation. In order for the individual scattering patterns of all 10^{17} molecules to reinforce each other so that the signal can be measured, a three-dimensional ordered array is needed. That is why a crystal is needed.

I. CRYSTALLIZATION

Protein Crystallography Requires Large, Well-Ordered Crystals

It is remarkable that objects as irregular in shape as proteins or tRNAs crystallize. Yet many of them do so readily. However, forming crystals is one thing, and forming what is called *diffraction-quality* crystals is quite another. In order to be suitable for a structure determination, protein crystals must be both large and well ordered. Large means at least a few tenths of a millimeter on a side (although if two dimensions are at least this big, the third can sometimes be smaller). All dimensions 0.5 to 1 mm would be ideal. Well-ordered really means two things: the crystal must be a single crystal, and it must scatter X-rays to high resolution in all directions. Not all crystals are single; some are *twinned*, which means two differently oriented lattices growing together. Severe twinning problems can stop a structure determination before it starts. And many protein crystals scatter X-rays weakly (some perfectly shaped crystals do not scatter them at all!) in one or more directions. Depending on the information desired, some of these crystals can still be used, but in general crystals should diffract to at least 3 Å resolution to be considered suitable.

Protein Crystals Have a Very High Solvent Content

Protein crystallographers are often asked what relationship the structure of a protein molecule in the solid state has to its structure in aqueous solution. The reply is that we have no idea, because we have never seen a protein molecule in the solid state! Protein crystals are not like crystals of sodium chloride or tartaric acid. When proteins crystallize, the large irregular shapes of these macromolecules cause them to make contact with one another at only a few points on their molecular surfaces. Large empty spaces are left in the crystal, and these spaces are filled with the solvent from which the crystals were grown. A typical protein or nucleic acid crystal will be at least 50% solvent by volume (Fig. 3). Protein crystals are more like a highly ordered gel than like a crystal of, say, quartz. Sometimes the solvent-filled regions of a protein crystal can be 30 or more Å in diameter, comparable to the size of the macromolecule itself. Protein and nucleic acid crystals are quite fragile as a result of this and must be maintained in sealed capillary tubes under high humidity so they do not dry out and lose their ordered arrangements.

The high solvent content of protein crystals has three very important consequences: first, it means that the structures of biopolymers determined in the "crystalline" state will be very similar, if not identical, to the structures these molecules have in aqueous solution. Second, it means that small molecules such as substrates or inhibitors can be diffused into the crystal lattice, where they will bind to, say, the active site unless it is obstructed by a protein–protein contact. Most crystalline enzymes are active in the "solid" state, although sometimes the constraints of the neighboring molecules will slow down or inhibit altogether a necessary protein conformational change. Finally, the high solvent content of protein crystals usually means that they are intrinsically less well ordered than crystals of smaller molecules. In severe cases, this may mean that the regularity of the crystal is only preserved for the overall shape of the molecule, so that no diffraction data at all can be observed beyond, say, 6 Å resolution. Fortunately, such cases are uncommon, but many protein crystals do not scatter X-rays strongly beyond about 3 Å resolution, and in such cases information about their structures is limited in both precision and accuracy (see below).

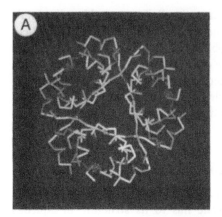
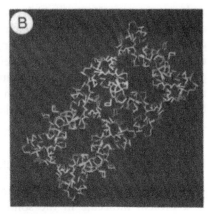
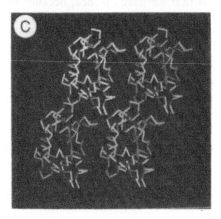

FIGURE 3 Packing of protein molecules in different unit cells. The typical protein crystal is 50% solvent by volume, and there are few contacts between molecules in the lattice. (a) Six insulin molecules pack together to form a hexagon. (b) Eight molecules of lysozyme are shown in this view of the tetragonal crystal form. (c) A different crystal form of lysozyme, showing how the same protein can crystallize in more than one type of lattice. The structure of this crystal form of lysozyme can be determined from the structure of lysozyme in the crystal form shown in (b) by the method of molecular replacement, and that is exactly how it was determined. Note the large solvent-filled channels between the molecules in each of these lattices.

Growth of Suitable Single Crystals Is Usually the Most Difficult Step in Protein Crystallography

The rate-limiting step in the determination of the structure of a biomolecule is usually the production of suitable single crystals. The operative word is "suitable." As we discussed, to be useful for diffraction, protein crystals must be at least 0.2 mm on a side, and must be well ordered.

Diffraction data observable beyond 2 Å resolution is desirable, but as the Rolling Stones sang, "You can't always get what you want." Another aspect of suitability that we have not yet discussed is the protein content of the basic repeating unit of the crystal, because that is what we must actually solve. For ease of structure determination, it is desirable that this unit contain the smallest possible number of molecules. For example, if the protein is tetrameric, it is ideal for the repeating unit in the crystal to contain only a single monomer. If it contains a dimer (or, even worse, one or more copies of the entire tetramer), twice (or four or more times) as many atoms must be located.

When reading a protein crystallography paper, the first thing to look for is the quality of the crystals that have been used. If they were large and diffracted strongly to high resolution, and if the repeating unit was not too big, then the structure is more likely to be of high precision and accuracy.

Although many different techniques have been developed to crystallize proteins, nucleic acids, and viruses, all of them are variations on the same general principle that is used to crystallize any molecule: a solution of the desired substance in some buffer conditions in which it is stable is slowly brought to supersaturation in the hope that crystals rather than amorphous precipitate will form. A variety of methods have been developed to carry out this operation, from microdialysis to vapor diffusion. Most of these methods are designed to use as little of the precious protein sample per experiment as possible, typically, 1–2 μl of protein solution. Unfortunately, protein crystals usually only grow from concentrated solutions. Typical values are 5–40 mg/ml, with 10 mg/ml being the most common. Two microliters of solution will then have about 20 μg of protein, so at least a milligram of protein will be used up for every 50 sets of conditions that are tried.

Many hundreds of trials may be needed before the right conditions of protein concentration, pH, temperature, and precipitant are found. Sometimes they are never found. It is our belief that, given enough time and effort, most highly purified proteins will crystallize. Unfortunately, this can be like the Eddington monkey problem, and most biochemists do not want to wait an infinite amount of time. If a protein has not crystallized after a year or so of honest effort, the outlook is not promising. But it pays to be patient: many protein crystals take weeks or months to grow, although some grow overnight; it depends on the protein. And enough material must be supplied: if several hundreds of experiments are going to be carried out and different protein concentrations are to be explored, the crystallographer will need tens of milligrams of pure protein. Uncrystallized, precipitated protein frequently cannot be recovered, which places further demands on the supply of pure material. We will start an investigation with 10 mg or so, but we expect to need at least 50 mg for a complete study, and often much more.

A word about purity: although crystallization is a purification step for small organic and inorganic molecules, it is usually not one for proteins. The protein must be pure to begin with. Protein preparations less than 90% pure have crystallized, but not often. Usually at least 95% purity is required and better than 98% is highly desirable. It is worth almost any amount of effort to attain this level of purity, because nothing else has so big an effect on whether the protein crystallizes (for dramatic examples, see Kuriyan *et al.*, 1991, and McPherson, 1985).

If a Protein Does Not Form Suitable Crystals, First Improve the Purity: If That Does Not Help, Change the Organism from Which It Is Isolated

Even highly pure protein samples may not be as pure as they appear. Purity is usually assessed by SDS–polyacrylamide gel electrophoresis. If the protein runs as a single band on the overloaded, Coomassie-stained gel it is believed to be at least 98% pure. But significant microheterogeneity can still exist even then, and often that will prevent crystallization. Two techniques are particularly good at detecting microheterogeneity: chromatofocusing and isoelectric focusing. Especially if the protein gives only very small or twinned crystals, it is important to use one of these methods to look at the sample. Sometimes three or more closely separated bands can be seen. A striking example of the power of this method is provided by the work of Kuriyan and associates on trypanothione reductase (Kuriyan *et al.*, 1991). The apparently pure sample, which only gave very poor crystals, was found on isoelectric focusing to consist of three components. When these were separated by preparative isoelectric focussing, two of the three produced beautiful single crystals.

If good crystals still cannot be grown, the next thing to do is to change the source of the material. If the *Escherichia coli* protein does not crystallize, the *Salmonella typhimurium* protein or the yeast protein or the rat liver protein may very well do so. Many biochemists balk at this step, partly because they are skeptical that a small change in amino acid sequence can make so big a difference in crystallization and partly because they dread the labor in developing a purification scheme for a new source of protein. Yet the literature abounds in cases where this has made all the difference. For example, *E. coli* mercuric ion reductase never gave suitable single crystals despite years of effort. The *Bacillus megatherium* enzyme crystallized overnight in large, well-diffracting prisms (Moore *et al.*, 1989).

Particularly valuable in this regard are proteins isolated from thermophilic organisms: the late Ian Harris once told one of the authors that thermophilic enzymes would always crystallize more readily than their mesophilic counterparts because they were more rigid and more resistant to denaturation. We have found this advice to be sound. Over the past 20 years we have repeatedly produced crystals of thermophilic enzymes after long and fruitless struggles with the same protein from mammalian or mesophilic bacterial sources (for examples, see Neidhart *et al.*, 1987, and Stoddard *et al.*, 1987). When all else fails, change the source. It is amazing how often this simple rule will work.

II. DATA COLLECTION

A Data Set in Crystallography Consists of the Measured Amplitudes of the Scattered X-Ray Waves, to Some Specified Resolution

Once suitable crystals are available, their scattering data must be recorded. Unlike the continuous scattering pattern from a single molecule or a randomly oriented ensemble of molecules, the pattern of X-rays scattered by a crystal consists of a series of discrete waves in specific directions (Fig. 4).

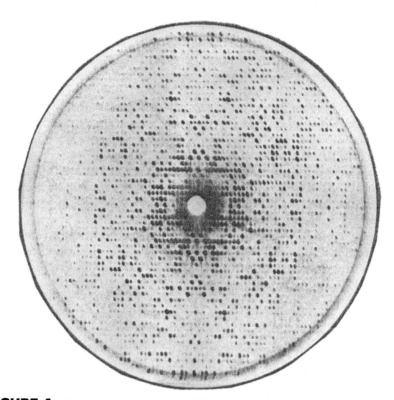

FIGURE 4   X-ray precession photograph of a protein crystal diffraction pattern. Note the symmetry in the photograph, which reflects the symmetry in the way the four protein molecules in each unit cell pack together in the crystal. Reflections closer to the center of the photo are at low resolution; those at the edges, high resolution (here, about 2.5 Å).

This effect is a consequence of the molecule having been incorporated into a crystal lattice: the individual molecules in the crystal each scatter, but the scattered waves interfere with each other destructively as well as constructively, giving rise to a series of spots arranged in regular positions in space. The spacing between the spots is related to the reciprocal of the dimensions of the lattice on which the molecules are organized; the distribution of intensities of the spots reflects the continuous scattering pattern of a single molecule. In other words, it is as though someone had placed a mask on top of the scattering pattern of a single molecule, only allowing the underlying Fourier transform to leak through and be observed at positions corresponding to the transform of the crystal lattice.

Since the pattern of spots resembles what would be observed if light were shined through a diffraction grating, the Fourier transform of a crystal is called the *diffraction pattern*. Because the crystal is three-dimensional, the diffraction pattern is a three-dimensional array of spots; Fig. 4 only shows a section through the complete pattern. The complete three-dimensional diffraction pattern of a protein crystal will have tens of thousands of spots, or more. Crystallographers call these spots *reflections* although nothing is really being reflected. The structure can be solved by taking the inverse Fourier transform of the diffraction pattern; the fundamental theorem states that such an operation will give an image of the electron density in the molecule.

Diffraction spots are waves, and a complete description of each wave must be fed into the computer before the inverse transformation can be carried out. How is a wave characterized? If a man were standing on a Pacific island and a friend came rushing up to tell him that a tidal wave was coming, he would want to know two things: when it was due to arrive (this is called the "phase" of the wave), so he would know how much time he had to get to high ground, and how big the wave was (its "amplitude") so he would know how high he had to climb! The same information is needed for each of the diffracted waves in a crystal structures determination. Phases cannot be measured directly (see below) but amplitudes can. Data collection in crystallography is just measurement of the amplitudes of all of the scattered waves (reflections) to a particular resolution.

The Number of Reflections that Must Be Collected for a Complete Data Set Depends on the Reciprocal of the Resolution, Cubed

The resolution of the final structure is determined by how many reflections are measured (Fig. 4). If only the strong central region of the diffraction pattern is measured, only a low resolution picture will be obtained. High resolution structure determination requires the time-consuming measurement of all observable structure amplitudes, even the weak ones at the limit of the pattern. And there are a lot of reflections to measure.

The number of reflections that must be measured to obtain a structure at a given resolution is inversely proportional to the cube of the resolution (we measure data in a sphere of a certain volume, and the resolution is related to the reciprocal of the radius of that sphere). The number of reflections is also directly proportional to the molecular weight of the basic repeating unit (called the asymmetric unit) of the crystal. Thus, if the number of reflections for a particular size protein to a particular resolution is known, we can easily calculate the approximate number that must be measured for any other protein to any desired resolution. For example, there are 8000 reflections in the diffraction pattern of bovine pancreatic ribonuclease A, a protein of 13,500 Da with one molecule in the asymmetric unit of its crystal form, to 2.0 Å resolution. Therefore, there will be 64,000 reflections to 1 Å resolution for ribonuclease A [$1/(1)^3$ is 8 times $1/(2)^3$, so the number of reflections increases eightfold]. Chymotrypsin, on the other hand, which is twice as big as ribonuclease A, will have 16,000 reflections to 2 Å resolution if there is a monomer in its asymmetric unit, and 32,000 reflections if there is a dimer. Large oligomeric proteins and viruses may have hundreds of thousands of reflections that must be measured if high resolution is desired.

Data Collection Is Now Rapid Thanks to Advances in Detector Technology

The amplitude of each reflection is easy to measure. It is related to the relative blackness of each spot on the film in Fig. 4, and can be quantified by optical scanning of the film. Film processing is laborious, and it means that data acquisition must occur in two discrete and time-consuming steps: recording the complete set of spots in the diffraction pattern on many films, and then extracting the amplitude information by scanning them. Normally, it takes weeks to obtain a protein data set by this method, and a large portion of that time is spent in the manual labor of scanning films.

This labor can be eliminated by counting the scattered photons directly. A machine called a *diffractometer* is designed to do that. It consists of an Eulerian cradle for positioning the crystal and a scintillation counter, similar to a Geiger counter, mounted on a movable arm. Motion of the crystal and detector is under computer control, and the crystallographer can program the diffractometer to measure the number of scattered X-rays in each reflection automatically, one at a time. Unfortunately, it may take a minute or so to count the scattered photons in one reflection, and if one has 32,000 reflections to collect, at least 32,000 min (over 3 weeks) will be required to measure a data set. Few protein crystals will survive in the X-ray beam for that length of time at ambient temperatures, so several crystals per data set are usually required, and scaling the partial data sets together often does not remove systematic differences between them.

A newer method that has greatly speeded up the process of amplitude measurement is the use of *area detectors*. Area detectors are just electronic film: a two-dimensional grid of wires that is sensitive to the positions and numbers of photons in each reflection. Since area detectors count the scattered X-rays directly, but over a larger area of space than a single counter on a diffractometer, an area detector interfaced to a computer can produce the amplitude information for hundreds of reflections per minute, while it is measuring. Area detectors are expensive: the cost for a simple one is well over $100,000 and some are more then twice that. But the money is well spent: with an area detector, a complete set of X-ray amplitudes can be measured from most crystalline biomolecules in only a few days.

There Is No Easy Way to Assess the Quality of a Data Set from Published Information

The amplitude of a reflection whose position in the diffraction pattern is given by the triple set of indices (h,k,l) is denoted F_{hkl}. Sometimes it is also written as F_{obs} to show that the quantity has been observed experimentally. The scattered intensity that gave rise to that amplitude is I_{hkl}. During data collection, a given I_{hkl} can be measured more than once, because of symmetry in the diffraction pattern. The precision of the data set can be estimated from the agreement among multiple measurements. The usual quantity reported is R_{sym}, which is the average percentage disagreement among symmetry-related measurements of the same I_{hkl}. For precise data, R_{sym} tends to be 5% or less. A related quantity, R_{merge}, reports the average percentage disagreement between common reflections measured either from different crystals or, more often, from different orientations of the same crystal. R_{merge} tends to be higher than R_{sym} by a few percentage points.

Unfortunately, although these R values can give a sense of the *reproducibility* of the data, they are not good measures of its *accuracy*. There are many sources of systematic error in X-ray data collection, and only comparison of complete data sets from different crystals measured on different types of detectors would allow their detection. In practice, such comparisons are almost never made, and when they are, the percentage disagreements tend to be much higher than the internal R values from one data set. Crystallographers will report R_{sym} and R_{merge} in publications, but a single number is never a good measure of the quality of a huge set of data. Readers of crystallographic papers should take some consolation, however, in the fact that modern data collection instruments and software are fairly sophisticated, so it is not difficult to measure good data from well-diffracting macromolecular crystals.

III. PHASE DETERMINATION

The Phases of the Scattered X-Ray Waves Cannot Be Measured Directly

Phase determination is much more difficult. X-rays travel at the speed of light, so there is no way to measure directly the relative time of arrival of each diffracted wave at the detector. This loss of phase information in the diffraction pattern leads to what is called *the phase problem* in crystallography. Somehow, the missing phases must be deduced from the only things that can be measured, the amplitudes.

In small molecule crystallography a number of very sophisticated mathematical and computational techniques have been developed that can produce phases directly from measured amplitudes. These *direct methods* of phase determination make small-molecule structure determination automatic in most cases. Thus far, technical obstacles have prevented their application to protein crystallography, so we are not yet able simply to take the measured F_{hkl} values, plug them into a computer program, and compute their phases from relationships between their amplitudes.

The Method of Isomorphous Replacement Uses Specific Heavy-Atom Substitution to Determine the Phases

However, if the macromolecules in the crystal can be labeled specifically with a heavy atom (such as might be accomplished by diffusing a solution of mercuric chloride into the crystal and letting it react with any cysteine residue that might be on the surface of the proteins), the diffracted intensities will change because the electron density distribution in the crystal will have changed. Some spots will increase in intensity, some will decrease, and some will remain unchanged. If a second derivative can be made that places a new heavy atom at a different position, the diffraction pattern will change as well, but the different reflections will increase and decrease in intensity. These three data sets (two derivative and one native protein data set) can be combined to give an estimate for the phase of each reflection. Phase determination in this way is somewhat analogous to positional location by triangulation from known distances. There will only be a single value for the phase of a reflection that allows the calculation of the correct effects (increase, decrease, or no change) on the reflection intensity of placing two heavy atoms at two known, but different, positions in the crystal lattice. At least two different derivatives are needed because phase determination from a single derivative has an ambiguity, just as position location from too few distances would be ambiguous. (Of course, to carry out this calculation it is essential that the positions of the heavy atoms in the crystal be known exactly. A number of methods have been developed to find the heavy atom locations directly from the intensity differences.)

This method of phase determination is called *multiple isomorphous replacement:* multiple, because more than one derivative is needed. Isomorphous means that the binding of the heavy atom must not change the protein structure, since the method depends on all of the intensity differences being due solely to the presence of a heavy atom at a particular place. Crystallographers call this phasing method MIR for short. In order for a derivative to be useful in phase determination, it must also give changes that are large enough to be measured accurately. Structure papers often present tables for each derivative of the ratio of the average contribution of the heavy atom to each reflection, abbreviated f_h, to the average error in the measurement and interpretation of that derivative, abbreviated E. The quantity f_h/E is sometimes called the "phasing power" but is really the signal-to-noise ratio for that derivative; a value less than one is supposed to mean that the derivative is providing little useful information, a value of 2 or greater signifies a very good derivative. Unfortunately, it is not straightforward to estimate the average error E, so the fact that a structure determination reports poor values for the phasing power should not be sole grounds to suspect the accuracy of the results.

The Method of Molecular Replacement Allows Phases to Be Determined Using Only a Native Data Set, Provided the Structure of a Related Molecule Is Already Known

Proteins are known to come in structural families, and it would seem that the knowledge that a protein of undetermined structure is homologous to a protein whose structure has already been solved could be used somehow in phasing. A second method of phase determination, called the method of *molecular replacement,* is based on this idea. Recall that the diffraction pattern is the Fourier transform of the molecule sampled at points corresponding to the lattice of the crystal. If two molecules are similar in structure, their molecular Fourier transforms will also be similar. Suppose we wish to determine the structure of the aspartic protease from the human immunodeficiency virus (HIV or the AIDS virus for short). Its structure is likely to be similar to that of the avian myoblastoma virus (AMV) protease, because their sequences are similar. An atomic model of HIV protease can be constructed from the known crystal structure of the AMV protease by replacing the corresponding amino acids in the known structure with their HIV counterparts. This model will be highly inaccurate, but if it is Fourier transformed to produce only low resolution data it should be a reasonable approximation to the actual HIV protease transform at low resolution. The model structure can then be moved in a computer through all possible orientations in the lattice of the actual HIV protease crystal until its calculated low-resolution diffraction pattern matches the observed intensity distribution from the actual measured data set for the HIV protease crystal. When a match is found, the rotated and translated model structure provides a set of crude atomic coordinates for the HIV protease, properly oriented in its lattice. Fourier transformation of the electron density distribution calculated from these coordinates provides a set of calculated phases for every reflection; these can be combined with the original measured amplitudes, yielding a new electron density map at higher resolution. If the process was successful, this map can be reinterpreted in terms of an improved atomic model.

Although the molecular replacement method is dependent on having a good starting model and often fails when the known and unknown structures are somewhat different, it is very convenient because it is purely computational, requiring no heavy-atom derivatives and only a single native data set. As the data base of known structures increases, and as protein engineering produces many mutant and hybrid proteins from well-studied old materials, it is likely that molecular replacement will become even more common a phasing tool than MIR.

IV. ELECTRON DENSITY MAP INTERPRETATION

Electron Density Map Interpretation Is Subjective

Once phased amplitudes are available, the diffraction data can be Fourier transformed to give an image of the electron density in the crystal. Depending on the resolution and the quality of phase determination, this map may be easy or difficult to interpret.

It is in the map interpretation stage that most of the errors in structure determination occur. If the phases are good, a crystallographically derived electron density map is a completely objective view of the distribution of electrons in the asymmetric unit of the crystal. However, when individual atoms are not resolved, their positions must be inferred from the size and shape of blobs of electron density, a subjective process carried out by fallible human beings. Refinement of the structure (see below) can in principle correct most, or even all, of the errors introduced during model building, but sometimes the mistakes are difficult to detect. The most common error in map interpretation is getting the connectivity of the secondary structural elements wrong. Individual α-helices and β-strands are usually easy to see, but the order in which they are joined together is sometimes ambiguous, especially if two distant portions of the polypeptide chain come together like a letter X. It is difficult to tell whether two diagonally running segments (×) are crossing or whether two U-shaped segments are in contact (⊃⊂). The wrong choice leads to the wrong connectivity.

The Molecular Boundary Should Be Clearly Visible

How can users determine that an error is likely, or not likely, to have been made? It helps to have some sense of the quality of the map. A picture of the initial electron density map, if available, gives information about the clarity of separation between the macromolecules and the solvent. The molecular boundary should be clear, since the solvent regions will have lower average density than that of the protein. Unfortunately, many crystallographers employ a computational procedure called solvent-flattening to improve their phases by enhancing the contrast between protein and solvent. A picture of a solvent-flattened map will always show a spectacularly clear molecular boundary, but this tells users nothing about the quality of the original map, which is a much better guide to the overall quality of the structure determination.

The Path of the Polypeptide Chain Should Be Unambiguous and Free from Breaks

The path of the polypeptide chain will be discernible as a ribbon of continuous density with branches (the side chains) every 3.8 Å. At low resolution, α-helices may be apparent as rods of high density, but chain connectivity is rarely obvious at 6 or 5 Å resolution. Structures determined at 3 Å resolution should show helices clearly and the strands of β-sheets will appear as twisted ropes of density with side chains pointing alternately in opposite directions. At 3 Å resolution the direction of the chain can be determined, since side-chain branching density in helices slants back toward the N-terminus of the helix.

Breaks in the continuity of the polypeptide chain electron density cause problems in the interpretation, since crystallographers must guess how the discontinuous pieces are joined. Serious errors in connectivity can occur from guessing. A map that presents no such ambiguities is much more likely to yield a correct final structure.

The Sequence Cannot Be Deduced Unambiguously from the Electron Density Map, but Errors in the Sequence Can Often Be Detected

In a 3-Å resolution map it may be possible to identify bulky side chains such as tryptophan, but in general even at high resolution it is impossible to derive the sequence of the polypeptide or nucleotide from the electron density map alone. Sequence information is essential in generating an accurate atomic model, even from the best maps.

Unfortunately, many sequences contain errors, even those determined by sequencing cloned DNA. (We have hardly ever determined the crystal structure of a protein without detecting at least one sequence error.) After the molecular boundary and chain direction and path have been determined, building the atomic model usually starts with an attempt to fit part of the known sequence into the map. A segment of polypeptide with a number of aromatic residues is often chosen first, as these large side chains give characteristic large flat electron density at medium to high resolution. Heavy-atom binding sites may provide markers that allow the sequence to be aligned with the electron density. For example, mercury has a high affinity for sulfur and reacts to form a covalent derivative, so mercury binding sites are often the locations of accessible cysteine residues. Computer graphics devices are used to display the map as a basket of lines, with the model superimposed in a different color in stick representation (Fig. 1). Fitting programs allow positioning of the model into the density and alter the conformational parameters of the main and side chain groups to achieve the optimal fit.

The process is highly subjective and mistakes are likely, especially if the map is poor in quality due to low resolution, badly measured data, imperfect model structure in the case of molecular replacement phasing, or lack of isomorphism in the derivatives in the case of MIR phasing. Even a good map can have confusing regions if the molecule is flexible and the density in that part of the map represents contributions from more than one conformation. In the end, crystallographers will try to generate a tentative atomic model with positions for most, if not all, of the residues in the protein or nucleic acid.

V. REFINEMENT

Careful Refinement Can Improve the Quality of a Structure but Refinement Is Not an Assurance of Accuracy

Once an atomic model has been built, some idea of its quality can be obtained by Fourier transforming its electron density to obtain calculated reflection amplitudes and phases. Comparison of the calculated amplitudes with those actually measured from the crystal by film or area detector yields a quantity denoted R, the crystallographic residual,

$$R = \sum_{hkl} |F_{obs} - F_{calc}| \Big/ \sum_{hkl} |F_{obs}|,$$

where F is the amplitude and the sums are over all measured reflections. If the model and data were both error-free, R should equal 0, but that never happens. A random ensemble of atoms that occupied the fraction of space as the true structure would give $R = 0.59$. It is somewhat disturbing that the crude atomic models obtained from the first fitting of an electron density map usually have $R = 0.45$ or so. An R factor of this magnitude implies that the average random coordinate error in the structure is close to 1 Å and that a number of errors of interpretation may exist. Nevertheless, these models are demonstrably better than random and they can be improved. The process of improvement is called *refinement*.

Structure refinement is based on the comparison of observed and calculated structure amplitudes used in the R factor. If F values from the model do not agree well with the measured data, it is possible to define a least-squares function,

$$\text{Function} = \sum_{hkl} (F_{\text{obs}} - F_{\text{calc}})^2,$$

which should be a minimum for the best model that the data can provide. The sum is taken over all measured reflections. Minimization of this function is a nonlinear least-squares problem and therefore cannot be achieved in one step, but some very sophisticated computational methods have been developed to make small changes in the model so that the value of the function is reduced, albeit slowly. Many cycles of refinement are required, and the computer algorithms cannot make very large corrections to the structure, so every so often the process must be interrupted and the model must be rebuilt manually. Monitoring the R factor is one way to tell when manual intervention is necessary; it will decrease as the function decreases and will level off when the least-squares approach cannot make any further improvements. Manual rebuilding is greatly aided by the refinement, however, since as the R factor decreases the quality of phases calculated from the refined model will increase. New electron density maps calculated with measured data and phases computed from the atomic model should be greatly improved over the original error-filled maps, and better interpretation of problem areas should be possible. Refinement is tedious and time-consuming, but can dramatically improve a structure: the final R factors for most well-refined protein crystal structures are usually 0.20 or lower. Zero is never achieved because the measured data have errors of at least a few percentage points, proteins have regions of conformational flexibility that are difficult to characterize, and the large regions of disordered solvent in the crystal are impossible to model accurately.

The *R* Factor Is, at Best, Only a Rough Guide to the Quality of a Structure

The precision of a protein crystal structure can be estimated from its R factor and resolution. Unrefined structures at medium resolution can have huge errors and should never be trusted quantitatively, although they may give a fairly accurate picture of the polypeptide chain fold and the relative arrangements in space of the side chains. Refined structures at near 3 Å resolution will still have average positional errors of 0.5 Å or more, and so they cannot be used to determine whether two groups are within hydrogen-bond distance of one another. Refined structures at 2 Å resolution will have coordinate precision of about 0.3 Å and can be used to infer most details of hydrogen bonding. Bound solvent molecules can also be located correctly at this resolution. Structures refined at 1.5 Å resolution or better will have positional errors of 0.2 Å or less and errors in bond angles of only a few degrees.

The Only Reliable Criterion for the Accuracy of a Protein Crystal Structure Is Based on Common Sense, Not Mathematics: Is the Structure Consistent with the Body of Biochemical Data on the Molecule?

It is important to remember that all these error ranges are in precision, not accuracy. Even high resolution structures with very low R factors may have a few wrongly placed atoms, although gross errors of interpretation are less likely. As protein crystallography becomes more and more a routine tool for structure/function investigation, experimental details are often too tersely reported for even an experienced referee to be certain that no mistakes have been made. We have tried in this chapter to give the user of structural information some indication of where problems can arise and what sort of warning signs to watch for, but it is our belief that no number, or set of numbers, can reliably indicate the *accuracy* of a protein structure determination.

Fortunately, there is a completely reliable indicator that does not depend on any number at all. It is biochemical common sense. The structure must explain and be consistent with the body of experimental data for the molecule in question. If it does, it is probably right in general and also in most of its details. If it does not, it is almost certainly wrong in whole or in part. Every incorrect protein structure, without exception, could have been suspected immediately by this criterion regardless of any of the published experimental details or "reliability" indices. Of course, if only one or two experiments out of many are inconsistent with the structure, they should probably be questioned. But incorrect structures nearly always fail to explain many things, not just one or two. We urge you to try this method of judging structures for yourself, on a structure that has been shown to be completely incorrect (Ghosh *et al.*, 1991, reports the original, incorrect structure and Antonio *et al.*, 1982, illustrates how the structure made no sense in terms of the spectroscopic data on the cofactor; Stout *et al.*, 1988, reports the correct structure determination, which fits the spectroscopic data beautifully).

Wise biochemists—and wise crystallographers—will therefore always check to see whether the structure makes sense in terms of what is known about the molecule from experimental data, and if the structure does not explain the results of, say, nearly all of the mutagenesis experiments, the structure should be questioned.

VI. EXPLOITING THE STRUCTURE

It Is Not Possible to Predict How Long It Will Take to Solve a Structure *de novo*, but Structures of Mutants or Structures of Protein—Ligand Complexes Can Be Obtained Very Rapidly

The best thing about crystallography is that once all of the work of solving a structure has been done, not only is there a huge amount of information about the protein, but there is also the ability to exploit that information very rapidly. If the structure of the protein does not change very much when ligands ((such as enzyme inhibitors) bind, it may be possible to diffuse them into the crystal lattice and let them form a complex with the active site without having the crystal fall apart. An alternative method is to cocrystallize the protein in the presence of the small molecule and hope that crystals of the complex will grow under much the same conditions as the native crystals. Cocrystallization is more problematic than diffusing ligands into native crystals, but it does obviate the concern that lattice forces in a pregrown native crystal might inhibit binding or block important conformational changes.

Ligand binding will cause the intensities in the diffraction pattern to change, but the phases will not change very much if the overall structure has not changed. Thus, for the structure of the protein—ligand complex, the phase problem has already been solved. To determine the structure of the complex, measure the amplitudes of the reflections from the crystal of the complex, and compute an electron density map with coefficients $(F_{p1} - F_p)$, where F_p is the amplitude of the reflection from the native crystal and F_{p1} is the amplitude from the crystal with the ligand bound. The phases for each reflection are the phases determined experimentally for the native structure, or calculated from the refined atomic model. Such a map is called a *difference electron density map* and it shows the location of the additional electron density due to the bound ligand and any changes induced in the structure of the protein.

Calculation of a difference map for any ligand complex requires only a few days, most of which time is spent collecting the data. If the map is easy to interpret, a good atomic model can be produced for, say, an enzyme—drug complex in less than a week (Mattos *et al.*, 1994). Mutant proteins produced by recombinant DNA technology can be analyzed equally rapidly (Joseph-

McCarthy *et al.*, 1994). This ease of exploitation of a structure is the reward for all of the hard work that went before.

Of course, the structure of the protein–ligand complex or of the mutant protein may be sufficiently different from the native structure that the crystals are not isomorphous or even in the same unit cell and space group. If that happens, the original phases cannot be used and the problem of solving a new structure arises. However, the path to such a solution is clear: molecular replacement using the refined native structure as the search model can usually be employed, and if that fails, the same heavy atom derivatives that worked the previous time can be used (Davenport *et al.*, 1991).

Protein crystallography is difficult and tedious, but by thinking of a protein crystal structure as the beginning of an investigation, not its end, the enormous power of the method is realized (Fitzpatrick *et al.*, 1993). In most cases, very important structures of, say, mutant proteins, can be determined to atomic resolution with ease in less time than it takes to sequence them. For protein design, this rapid turnaround means rapid feedback that makes the design process much more rational.

VII. SOME FINAL COMMENTS

Protein crystallography is a wonderful technique. It can reveal the locations of atoms in structures as large as viruses. It is also an often frustrating labor-intensive process. The protein that crystallographers care about most may never crystallize. Or, it may give beautiful crystals that do not diffract at all. It may give only low resolution diffraction when high resolution is essential to understanding the relationship between structure and function. It may give high resolution diffraction but may have 12 molecules in the basic repeating unit of the crystal. It may refuse to bind heavy atoms. Or, it may bind them in too many places and change its structure when it binds them. It may contain disordered regions that do not give well-defined electron density, and these may be the regions of greatest biological importance, and so on. The litany of possible woes is long. But as countless cell biologists, biochemists, and pharmacologists will tell you, the results are well worth the effort.

APPENDIX: SOME USEFUL ARTICLES AND ADDRESSES

1. Branden, C. I., and Tooze, J. (1991) *Introduction to Protein Structure*. Garland, New York. The modern successor to the classic book by Dickerson and Geis. A beautifully illustrated, clearly written summary of 20 years of structural information. Should be a required text in all biochemistry courses.

2. Moffat, K. (1984) Protein crystallography. In *Spectroscopic and Diffraction Methods for Biological Research.* (D. L. Rousseau, ed.), Academic Press, New York. A review article somewhat along the lines of this one but with more mathematics and physics. Well written and not too out of date.

Protein atomic coordinates are stored in the Brookhaven Protein Data Bank. It costs a few hundred dollars per year to obtain coordinates for all of the proteins whose structures have been deposited, which works out to a few dollars per structure, a lot less than it cost the crystallographers or NMR spectroscopists who did the structures! The address is Protein Data Bank, Brookhaven National Laboratory, Upton, New York 11973.

We urge anyone interested in structural information to think seriously about getting their own molecular graphics workstation. The cost of these devices is now comparable to that of the average spectrophotometer, and nothing can substitute for the sense they give of touching the molecules one is interested in. There are a number of commercial software packages that will allow display and manipulation of protein structures on a variety of workstations. For those interested in lower-end graphics, some packages are now available for personal computers that, although limited, do a surprisingly good job of visualization.

REFERENCES

Antonio, M. A., Averill, B. A., Moura, I., Moura, J. J. G., Orme-Johnson, W. H., Teo, B. K., and Xavier, A. V. (1982) *J. Biol. Chem.* 257, 6646–6649; Beinert, H., Emptage, M. H., Dreyer, J.-L., Scott, R. A., Hahn, J. E., Hodgson, K. O., and Thomson, A. J. (1983) *Proc. Natl. Acad. Sci. U.S.A.* 80, 393–396.

Bjorkman, P. J., Saper, M. A., Samraoui, B., Bennett, W. S., Strominger, J. L. and Wiley, D. C. (1987) *Nature (London)* 329, 506–512; Harrison, D. H., Bohren, K., Ringe, D., Petsko, G. A., and Gabbay, K. H. (1994) *Biochemistry* 33, 2011–2020.

Davenport, R. C., Bash, P. A., Seaton, B. A., Karplus, M., Petsko, G. A., and Ringe, D. (1991) *Biochemistry* 30, 5821–5826.

Fitzpatrick, P. A., Steinmetz, A. C. U., Ringe, D., and Klibanov, A. M. (1993) *Proc. Natl. Acad. Sci. U.S.A.* 90, 8653–8657.

Ghosh, D., Furey, W., Jr., O'Donnell, S., and Stout, C. D. (1981) *J. Biol. Chem.* 256, 4185–4192.

Joseph-McCarthy, D., Lolis, E., Komives, E. A., and Petsko, G. A. (1994) *Biochemistry* 33, 2815–2823.

Kuriyan, J., Kong, X. P., Krishna, T. S. R., Sweet, R. M., Murgolo, N. J., Field, H., Cerami, A., and Henderson, G. B. (1991) *Proc. Natl. Acad. Sci. U.S.A.* 88, 8769–8773.

Mattos, C., Rasmussen, B., Ding, X., Petsko, G. A., and Ringe, D. (1994) *Nature Struct. Biol.* 1, 55–58.

McPherson, A., (1985) *Methods Enzymol.* 114, 112–120.

Moore, M. J., Distefano, M. D., Walsh, C. T., Schiering, N., and Pai, E. F. (1989) *J. Biol. Chem.* 264, 14386–14388.

Neidhart, D. J., DiStefano, M. D., Tanizawa, K., Soda, K., Walsh, C. T., and Petsko, G. A. (1987) *J. Biol. Chem.* 262, 15323–15326.

Stock, A. M., Martinez-Hackert, E., Rasmussen, B. F., West, A. H., Stock, J. B., Ringe, D., and Petsko, G. A. (1993) *Biochemistry* **32**, 13375–13380.

Stoddard, B., Howell, L., Asano, S., Soda, K., Tanizawa, K., Ringe, D., and Petsko, G. A. (1987) *J. Mol. Biol.* **196**, 441–442.

Stout, G. H., Turley, S., Sieker, L., and Jensen, L. H. (1988) *Proc. Natl. Acad. Sci. U.S.A.* **85**, 1020–1022.

Tsernoglou, D., Petsko, G. A., and Tu, A. T. (1977) *Biochem. Biophys. Acta* **491**, 605–608.

Yonath, A., and Berkovitch-Yellin, Z. (1993) *Curr. Opinion Struct. Biol.* **3**, 175–181.

8

Spectroscopic and Calorimetric Methods for Characterizing Proteins and Peptides

PAUL R. CAREY*

WITOLD K. SUREWICZ†

*Department of Biochemistry, Case Western Reserve University, 10900 Euclid Avenue, Cleveland, Ohio 44106; and †Departments of Ophthamology and Biochemistry, University of Missouri, Columbia, Missouri 65212

I. INTRODUCTION

The investigator faced with characterizing a newly designed protein or peptide has to select the technique that meets his or her need. If a complete 3-D structure is required, crystallographic or NMR studies will have to be undertaken. However, many cases require a rapid assessment of changes in secondary structure or in thermal stability. For these there are a number of spectroscopic techniques available that can provide answers to critical questions with a much smaller investment of time and effort than is needed to undertake complete 3-D analysis. Often data can be collected and interpreted in hours or a few days rather than the months of intensive work needed to derive an accurate total structure by crystallography or NMR. Moreover, some of these techniques are able to address questions of, for example, variation in function in a family of protein-engineered enzymes. In this chapter we outline four approaches to probing protein and peptide properties, two of these involve vibrational spectroscopy—Raman and FTIR spectroscopies. For both of these, recent technical progress has made them more user friendly and generally applicable. The other methods, circular dichroism and calorimetry, continue to be widely used in studies of secondary structure and thermal stability, respectively. Table I outlines some of the

TABLE I Comparison of Spectroscopic Techniques and Calorimetry

Technique	Advantage	Atomic detail?	Secondary structure?	Rapid comparison of family members?	Measure conformat. transitions?
Circular dichroism		No	Yes	Yes	Yes
Raman		Some	Yes	No	Yes
Resonance Raman	Detailed info. on biological chromophores	Yes	No	Sometimes	Sometimes
FT infrared		Some	Yes	Yes	Yes
Calorimetry	Only technique that provides insight into thermodynamics of protein structure	No	No	Yes	Yes

uses and strengths of the methods treated here in protein and peptide characterization.

II. CIRCULAR DICHROISM

Circular dichroism (CD) spectroscopy is a branch of absorption spectroscopy, associated with electronic transitions, that is used to measure the interaction of polarized light with optically active (i.e., asymmetric) molecules. An introduction to the basic principles of CD may be found in a number of books and review articles (e.g., Cantor and Schimmel, 1980; Johnson, 1985). In this chapter, we will discuss mainly the practical aspects of the uses and limitations of CD spectroscopy in protein research. It should be emphasized that many individual chromophores present in proteins are symmetric and thus optically inactive, a feature prohibiting circular dichroism. However, optical activity of these chromophores may be induced by their interaction with an asymmetric environment provided by the rest of the protein molecule. In other words, circular dichroism of proteins and polypeptides results primarily from the spatial asymmetry of the constituent amino acids. This is the main reason why CD spectroscopy provides a sensitive tool to probe the conformational properties of these macromolecules.

CD spectra of proteins and peptides are usually measured in two spectral regions. Bands in the far-UV region (below 250 nm) represent primarily electronic transitions of the amide groups of the protein backbone. The sign, magnitude, and position of these bands are strongly dependent on the ϕ and ψ angles of the peptide bond. The spectrum in this region contains information about the secondary structure of the protein. CD bands in the near UV region (approximately 250–310 nm) originate primarily from the side chains of aromatic amino acids and from disulfide bonds. Circular dichroism of the aromatic amino acids is highly sensitive to the local environment of these residues and their orientation with respect to the backbone. The CD of disulfide bonds, on the other hand, depends on the overall geometry of C-S-S-C groups. Although very difficult to interpret in molecular terms, spectra in the near-UV region provide a fingerprint of the tertiary structure of the protein molecule.

Circular dichroism is defined as the difference between the extinction coefficients for left and right circularly polarized light,

$$\Delta\epsilon(\lambda) = \epsilon_L(\lambda) - \epsilon_R(\lambda), \tag{1}$$

where $\epsilon_L(\lambda)$ and $\epsilon_R(\lambda)$ represent the extinction coefficients for left and right polarized light, respectively, at wavelength λ. Since, according to the Beer–Lambert law extinction coefficients, $\epsilon_{L,R}$ are related to the ordinary absorption $A_{L,R}$ by the formula $\epsilon_{L,R} = A_{L,R}(\lambda)/lc$, where l is the pathlength of the cell (in cm) and c is the concentration (in mol/liter), one can express circular dichroism as

$$\Delta\epsilon(\lambda) = \Delta A_{L,R}(\lambda)/lc, \tag{2}$$

where $\Delta A_{L,R}(\lambda) = A_L(\lambda) - A_R(\lambda)$.

For mostly historical and instrumental reasons, the CD data are often expressed not as $\Delta\epsilon(\lambda)$ but in terms of molar ellipticity, $\Theta(\lambda)$, which has units of deg cm^2 decimol^{-1} and is related to $\Delta\epsilon(\lambda)$ by the formula $\Theta(\lambda) = 3298\Delta\epsilon(\lambda)$. One should note certain inconsistencies in the literature regarding the molar units used to normalize CD spectra. In the far-UV region, the principal chromophore of interest is the peptide bond and the molar unit is based on the mean residue weight (i.e., the molecular weight of the protein divided by the number of amino acids in the protein molecule). On the other hand, near-UV spectra measure the signal from specific chromophoric groups (aromatic side chains, prosthetic groups). The molar unit used to analyze these spectra is often based on the concentration of the whole macromolecule.

Since circular dichroism measures a difference in absorption by the sample of left and right circularly polarized light, the CD signal is relatively very weak and usually represents only approximately 10^{-4} of the absorption. The need to measure such weak signals imposes specific conditions on the design of circular dichroism instrumentation (e.g., the use of very bright light sources) as well as on the preparation of samples for CD spectroscopy. In general, the intensity of the CD signal will increase as the optical density (and thus concentration of the sample) is increased. However, since the noise in circular dichroism spectroscopy is statistical in nature, an increase in optical density will also result in a deterioration of the signal-to-noise ratio.

In theory, the best compromise between the sensitivity requirements and the signal-to-noise ratio is obtained at the absorbance of 0.869 (Johnson, 1985). For practical reasons, it is very difficult to make accurate CD measurements when the total absorbance of the sample (i.e., that of the protein and the solvent) exceeds one. High absorption of the solvent presents a particularly serious problem when the measurements are made in far-UV region. This problem can be partially circumvented by using cells with a very short pathlength and/or reducing the concentration of buffer salts. Typically, cells with a 0.05 to 0.2-mm pathlength are used for far-UV measurements, whereas longer pathlength cells (5–10 mm) are adequate for near-UV experiments. Even with short pathlength cells, measurements below approximately 200 nm are often hampered by a poor signal-to-noise ratio and require the use of multiple scans and computerized signal averaging.

A. Far-UV Circular Dichroism Spectra and Protein Secondary Structure

The CD spectrum of proteins and peptides below 250 nm originates largely from the contributions from electronic transitions within the backbone amide groups. The "isolated" amide chromophore has a plane of symmetry and by itself is optically inactive. Optical activity of this chromophore in the peptide or protein may be induced by (i) the presence of an asymmetric α-carbon and (ii) the interaction with the asymmetric environment of the macromolecule. Whereas the first effect is relatively minor, the circular dichroism induced by the second mechanism is expected to be large. Thus, since the optical activity of the amide chromophore is determined largely by its immediate environment, CD spectra in the far-UV region are very sensitive to the secondary structure of polypeptide chains.

Early measurements of CD spectra were focused on simple homopolypeptides, which are known to adopt well-defined secondary structures (Greenfield and Fasman, 1969). A good model system for studying spectra-structure correlations is poly-L-lysine since the conformation of this polypeptide is strongly dependent on environmental conditions such as pH and temperature. The spectra of poly-L-lysine that are representative of three most common elements of the secondary structure, i.e., α-helix, β-pleated sheet and "random structure" are shown in Fig. 1. The α-helical spectrum shows a typical double minimum at 222 and 208–210 nm and a maximum at 190–195 nm. These extrema represent the n-π^*, π-$\pi^*_{\parallel}$, and π-$\pi^*_{\perp}$ transitions, respectively. The β-pleated sheet gives rise to a spectrum with a minimum at 217 nm and a maximum around 195 nm. Finally, the spectrum of poly-L-lysine in a random conformation is characterized by a negative band at around 197 nm.

CD spectra of heteropolypeptides or proteins usually show a more com-

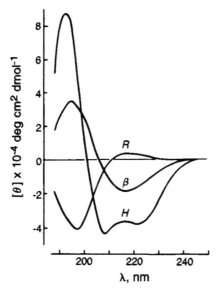

FIGURE 1 Circular dichroism spectra of polylysine in different conformations. R, unordered form at neutral pH; H, α-helix at pH 10.8; β, β-structure at pH 11.1 after heating for 15 min at 52°C and cooling back to room temperature. Reproduced from Yang *et al.*, (1986), with permission.

plex pattern, as these spectra contain contributions from various elements of the secondary structure. A number of different methods have been developed for quantitative analysis of the CD spectra in terms of protein secondary structure. Most of these approaches assume that the experimental spectrum can be expressed as a superposition of a number of "pure" components that represent different classes of the secondary structure, i.e.,

$$\Theta(\lambda) = f_k \, \Theta_k(\lambda), \tag{3}$$

where Θ_k denotes the basis spectrum of a pure component k and f_k denotes the fraction of this component. Coefficients f_k (and thus the secondary structure content) may be obtained by fitting the experimental spectrum with the reference spectra believed to represent pure components of the secondary structure. Early work was based on the use of reference spectra of model peptides. Greenfield and Fasman (1969) pioneered these efforts by proposing that the CD spectrum of a protein may be estimated by utilizing reference spectra of polylysine measured under conditions that favor formation of α-helix, β-sheet, and random structure. Somewhat later, the reference set has been expanded to include the spectra of β-turn forming peptides (Brahms and Brahms, 1980).

The analysis of the CD data based on the use of reference spectra of model peptides played a historically fundamental role. Today, this method is still used with good results for determining the secondary structure of peptides. However, when applied to proteins, the use of reference peptide spectra presents a number of serious disadvantages. Although the spectra of long polypeptides may provide a good model for infinitely long α-helices or β-sheets, these structural segments in proteins are often very short and their spectral properties may differ from those of indefinitely long elements. Furthermore, there is a substantial controversy regarding the suitable peptide models for various types of β-turns and a "random" structure. CD of unordered segments in proteins is not necessarily well represented by a "random coil" spectrum of model peptides. An additional problem is presented by the fact that there is more than one type of β-sheet structure and substantial differences may be seen in the CD spectra of various model β-sheet peptides. The above difficulties have stimulated efforts aimed at developing improved methods for the analysis of protein far-UV CD spectra. Generally, the more recent approaches do not rely on peptide reference spectra but rather make use of the spectra of proteins for which the structure has been determined by X-ray crystallography. For example, Saxena and Wetlaufer (1971) used CD spectra of three known proteins to derive basis spectra for α-helix, β-sheet, and unordered structure. A somewhat improved accuracy was obtained later by introducing an empirical parameter accounting for the chain-length dependency of the α-helical spectrum (Chen *et al.*, 1974), including in the analysis β-turns and enlarging the number of reference proteins (Chang *et al.*, 1978). However, even with these developments the reliability of the secondary structure estimates was often less than satisfactory. A major difficulty associated with this type of analysis is related to the introduction of a large number of parameters. This leads to instability of the solution and to experimental error. Provencher and Glockner (1981) have overcome this problem by using a constrained regularization procedure. This mathematical procedure stabilizes the solution by penalizing fits that favor a particular protein in a reference set, unless the emphasis of this protein is necessary for obtaining a satisfactory fit. The above method allowed the authors to analyze an experimental CD spectrum as a linear combination of the spectra of 16 proteins of known secondary structure. This analysis provided a significant improvement in the accuracy of the secondary structure estimation. The correlation coefficients between the computed fractions and the fractions obtained from crystallographic analysis were found to be very good for α-helix and β-sheet (0.96 and 0.94, respectively), whereas those for β-turns and the "remainder" were significantly less satisfactory (0.31 and 0.49, respectively). The computer program CONTIN has been written by Provencher in Fortran IV to employ this method. This program is now used by many laboratories worldwide. In its original version, CONTIN accepts experimental data between 190 and 240 nm.

A somewhat different methodology has been developed by Hennessey and Johnson (1981), who have made use of a singular value decomposition method. More recently, the singular value decomposition approach has been combined with the statistical procedure called "variable selectivity" (Manavalan and Johnson, 1987). This combined methodology has been reported to be very flexible and give accurate estimates of the secondary structure (Johnson, 1988; Johnson, 1990). However, a certain disadvantage is caused by the fact that the latter method was designed for a wavelength range extending to 178 nm. Although this undoubtedly increases the information content, collection of the CD data below 190 nm often present serious practical difficulties and, in many cases, is precluded by the absorption of the solvent.

A new and important development in CD spectroscopy of proteins is the introduction of a deconvolution method called convex constraint analysis (Perczel *et al.*, 1991). Unlike other approaches, this method uses only experimental CD curves, *without making use of any reference spectra originating from model peptides or proteins of known secondary structure*. The convex constraint analysis algorithm allows decomposition of CD spectra of globular proteins into at least four pure components that represent independent sources of chirality. Importantly, these components show remarkable resemblance to the CD spectra of known secondary structures (Fig. 2). Since the convex constraint analysis does not make use of crystal structures or any other reference data, this method is highly unbiased. Furthermore, it should

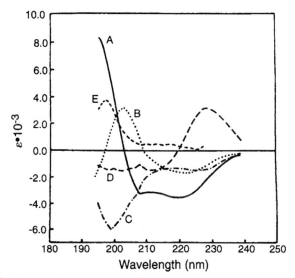

FIGURE 2 The weighted pure component curves that yield the best fit of the circular dichroism spectrum of pepsinogen. A, α-type; B, β-form; C, unordered form; D, β-turn; E, additional chiral contribution. Reproduced from Perczel *et al.*, (1991), with permission.

be applicable equally well to analyze the CD spectra of globular proteins as well as those of integral membrane proteins. For the latter proteins, the structural basis (i.e., reference spectra) required in other types of analysis is still lacking.

A critical evaluation of various methods used in the analysis of far-UV protein spectra and the discussion of possible sources of error may be found in the papers by Woody (1985) and Van Stokkum et al., (1990). Here, we emphasize that regardless of the method used, the estimations of the secondary structure derived from CD spectra should be always viewed critically. If at all possible, the CD estimates should be verified by another spectroscopic technique. Some proteins are particularly prone to error and their analysis by CD spectroscopy yields notoriously poor results. This statement applies especially to proteins that are rich in aromatic residues. Aromatic side chains can make significant contributions in the far-UV region (Manning and Woody, 1989) and, depending on their spatial distribution, local environment, and interactions, may lead to anomalous features such as positive maxima around 230 nm (Khan et al., 1989) or unusually weak negative ellipticity at 208 nm in the CD spectra of helical proteins (Arnold et al., 1992). CD spectra in the far-UV region may also be distorted by the presence of disulfide bonds and certain prosthetic groups. A special problem is encountered in CD studies of larger assemblies such as viruses or proteins in a membrane environment. Circular differential scattering can be an important part of the circular dichroism of these systems (Bustamante et al., 1983) and, if unaccounted for, may introduce serious errors in the analysis of spectroscopic data.

B. Near-UV Circular Dichroism of Proteins

Circular dichroism spectra in the near-UV region originate largely from aromatic side-chain groups and from disulfide bonds. Although there is a relatively solid theoretical and experimental basis for understanding the spectral characteristic of isolated aromatic amino acids (Kahn, 1979), detailed interpretation of the near-UV spectra of these residues in proteins is generally very difficult. Circular dichroism of aromatic residues is influenced in a rather complex and not fully understood manner by environmental conditions such as local hydrophobicity, temperature, and pH. The situation is further complicated by the fact that near-UV CD bands of various aromatic groups may overlap. Since many proteins contain multiple aromatic residues, the assignment of individual spectral feature of these proteins is very difficult, if not impossible.

Detailed analysis of near-UV CD spectra of various aromatic chromophores may be found in the review paper by Kahn (1979). Of all three aromatic amino acid side chains, the most diagnostic appears to be tyrosine. The position of the major near-UV CD band of this residue shifts in a regular way from 282 nm in water to approximately 290 nm in a hydrophobic environment. Furthermore, certain transitions giving rise to the near-UV CD spectrum of tyrosine are very sensitive (both in magnitude and sign) to the position of the aromatic group with respect to the main chain. In the absence of other aromatic residues in the protein (especially tryptophans), the near-UV CD spectra of tyrosines may provide fairly specific information about local interactions and/or conformational changes in proteins. An illustrative example of this are the studies with ribonuclease that allowed spectroscopic identification of buried and exposed tyrosines as well as provided insight into local conformational changes in the vicinity of Tyr-25 upon conversion of ribonuclease A to ribonuclease S (Strickland, 1972; Horwitz and Strickland, 1971). Experiments with another tryptophan-free protein, insulin, have demonstrated the potential of tyrosine CD in probing subunit–subunit interactions in the assembly of multimeric proteins (Strickland and Mercola, 1976).

The intrinsic optical activity of the disulfide bond is strongly dependent on the C—S—S angle and the dihedral angle of the bond as a whole. Therefore, disulfide CD spectra are potentially of considerable diagnostic value. Unfortunately, in protein spectra the disulfied bands are usually obscured by the contributions from aromatic residues; extraction of specific information regarding the geometry of the CSSC group is very difficult and has been accomplished only in few special cases (Kahn, 1979; Reed, 1990).

It is evident from the brief discussion above that in most cases the near-UV CD spectrum of proteins is very complex and individual peaks usually cannot be unambiguously assigned to transitions of specific amino acid side chains. However, despite this fundamental difficulty with the interpretation at a molecular level, the near-UV CD spectra provide an extremely sensitive probe (or a fingerprint) of a global tertiary structure of proteins. The technique is routinely used to monitor folding–unfolding transitions induced by temperature or chemical denaturants as well as to probe protein conformational changes induced by ligand binding. Recently, circular dichroism measurements contributed greatly to the identification of an important folding intermediate of proteins known as a "molten globule" state. In this conformational state the protein is still compact and retains a large proportion of its native secondary structure. However, the tertiary structure of a molten globule is significantly loosened, leading to the collapse of the near-UV CD spectrum. As an illustration, Fig. 3 shows the changes in far- and near-UV CD spectra of colicin A upon pH-induced transition of this protein from a native conformation to a molten globule state (van der Goot *et al.*, 1991).

CD spectroscopy has been also used to study the induced dichroism in the visible region of certain chromophoric proteins, especially those containing a porphyrin prosthetic group. An overview of this field may be found in more specialized reviews (e.g., Myer, 1985).

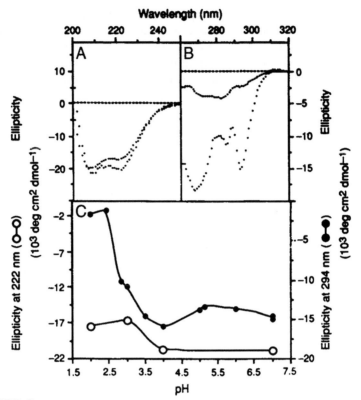

FIGURE 3   Circular dichroism spectra of the pore-forming fragment of colicin A in far ultraviolet (A) and near ultraviolet (B) at pH 7 (lower spectra) and pH 2 (upper spectra). C, influence of pH on the ellipticity at 222 nm and 294 nm. Reproduced from van der Goot *et al.*, (1991), with permission.

III. FOURIER-TRANSFORM INFRARED SPECTROSCOPY

Infrared spectroscopy measures the transitions between vibrational levels of the ground electronic state of a molecule. Typically, vibrational bands arise from transitions that are somewhat localized on a specific chemical group of a molecule. This has led to a concept of group vibrational modes or group frequencies such as, for example, C–H stretching, C=O stretching or N—H stretching. However, in a larger molecule the vibrations of individual groups are often nonindependent and may be strongly coupled. This coupling may, in turn, affect the positions of infrared bands. Theoretical calculations of vibrational spectra must thus include numerous interactions; in the case of biopolymers these calculations are extremely complex and present a very difficult mathematical problem.

Infrared spectroscopy was among first techniques that were recognized as potentially useful in conformational studies of biopolymers. However, for many years the application of this method to protein research was limited in scope. The major factors that contributed to this limitation include poor sensitivity of traditional dispersive instruments and a strong absorption of water, which often obscures the signal of interest. An additional problem is presented by the intrinsic broadness of individual infrared bands, which leads to instrumentally unresolvable multicomponent band contours. A major breakthrough in biological infrared spectroscopy occurred with the development of computerized Fourier-transform infrared (FT-IR) techniques. The contemporary FT-IR instrumentation has dramatically improved the accuracy, reproducibility, and sensitivity of spectroscopic measurements. Furthermore, it allows routine measurements with strongly absorbing (low-throughput) samples and thus permits accurate detection of weak protein bands that are superimposed on much stronger absorption bands of water. Rapid advances in instrumentation were matched by the development of increasingly sophisticated methods for numerical data analysis. These methods made possible accurate subtraction of the background, separation of overlapping bands by Fourier self-deconvolution procedures, or analysis of complex multicomponent spectra by the "pattern recognition" procedures.

A good introduction to both theoretical and experimental aspects of modern FT-IR spectroscopy may be found in the monograph by Griffiths and de Haseth (1986). Here, we will discussed major applications of the FT-IR method in protein research. These applications include analysis of the amide bands in terms of the secondary structure, assessment of the accessibility of the protein backbone to hydrogen–deuterium exchange, and studies at the level of individual chemical groups of proteins or protein-bound ligands.

A. Infrared Amide Bands and Protein Secondary Structure

Infrared spectra of proteins and peptides exhibit nine so-called amide bands that represent different vibrations of the peptide linkage. Theoretical grounds for understanding these vibrational modes has been provided by normal coordinate analysis (for a review see Krimm and Bandekar, 1986). Although remarkably successful in predicting vibrational spectra of small peptides, rigorous normal mode analysis for more complex systems is extremely difficult and does not seem to be practical for large proteins. Nevertheless, normal mode calculations have played a fundamentally important role by providing a theoretical basis for understanding the empirical correlations between protein infrared spectra and the conformation of the polypeptide backbone.

The vibrational mode most useful for studying protein secondary structure is the amide 1. This mode represents primarily stretching vibrations of the $C{=}O$ groups of the peptide linkages (coupled to the in-phase bending of the N—H bond and stretching of the C—N bond) and gives rise to an infrared band(s) between approximately 1600 and 1700 cm^{-1}. The conformational sensitivity of the amide I band may be illustrated by inspecting the spectra of poly-L-lysine, which, depending on environmental conditions, can refold from random coil to α-helix or to β-sheet structure. Figure 4 shows that each of these conformational states has a distinct spectroscopic signature in the amide I region. However, in contrast to simple homopolypeptides that often fold into well-defined and homogenous structures, proteins are complex three-dimensional entities that usually contain a mixture of α-helices, parallel and antiparallel β-sheets, various turns, and unordered segments. Each of these conformers contributes to the infrared spectrum. The observed amide I band contour is thus a complex composite of many overlapping component bands that represent various elements of the secondary structure. Since the widths of these contributing bands are usually greater than the separation between their maxima, the individual components cannot be directly resolved and/or identified in the experimentally measured spectra. Two types of methods are currently used to extract the structural information encoded in amide I band contours of proteins. One type of analysis, which may be referred to as "frequency-based" approach, makes

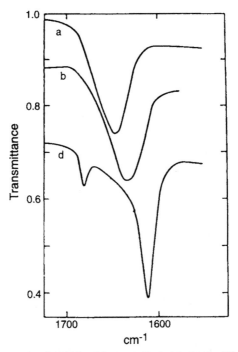

FIGURE 4   The amide I band of polylysine in D$_2$O solution in different conformational states. a, random conformation; b, α-helix; d, β-sheet structure. Reproduced from Susi, H., Timasheff, S. N., and Stevens, L. (1967) *J. Biol. Chem.* **242**, 5460–5466, with permission.

use of computational procedures for band narrowing that allow for the separation of broad band contours into underlying components. An alternative approach is based on the principle of "pattern recognition."

The mathematical procedure of Fourier self-deconvolution was first introduced to infrared spectroscopy by Kaupinnen *et al.*, (1981). The above procedure, often also referred to as "resolution enhancement," reduces the widths of infrared bands. This allows for increased separation (and thus better visualization) of the overlapping components under broad infrared band contours. Algorithms for Fourier self-deconvolution are now often included in the software packages of commercial FT-IR instrumentation. Increased band separation may be also achieved by calculating second or higher order derivative of the absorption spectra (Susi and Byler, 1986; Dong *et al.*, 1990). However, a distinct advantage of the deconvolution method is that it introduces less spectral distortion and, especially, does not affect the integrated intensities of the individual component bands. Despite the apparent simplicity of the "resolution enhancement" procedures, great care is recommended in applying these methods to the spectra of proteins. In particular, one should be aware that the application of Fourier self-deconvolution (or derivation) will result in disproportionally strong amplification of all sharp features such as spectral noise or bands originating from any residual water vapor. Although often "invisible" in original spectra, these features may appear in resolution-enhanced spectra as artifacts that are indistinguishable from the real components of the protein amide band. A very high signal-to-noise ratio and virtually complete elimination of water vapor is thus a prerequisite of meaningful deconvolution. It is recommended that the absence of artifacts be verified by careful examination of the resolution-enhanced spectrum in the region where no protein bands are expected (e.g., between 1720 and 1780 cm^{-1}).

The structural assignment of bands seen in deconvolved spectra of proteins is guided by theoretical calculations and by spectra–structure correlations established for model proteins of known three-dimensional structure. Thus, amide I bands centered between approximately 1650 and 1658 cm^{-1} are usually assigned to α-helices, wherease those between 1620 and 1640 cm^{-1} are considered to be highly characteristic of a β-pleated sheet structure (Byler and Susi, 1986; Surewicz and Mantsch, 1988). For many proteins more than one "β-band" is seen in the 1620–1640 cm^{-1} region (Susi and Byler, 1987) and, in same cases, the low frequency band may be shifted even below 1620 cm^{-1} (Surewicz and Mantsch, 1988). Although this multiplicity is of potential diagnostic value, no correspondence has been established between various "β-bands" and specific types of β-sheet structure. Antiparallel β-sheets are also characterized by a high frequency component in the 1670–1690 cm^{-1} region. This component results from the large splitting of the amide mode caused by interstrand interaction (Krimm and Bandekar, 1986). However, identification of the high frequency "β component" in the spectra of proteins is often hampered by its overlapping with the spectral contributions from various types of turns (Torii and Tasumi, 1992). The assignment of the amide I bands discussed above should be viewed as a general guide only. Theoretical calculations and a growing number of experimental observations indicate that the spectral contributions of unordered structures and turns or loops may spread over the wide wavenumber region (Krimm and Bandekar, 1986, Torii and Tasumi, 1992; Wilder *et al.*, 1992; Prestelski *et al.*, 1991). Furthermore, under certain conditions the helical bands may be shifted below 1650 cm^{-1} (Jackson *et al.*, 1991). Detailed discussion of the problems associated with the structural assignment of amide I bands may be found in a review by Surewicz *et al.* (1993).

Fourier self-deconvolution of amide I bands provided also a basis for quantitative estimation of protein secondary structure. Byler and Susi (1986) proposed a procedure that involves curve fitting of deconvolved amide I band contours as a linear combination of individual component bands. The resulting fractional areas of the bands assigned to different types of secondary structure were assumed to represent percentages of these structures in a given protein. This procedure was claimed to provide estimates of secondary structure that correlate well with crystallographic data and was adopted, with different modifications, by many laboratories (e.g., Surewicz and Mantsch, *et al.*, 1988; Goormaghtigh *et al.*, 1990, Prestelski *et al.*, 1991; Dong *et al.*, 1990; Arrondo *et al.*, 1993). However, critical analysis of various assumptions inherent in this method (Surewicz *et al.*, 1993) calls for caution in using curve-fitting analysis as a generally valid method to assess quantitatively the *absolute* content of protein secondary structure. On the other hand, if used cautiously, the methods based on band narrowing provide a very sensitive tool for monitoring, *in relative terms*, the nature and extent of changes in the conformation of the protein backbone.

Approaches based on "pattern recognition" avoid Fourier self-deconvolution of the spectra and are based on the use of a calibration matrix of the amide bands of proteins whose three-dimensional structure has been determined by X-ray crystallography. Conceptually, these approaches are similar to those used in the analysis of circular dichroism spectra. Three different variants of this methodology have been published in recent years: the partial least-squares method of Dousseau and Pézolet (1990), the factor analysis method of Lee *et al.* (1990), and the matrix method of Sarver and Kruger (1991). Depending on the particular version, either amide I alone or both amide I and amide II regions of the reference spectra are used to construct the calibration matrix. Although these methods eliminate the element of subjectivity implicit in resolution enhancement procedures, they are not without shortcomings (Surewicz *et al.*, 1993). In particular, the lack of uniqueness in spectra–structure relationships (see above) makes the quantitative analysis of amide bands somewhat vulnerable to unpredictable errors. Furthermore, the methods based on pattern recognition are much less reliable when applied to the spectra measured in D_2O. The need to subtract a very strong H_2O band that overlaps the amide I region may introduce an additional source of error. Finally, regardless of the method used, the results of a quantitative analysis of amide bands may be significantly distorted by spectral contributions of the amino acid side-chain groups (Chirgadze et al., 1975; Venyaminov and Kalninin, 1990). This presents a particularly serious problem in studying proteins rich in amino acids such as asparagine and glutamine. Ideally, the side-chain absorption should be subtracted prior to analysis of the spectrum in terms of protein secondary structure.

FT-IR spectroscopy has proven particularly valuable in probing the secondary structure of membrane proteins (for recent reviews see Surewicz and Mantsch, 1990; Haris and Chapman, 1992) and large protein aggregates (e.g., Fraser et al., 1991; Lansbury, 1992). These systems are notoriously difficult to study by other spectroscopic techniques.

Important recent advances are linked to the development of the methodology of isotope-edited FT-IR spectroscopy. Especially, this technology is of potentially great value in studying protein–protein, or protein–nucleic acid, or protein–ligand (Tonge et al., 1991) interactions. Conformational studies of proteins (or peptides) in complexes with other biomolecules by methods such as CD or conventional FT-IR spectroscopy are severely limited by the overlapping of the spectroscopic signals from the two macromolecules. As shown in a recent study on calmodulin–target peptide interaction (Zhang et al., 1994), uniform ^{13}C labeling of the protein may downshift the conformation-sensitive amide I band by as much as 55 cm^{-1} (Fig. 5). This results in the separation of the amide bands originating from the labeled and unlabeled molecules, allowing thus the assessment of the secondary structure of each component of the complex. In addition to uniform labeling, one can make use of specific ^{13}C labeling of individual amino acids. This promising approach has been used to locate different elements of secondary structure in conformationally heterogeneous model peptides (Tadesse et al., 1991), and to study the local conformation of amyloid-forming peptides (Lansbury, 1992).

B. Hydrogen–Deuterium Exchange

Hydrogen–deuterium exchange of the backbone amide groups may be followed by measuring the decrease in the intensity of the amide II band at around 1545–1555 cm^{-1}. Upon deuteration this band shifts downfield by about 100 cm^{-1}. The corresponding shift of the amide I band is much smaller (5–10 cm^{-1}) and the latter band may provide an internal standard for kinetic measurements. Studies of the kinetics of hydrogen–deuterium exchange may provide useful information regarding the compactness of folding and conformational dynamics of the protein molecule (Gregory and Rosenberg, 1986). Although the multiple-site exchange techniques such as

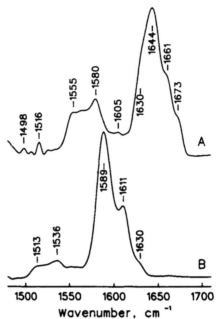

FIGURE 5   Infrared spectra of (A) unlabeled and (B) uniformly ^{13}C-labeled calmodulin in D_2O buffer. Spectra were deconvolved using a Lorentzian line shape with a half-bandwidth of 16 cm^{-1} and a resolution enhancement factor of 2. Reproduced from Zhang *et al.*, (1994), with permission.

FT-IR have been recently surpassed by high resolution approaches based on NMR spectroscopy or neutron diffraction, infrared spectroscopy still provides a viable option in some studies, especially those dealing with proteins in a membrane environment (e.g., Osborne and Nabedryk-Viala, 1986; Haris *et al.*, 1989; Jung *et al.*, 1986).

C. Infrared Spectroscopic Studies at the Level of Individual Chemical Groups

In addition to its widespread use in studying global secondary structure of proteins, FT-IR spectroscopy has found numerous applications as a tool to probe structural and/or conformational changes in proteins or their ligands at the level of individual chemical groups. Since infrared bands representing individual groups are usually obscured by the strong absorption of the bulk protein, such studies are technically very demanding and often require the use of a special strategy known as "reaction modulated difference spectroscopy." In this approach spectra of a protein in two functional states are measured and subtracted, yielding a difference spectrum that contains only those bands that are affected by the reaction of interest. Since the difference spectrum usually represents only a very small fraction of the bulk protein signal, reproducible results can often be obtained only if the spectra are recorded in rapid succession, with minimal sample manipulation. This requirement has led many investigators to focus on chromophoric proteins that can be triggered in a repetitive manner by light. Light-modulated difference IR spectroscopy has been particularly successful in probing molecular events associated with photochemical cycles in bacteriorhodopsin, rhodopsin, and photosynthetic reaction center complexes (e.g., Braiman and Rothschild, 1988; Bagley *et al.*, 1989; Rothschild *et al.*, 1990; Jager *et al.*, 1994). The assignment of individual bands in the difference spectra of these protein has been greatly aided by isotopic labeling and site-directed mutagenesis. Recently, the strategy of reaction modulated difference spectroscopy has been extended to include other noninvasive triggering methods such as application of electrochemical potential (Moss *et al.*, 1990) or photochemical release of "caged ligands" (Barth *et al.*, 1991; Buchet *et al.*, 1991). This topic has been recently reviewed by Mantele (1993).

One of very few functional groups in proteins whose infrared bands are relatively well separated from the bulk protein spectrum are thiols of cysteine residues. The SH stretching bands occur in the spectral region around 2550 cm^{-1}. These bands are highly sensitive to hydrogen-bonding and may be used as a local probe of the conformation in the vicinity of cysteine residues. For example, spectroscopic properties of thiol groups have been explored extensively by Alben and co-workers (Alben and Bare, 1989; Moh *et al.*, 1987) to study the local environment of -SH groups in hemoglobins.

FT-IR spectroscopy also shows a considerable potential as a tool to explore molecular aspects of protein–ligand interactions. The technique has been used to study carbonyl groups of acyl intermediates of various enzymes (Belasco and Knowles, 1983; Kurz and Drysdale, 1987; Tonge *et al.*, 1991), and to probe local interactions in carbon monoxide binding proteins (e.g., Einarsdottir *et al.*, 1988; Potter *et al.*, 1990; Fiamingo *et al.*, 1986).

IV. RAMAN AND RESONANCE RAMAN SPECTROSCOPIES

Raman spectroscopy provides vibrational spectroscopic information on peptides and proteins (Carey, 1982). A beam of monochromatic light from a laser is focused into a sample and the scattered light is examined by a dispersive spectrometer and a photon detection device. Analysis of the scattered light reveals that a tiny percentage of the incident photons have undergone inelastic scattering and in the present context have exchanged energy with the vibrational energy levels of the sample. Thus, like IR spectroscopy Raman spectroscopy provides a vibrational spectrum of molecules. For proteins, analysis of the vibrational spectra reveals molecular detail of peptide bonds, side-chain conformations, and environment. One advantage of Raman spectroscopy vis à vis IR is that the Raman spectrum of water is weak and is relatively easy to exclude from the Raman spectrum of a protein in solution. Another unique aspect found in Raman is the so-called resonance effect. When the wavelength of the laser used to generate the Raman spectrum lies within an electronic absorption band of the sample, greatly enhanced scattering from the chromophoric portion can occur, leading to a relatively intense resonance Raman (RR) spectrum. The resonance Raman effect introduces the potential of specificity and selectivity—the intense RR spectrum of a biologically important chromophore can be selectively derived when it is bound at the active site of a complex biological system. In a sense RR spectroscopy can be considered the high information vibrational counterpart of absorption spectroscopy—because vibrational spectra are information-rich compared to absorption spectra. Thus, RR spectra can provide very detailed molecular information concerning chromophores and chromophore–protein interactions.

An important consideration is that recent technical innovations, including the use of Fourier transform Raman instruments (Chen *et al.*, 1991) and red-light sensitive charge coupled photon-detectors (Kim *et al.*, 1993), are making Raman spectroscopy a more user-friendly "routine" analytical technique. The technical innovations include the extensive use of computer power and the use of high efficiency optical filters based on holographic principles and on the advent of red-sensitive multiplex photon detectors. Thus, Raman spectroscopy will be used more extensively in the characterization of peptides and proteins for both natural and redesigned materials. An advantage of Raman analysis that remains to be exploited is that Raman data can be collected for protein solutions, solids, and single crystals. Thus, Raman spectroscopy may serve as a useful bridge between the X-ray analysis of crystalline material and the structural analysis of the same material in solution by NMR.

A. Raman Spectra of Proteins and Peptides

Information on the secondary structure of the peptide backbone can be gleaned from two regions in the Raman spectrum known as amide I and amide III (Carey, 1982; Harada and Takeuchi, 1986). The amide I mode is described in the section in this chapter on FTIR. Amide III modes have a large contribution from a bending motion of the peptide N—H atoms. Table II summarizes the approximate positions and profiles of amide I and III features for α, β and unordered secondary structural elements. The bands can be used to form a qualitative picture of secondary structure and this can be put on a more quantitative footing if small changes occur among a series of related proteins or when, for example, small changes occur as a protein binds to a nucleic acid in a virus structure (Li *et al.*, 1990). Band fitting using a library of known structures and their Raman spectra has also been used to

TABLE II Approximate Positions (cm⁻¹) of Amide I and III Bands in Raman Spectra for Various Polypeptide Conformations

	Amide		Amide III	
α-Helix	1645–1660	Strong	1265–1300[a]	Weak
β-Sheet	1665–1680	Strong	1230–1240	Strong
Unordered	1660–1670	Strong and broad	1240–1260	Medium and broad

[a] May be confused with side-chain modes.

quantitate the amount of α, β and unordered structure in an uncharacterized protein (Carey, 1988). This carries uncertainties similar to those found in the FTIR approach (elsewhere) and suffers from an additional drawback in that the signal-to-noise ratio of Raman data is usually inferior to that obtained by FTIR.

Raman spectra can be informative for characterizing cysteine (-SH), tryptophan, and tyrosine side chains as well as disulfide (-S-S-) linkages. The information, with key references, is summarized in Table III. Many of the features listed in Table III are illustrated in Fig. 6. For a recent in-depth review of the application of Raman spectroscopy to proteins the reader is referred to an article by Miura and Thomas (1995).

The use of Raman spectroscopy to characterize peptides and proteins has been hampered by the technical difficulties associated with obtaining high quality experimental data; in particular fluorescence from the sample itself, or from minor impurities, has often obscured the Raman data. These problems have been largely removed with the advent of FT Raman and, also, charge coupled photon detectors (CCDs). Both those technologies can operate with deep red excitation sources (between 600 and 1000 nm). Since there are very few fluorophores in this region, the main impediment to obtaining Raman data is removed. Moreover, both FT Raman and CCD-based Raman operate in the "multiplex" mode enabling the data to be collected quickly. The impact of CCD detection is illustrated in Fig. 1; a high quality Raman spectrum of aqueous lysozyme was obtained in 10 min. Not only is the signal-to-noise ratio high but now spectra can be obtained routinely without days or weeks of effort being required to remove fluorescent impurities.

B. Resonance Raman Spectroscopy

When the laser wavelength used to excite the Raman spectrum is coincident with an absorption band of a chromophore, greatly enhanced Raman scattering is observed from the chromophore. This is termed the resonance Raman effect and, as mentioned before, has the advantage of selectivity and specificity. It enables us to observe selectively the Raman spectrum of a

TABLE III   Most Important Raman Bands
for Establishing Side-Chain Properties

Side chain	Approximate position of Raman feature (cm^{-1})	Information	Reference
–SH	2575	Environment about –SH, e.g., H-bonding	Li et al., 1992
		Conversion –SH to S–S	East et al., 1978
Tyrosine	850 and 830 "tyrosine doublet"	Intensity ratio 850/830 = R used to define H-bonding. R is small if –OH acts as H-bond donor; R is large if –OH acts as H-bond acceptor	Carey, 1988 / Harada and Takeuchi, 1986
Tryptophan	1361	Intensity profile may indicate buried or exposed side chain	Carey, 1988
	1386	From deuterated Trp. Used to follow H/D exchange	Takesada et al., 1986
	1550	Used to follow Trp destruction mediated by light	Pozsgay et al., 1987
	1550	Exact position provides C_2C_3–$C_\beta C_\alpha$ torsional angle	Miura et al., 1989
–S–S–	500–550	Torsional angles for bonds near –S–S– linkages	Harada and Takeuchi, 1986 / Qian and Krimm, 1992
C–S	630–745	Combined with –S–S– data to provide torsional angles	Qian and Krimm, 1992

chromophore in a complex biological milieu. In the present context it can be used in two ways. With excitation below 300 nm the chromophores due to aromatic side chains and the peptide bonds themselves can be probed (Harada and Takeuchi, 1986; Asher, 1993; Austin et al., 1993). At present this is a technique under rapid development and is not suitable for routine analysis. However, many biological chromophores exist in the visible region of the spectrum and in the past 25 years there have been a myriad of studies yielding very detailed information on, e.g., heme proteins, visual pigments, photosynthetic pigments, and metallo- and flavo-proteins. Much of the early work is summarized in the three-volume work edited by Spiro (1987). Some more recent reviews on heme systems are by Kitagawa and Ogura (1993) on the time-resolved resonance Raman spectra of heme proteins, and by Woodruff and co-workers (1993) on cytochrome oxidase. Advances in vibrational spectroscopic studies of bacteriorhodopsin and rhodopsin proteins are outlined in reviews by Mathies and colleagues (1991) and by Siebert (1993). A

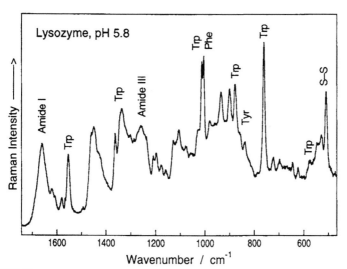

FIGURE 6 Raman spectrum of a 12% solution of lysozyme, with the spectrum of water subtracted. This spectrum was recorded in 1 min using a single monochromator, CCD detection, and a super notch filter to reject Rayleigh scattering (Kim *et al.*, 1993). The light used to excite the spectrum was 250 mW 752 nm from a krypton laser. M. Kim and P. R. Carey, unpublished work.

recent summary of the value of resonance Raman spectroscopy in characterizing photosynthetic systems is given by Robert (1995).

In many recent studies, the use of site-selected mutants has enhanced the scope of resonance Raman analysis and has demonstrated the value of the Raman method for characterizing redesigned proteins. The "oxyanion hole mutants" of subtilisin are a case in point. For subtilisin BPN', Asn 155 provides a H-bonding donor that binds the substrate's labile C=O moiety. In pioneering studies using protein engineering the kinetic consequences of changing Asn 155 to, e.g., alanine or glutamine, were explored extensively (Bryan *et al.*, 1986) but recent work using resonance Raman spectroscopy has provided novel and detailed information on the geometry and bonding of the substrate's C=O group during catalysis. Acyl enzyme intermediates of the kind

$$CH_3 - \text{(thiophene ring)} - CH=CH - C(=O) - O - \text{subtilisin}$$

have a λ_{max} near 350 nm due to the acyl group. Utilizing this absorption to obtain the RR spectrum permits the position of the carbonyl stretching frequency, $\nu_{C=O}$, to be monitored selectively under turnover conditions. The value of $\nu_{C=O}$ yields estimates of $r_{C=O}$, the carbonyl bond length, and ΔH, the change in enthalpy of H-bonding to the C=O in the active site. These studies were undertaken for a series of 12 acyl-serine proteases and a free energy-type relationship between $\nu_{C=O}$ (or $r_{C=O}$) and log k_3 (where k_3 is the rate constant for hydrolysis of the acyl enzyme) is illustrated in Fig. 7 (Tonge and Carey, 1992).

C. Raman Difference Spectroscopy

The resonance Raman technique has provided exquisite molecular detail on protein bound ligands both for native and redesigned proteins. Of course, it is limited to chromophoric ligands in order to meet the resonance Raman condition. However, the same advances in technology mentioned earlier, red-sensitive CCDs as photon detectors and holographic optical filters, are facilitating the obtention of Raman data from a protein-bound ligand under nonresonance conditions. Again this is due in large part to an increase in the sensitivity of detection coupled with elimination of unwanted fluorescence signals.

The principle is simple, using computer methods the Raman spectrum of parent protein is subtracted from the Raman spectrum of the protein–ligand complex. The method works far from resonance conditions but is most powerful when the ligand is a strong Raman scatterer, e.g., for groups like C=O, or a conjugated system such as -C=C-C=O, which has polariz-

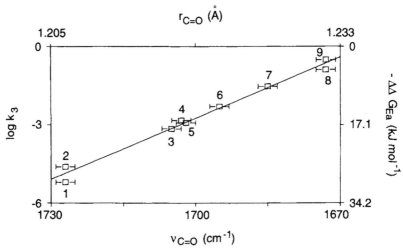

FIGURE 7 A relationship derived for $\nu_{C=O}$, the acyl group carbonyl stretching frequency, and reactivity, where k_3 is the deacylation kinetic rate constant. The carbonyl frequency was measured from the resonance Raman spectrum. The frequency can be converted to bond length and this figure demonstrates the extremely accurate conformational data that can be derived from Raman measurements (Tonge and Carey, 1992).

able electron clouds. The reasons for this are twofold. When the ligand is a strong scatterer it results in a higher signal-to-noise ratio in the spectral data after subtraction. Secondly, in principle, both ligand and protein bands can occur in the difference spectrum since binding the ligand can perturb the protein structure. However, if the ligand is a strong scatterer the protein features in the difference spectrum will be weak and possibly undetected.

The approach will have wide application in, for example, drug design where molecular information is being sought for a series of drugs binding to an enzyme or receptor. Alternatively the consequences of redesigning a binding site on the chemistry of binding of a certain ligand can be probed. The power of the approach is illustrated in Fig. 8, which shows a high quality Raman spectrum of a flavin bound to a protein. Historically flavin Raman

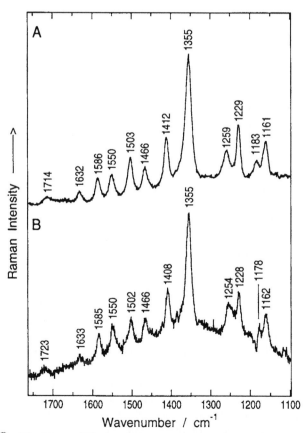

FIGURE 8 The Raman difference spectroscopic approach is used to obtain the Raman spectrum of (B) riboflavin bound to riboflavin-binding protein and (A) free, unbound ligand. Carbonyl features can be seen for the first time. The data were obtained using 400 mW, 647.1 nm excitation and 10 min acquisition time (Kim and Carey, 1993).

spectra have been very hard to obtain due to competing fluorescence. Figure 8 shows that using 647.1 nm of excitation and 600 s of data collection time the Raman difference approach yields excellent data. The weak feature near 1720 cm^{-1} is due to the -C_4=O of the riboflavin's isoalloxazine ring and the data show, for the first time, that hydrogen bonding interactions involving the C=O decrease upon going from water to the active site.

The advances in technology mentioned above have added another dimension to the Raman analysis of proteins, namely Raman optical activity (ROA). The experimental gains in spectral sensitivity have permitted the measurement of the very small differences in the scattering of left-handed and right-handed polarized light from chiral centers. In principle ROA has potential as a highly sensitive conformational probe and for early applications to proteins and peptides the reader is referred to a review by Barron and Hecht (1993).

V. DIFFERENTIAL SCANNING CALORIMETRY

Various spectroscopic techniques provide important information about the structural properties of protein molecules. However, spectroscopy is generally of very limited value in studying thermodynamic aspects of protein structure. Insight into protein thermodynamics is essential for understanding the mechanisms of protein folding as well as for elucidating the nature of forces that stabilize the protein molecule in its unique three-dimensional structure.

The technique that is particularly well suited for studying the thermodynamic stability of proteins is differential scanning calorimetry (DSC). The unique role of DSC is linked to the fact that it is the only method that can provide information about the enthalpy of the system as a function of temperature. The functional dependence between these two parameters gives access to all thermodynamic information about the macroscopic system in a given temperature range.

In a differential scanning calorimetry experiment one measures the heat capacity of a system as a function of temperature. The calorimeter consists of two cells: one filled with the solvent (reference cell) and one containing a biopolymer of interest (sample cell). The cells are heated at a constant rate. When an equilibrium process such as protein unfolding is triggered by the temperature, a certain amount of heat is absorbed by the sample, and extra electric power is supplied to the sample cell in order to maintain it at the same temperature as the reference cell. This extra power is proportional to the excess heat capacity of the sample, i.e., the amount by which the specific heat during the transition exceeds the baseline specific heat. The enthalpy change associated with the transition, ΔH, is equal to the area under the peak of the excess heat capacity versus temperature curve,

$$\Delta H = \int_{T_0}^{T_F} \Delta C_p dT, \tag{4}$$

where ΔC_p is the excess heat capacity associated with the transition, and T_0 and T_F are the lower and upper temperature limits of the transition, respectively.

In order to minimize the possible heat effects due to the interaction between individual protein molecules, DSC experiments should be performed using dilute protein solutions. However, in such solutions the heat effects associated with protein transitions are very small compared to the background of strong heat absorption by the solvent. Therefore, studies of protein thermodynamics impose special requirements on the sensitivity of scanning calorimeters. Routine studies of dilute proteins solutions have become feasible with the development of a new generation of instruments known as ultrasensitive scanning calorimeters or microcalorimeters. The DSC instruments most frequently used in protein research include the Microcal 2 (Microcal, Inc.) and the successors of the DASM-1M calorimeter originally described by Privalov *et al.*, (1975).

A. Reversible Transitions

Rigorous thermodynamic analysis of DSC data is possible only for reversible transitions. Such an analysis is based on the van't Hoff equation,

$$\left(\frac{dlnK}{dT} \right)_p = \frac{\Delta H_{vH}}{RT^2}, \tag{5}$$

where T denotes the absolute temperature, K is the equilibrium constant of the transition process, R denotes gas constant and ΔH_{vH} is van't Hoff enthalpy. The above equilibrium thermodynamics equation is directly applicable only to two-state (or "all-or-none") processes in which intermediate states between the initial and the final states are not populated at equilibrium. Furthermore, for strictly two-state processes carried under essentially equilibrium conditions

$$\Delta H_{vH} = \Delta H_{cal} = M\Delta h_{cal}, \tag{6}$$

where ΔH_{cal} is the total calorimetric enthalpy, M denotes the molecular weight of the molecule and Δh_{cal} is the calorimetric specific enthalpy. Since the heat absorbed by the system is proportional to the degree of progress of the reaction, the van't Hoff enthalpy can be determined directly from the calorimetric curve. *The fact that both ΔH_{cal} and ΔH_{vH} can be obtained from the same experimental curve constitutes one of the greatest advantages of the calorimetric method. Comparison of these two quantities allows the investi-*

gator to assess whether the transition under study can be approximated by a two-state model.

It can be shown that for the simplest possible reversible process $A \rightleftharpoons B$ the van't Hoff enthalpy can be expressed as

$$\Delta H = ART^2 C_{ex.1/2}/\Delta h_{cal}, \tag{7}$$

where $T_{1/2}$ is the absolute temperature at which the transition is half completed and $C_{ex.1/2}$ is the excess specific heat at $T_{1/2}$ and the factor A has a value of 4. In a pioneering study Privalov and co-workers have examined differential scanning calorimetric curves of a number of small globular proteins under the conditions of various pH (e.g., Privalov and Khechinashvili, 1974). Analysis of the DSC data revealed that most of these proteins undergo reversible, two-state thermally induced folding–unfolding transitions that can be described by Eqs. (6) and (7). Another important conclusion of these studies was that the temperature dependence of the denaturation enthalpy is largely due to nonpolar groups coming into contact with water during the transition from the native to the denatured states, whereas the enthalpy of the rupture of hydrogen bonds in proteins depends little on temperature. A comprehensive review of the early literature on DSC studies of small globular proteins has been published by Privalov (1979).

More recently, Sturtevant and co-workers have extended the thermodynamic treatment of DSC data to include the frequently encountered cases of two-state denaturation of oligomeric proteins with self-dissociation into monomers (reactions of the general scheme $A_n \rightleftharpoons nB$) and two-state denaturation with ligand dissociation (for review see Sturtevant, 1987). Analysis of the DSC curves for *Streptomyces* subtilisin inhibitor has revealed that thermal denaturation of this protein involves dissociation of the native dimeric structure into monomers (Takahashi and Sturtevant, 1981). Other examples of the application of this extended formalism include calorimetric studies of the tetrameric core protein of *lac* repressor (Manly *et al.*, 1985) and the studies on thermal denaturation of the arabinose binding protein of *Escherichia coli* (Fukada *et al.*, 1983).

Although thermal denaturation of many small proteins can be described as a simple two-state process (i.e., $\Delta H_{vH} = \Delta H_{cal}$), clearly not all proteins meet this criterion. When the van't Hoff enthalpy exceeds the calorimetric enthalpy ($\Delta H_{vH} > \Delta H_{cal}$), it may be concluded that the process under study involves intermolecular cooperation. A more interesting situation, frequently encountered in protein studies is when $\Delta H_{vH} < \Delta H_{cal}$. In the latter case, one can conclude that the denaturation of the protein molecule involves a multistate process or, in other words, it consists of several stages. Analysis of calorimetric curves representing multistate transitions is not straightforward and usually requires the use of deconvolution procedures. Several computer programs are now available that can be used to decompose complex calorimetric curves into simple constituents corresponding to the heat effects of individual transitions between macroscopic states (e.g., Freire and Biltonen, 1978; Filimonov *et al.*, 1982; Gill *et al.*, 1985; Privalov and Potekhin, 1986). In general, the presence of a number of discrete steps in the thermal unfolding of protein may indicate (i) that the whole molecule unfolds in discrete steps or (ii) that the protein consists of several domains and each of these domains unfolds independently in a two-state manner. Important lessons in this regard have been learned from the calorimetric study of plasminogen and its fragments (Novokhotny *et al.*, 1984). The results of this study are summarized in Fig. 9. Deconvolution analysis of the excess heat capacity profile of this protein has revealed that the thermal denaturation consists of seven two-state transitions (Fig. 9a). DSC curves of plasminogen fragments (obtained by treatment of the protein with various proteolytic enzymes) clearly show that each heat absorption peak identified by the deconvolution analysis corresponds to the melting of a certain part of the molecule. In other words, individual parts of plasminogen represent cooperative units that unfold independently of the rest of the protein molecule. From this example, as well as from other DSC studies with complex proteins it appears that proteins generally consist of independent cooperative domains. Whereas small proteins usually consist of a single domain, larger proteins may comprise as much as a dozen or more of different domains. Detailed discussion of calorimetric studies of multidomain proteins may be found in excellent reviews by Privalov (1982, 1989). *It is remarkable that insight into the domain structure of proteins can be obtained solely from the thermodynamic analysis of the calorimetric data, even in the absence of any direct structural information.*

B. Irreversible Transitions

A prerequisite for the validity of the thermodynamic analysis of calorimetric data is that the system under study remains in thermodynamic equilibrium throughout the temperature range in which the transition takes place. The most direct equilibrium criterion is the reproducibility of the calorimetric trace in the second heating cycle, the so-called calorimetric reversibility. Unfortunately, many proteins show no thermal effect in the second thermogram. The potential factors responsible for this irreversibility have been discussed recently (Freire *et al.*, 1990). They include protein aggregation, isomerization of proline residues, deamination of asparagine and glutamine residues, and destruction of disulfide bonds. Despite the lack of calorimetric reversibility, DSC curves for a number of proteins have been analyzed in terms of equilibrium thermodynamics (e.g., Manly *et al.*, 1985; Edge *et al.*, 1985; Hu and Sturtevant, 1987). The rationale for this has been based on the assumption that the irreversible step does not take place significantly in the temperature range of the transition but occurs, with negligible heat effect, at postdenaturational temperatures. An additional equilibrium

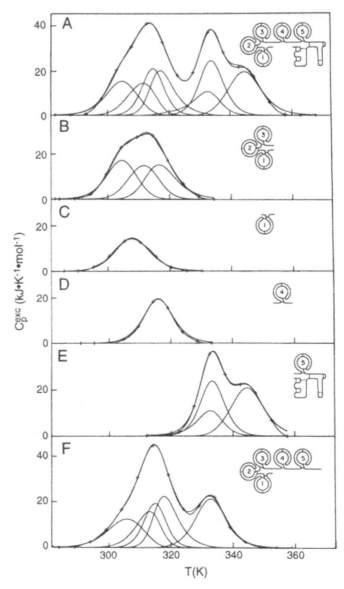

FIGURE 9 Deconvolution of the excess heat capacity function of Lys-plasminogen (A) and its various fragments (B–F). Crosses indicate experimental curves to distinguish them from the curves calculated using the deconvolution analysis. The structures of plasminogen fragments are given schematically in the upper corner of each panel. Reproduced from Novokhatny et al., (1984), with permission.

criterion applied in such cases was the agreement between the van't Hoff enthalpy values calculated from the calorimetric traces and those calculated from the effect of protein concentration or concentration of the ligand on the transition temperature. Although this approach may be justified in some instances, its general validity is questionable. In fact, recent work has shown that for many proteins the irreversible step takes place during the time the protein spends in the transition region (e.g., Sanchez-Ruiz *et al.*, 1988a,b; Guzman-Casado *et al.*, 1990; Freire *et al.*, 1990; Galisteo *et al.*, 1991). The DSC transitions are thus strongly rate limited and are amenable to kinetic, but not thermodynamic, analysis.

The simplest model of irreversible denaturation is a two-state process $N \xrightarrow{k} D$, in which the protein undergoes a transition from the native state N to a final state D. The transition is a first-order reaction characterized by the rate constant k, which changes with temperature according to the Arrhenius equation. Recently Sanchez-Ruiz *et al.* (1988a) have developed a rigorous mathematical treatment for the irreversible two-state calorimetric transitions, describing them in terms of kinetic parameters such as rate constants and energy of activation. Clearly, no thermodynamic information can be obtained from such transitions. The calorimetric traces for the two-state irreversible model are asymmetric toward the low temperature side of the transition and show pronounced scan-rate dependence, being shifted to higher temperatures at increasing scan rates (Fig. 10). The two-state irreversible model may seem unrealistically simple. Nevertheless, it was shown to provide good approximation of thermal denaturation of a number of pro-

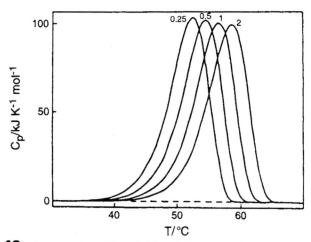

FIGURE 10 Scan-rate dependence of differential scanning calorimetry curves corresponding to two-state irreversible transition characterized by a ΔH of 800 kJ/M and an activation energy of 300 kJ/M. The number at each curve indicates scanning rate (in K/min). Reproduced from Freire *et al.*, (1990), with permission.

teins (e.g., Sanchez-Ruiz, 1988a,b; Guzman-Casado *et al.*, 1990). Theoretical aspects of more general situations have been discussed by several authors (Freire *et al.*, 1990; Sanchez-Ruiz, 1992; Lepock *et al.*, 1992). An important practical conclusion from the analysis of kinetically controlled transitions is that great caution should be exercised in using thermodynamic approaches to analyze DSC data for proteins that show no thermal effect in the second heating cycle. Sanchez-Ruiz *et al.*, (1988a) have proposed to use scan-rate dependence of thermograms as an independent equilibrium criterion in differential scanning calorimetry: *the detection of a clear scan-rate dependence of calorimetric traces should preclude the analysis of DSC data in terms of equilibrium thermodynamics.*

REFERENCES

Alben, J. O., and Bare, G. H. (1989) *J. Biol. Chem.* 255, 3892–3897.
Arnold, G. E., Day, L. A., and Dunker, A. K. (1992) *Biochemistry* 31, 7948–7956.
Arrondo, J. L. R., Muga, A., Castresana, J., and Goni, F. M. (1993) *Prog. Biophys. Mol. Biol.* 59, 23–56.
Asher, S. A. (1993) *Anal. Chem.* 65, 59A–66A.
Austin, J. C., Jordon, T., and Spiro, T. G. (1993) In *Advances in Spectroscopy* (R. J. H. Clark and R. E. Hester, Eds.), Vol. 20, Chap. 2. Wiley, Chichester, UK.
Bagley, K. A., Eisenstein, L., and Ebrey, T. G. (1989) *Biochemistry* 28, 3366–3373.
Barron, L. D., and Hecht, L. (1993) In *Advances in Spectroscopy* (R. J. H. Clark and R. E. Hester, Eds.) Vol. 21, Chap. 5. Wiley, Chichester, UK.
Barth, A., Mantele, W., and Kreutz, W. (1991 *Biochim. Biophys. Acta* 1120, 123–143.
Belasco, J. G., and Knowles, J. R. (1983) *Biochemistry* 22, 122–129.
Brahms, S., and Brahms, J. (1980) *J. Mol. Biol.* 138, 149–178.
Braiman, M. S., and Rothschild, K. J. (1988) *Annu. Rev. Biophys. Biophys. Chem.* 17, 541–570.
Bryan, P., Pantoliano, M. N., Quill, S. G., Hsiao, H.-Y., and Poulos, T. (1986) *Proc. Natl. Acad. Sci. U.S.A.* 83, 3743–3745.
Buchet, R., Jona, I., and Martonosi, A. (1991) *Biochim. Biophys. Acta* 1069, 209–217.
Bustamante, C., Tinoco, I., and Maestre, M. F. (1983) *Proc. Natl. Acad. Sci. U.S.A.* 80, 3568–3572.
Byler, M., and Susi, H. (1986) *Biopolymers* 25, 469–487.
Cantor, C. R., and Schimmel, P. R. (1980) *Biophysical Chemistry, Part II: Techniques for the Study of Biological Structure and Function.* Freeman, New York.
Carey, P. R. (1982) *Biochemical Applications of Raman and Resonance Raman Spectroscopies,* pp. 71–98. Academic Press, New York.
Carey, P. R. (1988). In *Modern Physical Methods of Biochemistry: Part B* (A. Neuberger and L. L. M. Van Deemen, Eds.), pp. 27–64. Elseveier, B. V.
Chang, C. T., Wu, C.-S. C., and Yang, J. T. (1978) *Anal. Biochem.* 91, 13–31.
Chen, W., Nie, S., Kuck, J. F. R., Jr., and Yu, N.-T. (1991) *Biophys. J.* 60, 447–455.
Chen, Y.-H., Yang, J. T., and Chau, K. H. (1974) *Biochemistry* 13, 3350–3359.
Chirgadze, Y. N., Fedorov, O. V., and Trushina, N. P. (1975) *Biopolymers* 14, 679–694.
Dong, A., Huang, P., and Caughey, W. S. (1990) *Biochemistry* 29, 3303–3308.
Dousseau, F., and Pézolet, M. (1990) *Biochemistry* 29, 8771–8779.
East, E. J., Chang, R. C. C., Yu, N.-T., and Kuck, J. F. R. (1978) *J. Biol. Chem.* 253, 1436–1441.

Edge, V., Allewell, N. M., and Sturtevant, J. M. (1985) *Biochemistry* 24, 5899–5906.

Einarsdottir, O., Choc, M. G., Weldon, S., and Caughey, W. S. (1988) *J. Biol. Chem.* 263, 13641–13654.

Fiamingo, F. G., Altschuld, R. A., and Alben, J. O. (1986) *J. Biol. Chem.* 261, 12976–12987.

Filimonov, V. V., Potekhin, S. A., Matveev, S. V., and Privalov, P. L. (1982) *Mol. Biol. USSR* 16, 551–562.

Fraser, P. E., Nguyen, J. T., Surewicz, W. K., and Kirschner, D. (1991) *Biophys. J.* 60, 1190–1201.

Freire, E., and Biltonen, R. L. (1978) *Biopolymers* 17, 463–479.

Freire, E., van Osdol, W. W., Mayorga, O. L., and Sanchez-Ruiz, J. M. (1990) *Annu. Rev. Biophys. Biophys. Chem.* 19, 159–188.

Fukada, H., Sturtevant, J. M., and Quiocho, F. A. (1983) *J. Biol. Chem.* 258, 13193–13198.

Galisteo, M. L., Mateo, P. L., and Sanchez-Ruiz, J. M. (1991) *Biochemistry* 30, 2061–2066.

Gill, S. J., Richey, B., Bishop, G., and Wyman, J. (1985) *Biophys. Chem.* 21, 1–14.

Goormaghtigh, E., Cabiaux, V., and Ruysschaert, J.-M. (1990) *Eur. J. Biochem.* 193, 409–420.

Greenfield, N., and Fasman, G. D. (1969) *Biochemistry* 8, 4108–4116.

Gregory, R. B., and Rosenberg, A. (1986) *Methods Enzymol.* 131, 448–508.

Griffiths, P. R., and de Haseth, J. A. (1986) *Fourier Transform Infrared Spectrometry.* Wiley, New York.

Guzman-Casado, M., Parody-Morreale, A., Mateo, P. L., and Sanchez-Ruiz, J. M. (1990) *Eur. J. Biochem.* 188, 181–185.

Harada, I., and Takeuchi, H. (1986) In *Spectroscopy of Biological Systems* (R. J. H. Clark and R. E. Hester, Eds.), pp. 113–175. Wiley, Chichester, U.K.

Haris, P. I., and Chapman, D. (1992) *Trends Biochem. Sci.* 17, 328–333.

Haris, P. I., Coke, M., and Chapman, D. (1989) *Biochim. Biophys. Acta* 995, 160–167.

Hennessey, J. P., and Johnson, W. C., Jr. (1981) *Biochemistry* 20, 1085–1094.

Horwitz, J., and Strickland, E. H. (1971) *J. Biol. Chem.* 246, 3749–3752.

Hu, C. Q., and Sturtevant, J. M. (1987) *Biochemistry* 26, 178–182.

Jackson, M., Haris, P. I., and Chapman, D. (1991) *Biochemistry* 30, 9681–9686.

Jager, F., Fahmy, K., Sakmar, T. P., and Siebert, F. (1994) *Biochemistry* 33, 10878–10882.

Johnson, W. C., Jr. (1985) In *Methods of Biochemical Analysis* (D. Glick, Ed.), Vol. 31, pp. 61–163. Wiley, New York.

Johnson, W. C., Jr. (1988) *Annu. Rev. Biophys. Biophys. Chem.* 17, 145–166.

Johnson, W. C., Jr. (1990) *Proteins* 7, 205–214.

Jung, E. K. Y., Chin, J. J., and Jung, C. Y. (1986) *J. Biol. Chem.* 261, 9155–9160.

Khan, P. C. (1979) *Methods Enzymol.* 61, 339–378.

Kauppinen, J. K., Moffatt, D. J., Mantsch, H. H., and Cameron, D. (1981) *Appl. Spectrosc.* 35, 271–276.

Kahn, M. Y., Villanueva, G., and Newman, S. A. (1989) *J. Biol. Chem.* 264, 2139–2142.

Kim, M., and Carey, P. R. (1993) *J. Amer. Chem. Soc.* 115, 7015–7016.

Kim, M., Owen, H., and Carey, P. R. (1993) *Appl. Spectrosc.* 49, 1780–1783.

Kitagawa, T., and Ogura, T. (1993) In *Advances in Spectroscopy* (R. J. H. Clark and R. E. Hester, Eds.), Vol. 21, Chap. 3. Wiley, Chichester, UK.

Krimm, S., and Bandekar, J. (1986) *Adv. Protein Chem.* 38, 181–364.

Kurz, L. C., and Drysdale, G. R. (1987) *Biochemistry* 26, 2623–2627.

Lansbury, P. T., Jr. (1992) *Biochemistry* 31, 6865–6870.

Lee, D. C., Haris, P. I., Chapman, D., and Mitchell, R. C. (1990) *Biochemistry,* 29, 9185–9193.

Lepock, J. R., Ritchie, K. P., Kolios, M. C., Rodahl, A. M., Heinz, K. A., and Kruuv, J. (1992) *Biochemistry* 31, 12706–12712.

Li, H., Wurrey, C. J., and Thomas, G. J., Jr. (1992) *J. Amer. Chem. Soc.* 114, 7463–7469.

Li, T., Chen, Z., Johnson, J. E., and Thomas, G. J., Jr. (1990) *Biochemistry* 29, 5018–5026.

Manavalan, P., and Johnson, W. C., Jr. (1987) *Anal. Biochem.* 67, 76–85.

Manly, S. P., Matthews, K. S., and Sturtevant, J. M. (1985) *Biochemistry* **24**, 3842–3846.
Manning, M. C., and Woody, R. W. (1989) *Biochemistry* **28**, 8609–8613.
Mantele, W. (1993) *Trends Biochem. Sci.* **18**, 197–202.
Mathies, R. A., Lin, S. W., Ames, J. B., and Pollard, W. T. (1991) *Annu. Rev. Biophys. Biophys. Chem.* **20**, 491–518.
Miura, T., Takeuchi, H., and Harada, I. (1989) *J. Raman Spectrosc.* **20**, 667–671.
Miura, T., and Thomas, G. J., Jr. (1995) In *Subcellular Biochemistry: Structure, Function and Protein Engineering* (B. B. Biswas and S. Roy, Eds.). Plenum, New York.
Moh, P. P., Fiamingo, F. G., and Alben, J. O. (1987) *Biochemistry* **26**, 6243–6249.
Moss, D. A., Nabedryk, E., Breton, J., and Mantele, W. (1990) *Eur. J. Biochem.* **187**, 565–572.
Myer, Y. P. (1985) *Curr. Top. Bioenerg.* **14**, 149–188.
Novokhatny, V. V., Kudinov, S. A., and Privalov, P. L. (1984) *J. Mol. Biol.* **179**, 215–232.
Osborne, H. B., and Nabedryk-Viala, E. (1986) *Methods Enzymol.* **131**, 676–681.
Perczel, A., Hollosi, M., Tusnady, G., and Fasman, G. D. (1991) *Protein Eng.* **4**, 669–679.
Potter, W. T., Hazzard, J. H., Choc, M. G., Tucker, M. P., and Caughey, W. S. (1990) *Biochemistry* **29**, 6283–6295.
Pozsgay, M., Fast, P., Kaplan, H., and Carey, P. R. (1987) *J. Invertebr. Pathol.* **50**, 246–253.
Prestelski, S. J., Byler, D. M., and Liebman, M. N. (1991) *Biochemistry* **30**, 133–143.
Privalov, P. L. (1979) *Adv. Protein Chem.* **33**, 167–241.
Privalov, P. L. (1982) *Adv. Protein Chem.* **35**, 1–104.
Privalov, P. L. (1989) *Annu. Rev. Biophys. Biophys. Chem.* **18**, 47–69.
Privalov, P. L., and Khechinashvili, N. N. (1974) *J. Mol. Biol.* **86**, 665–684.
Privalov, P. L., Plotnikov, V. V., and Filimonov, V. V. (1975) *J. Chem. Thermodyn.* **7**, 41–47.
Privalov, P. L., and Potekhin, S. A. (1986) *Methods Enzymol.* **131**, 4–51.
Provencher, S. W., and Glockner, J. (1981) *Biochemistry* **20**, 33–37.
Qian, W., and Krimm, S. (1992) *Biopolymers* **32**, 1025–1033.
Reed, J. (1990) In *Modern Methods in Protein and Nucleic Acid Research* (H. Tschesche, Ed.), pp. 367–394, de Gruyter, New York.
Robert, B. (1995) In *Advances in Photosynthesis* (J. Amesz and A. J. Hoff, Eds.), Chap. 5. Klawer, Dordrecht.
Rothschild, K. J., Braiman, M. S., He, Y.-W., Marti, T., and Khorana, H. G. (1990) *J. Biol. Chem.* **265**, 16985–16991.
Sanchez-Ruiz, J. M. (1992) *Biophys. J.* **61**, 921–935.
Sanchez-Ruiz, J. M., Lopez-Lacomba, J. L., Cortijo, M., and Mateo, P. L. (1988a) *Biochemistry* **27**, 1648–1652.
Sanchez-Ruiz, J. M., Lopez-Lacomba, J. L., Mateo, P. L., Vilanova, M., Serra, M. A., and Aviles, F. X. (1988b) *Eur. J. Biochem.* **176**, 225–230.
Sarver, R. W., and Krueger, W. C. (1991) *Anal. Biochem.* **194**, 89–100.
Saxena, V. P., and Wetlaufer, D. B. A. (1971) *Proc. Natl. Acad. Sci. U.S.A.* **68**, 969–972.
Siebert, F. (1993) In *Advances in Spectroscopy* (R. J. H. Clark and R. E. Hester, Eds.), Vol. 20, Chap. 1. Wiley, Chichester, UK.
Spiro, T. G., Ed. (1987) *Biological Applications of Raman Spectroscopy,* Vols. 1–3. Wiley, New York.
Strickland, E. H. (1972) *Biochemistry* **11**, 3465–3474.
Strickland, E. H., and Mercola, D. (1976) *Biochemistry* **15**, 3875–3884.
Sturtevant, J. M. (1987) *Annu. Rev. Phys. Chem.* **38**, 463–488.
Surewicz, W. K., and Mantsch, H. H. (1988) *Biochem. Biophys. Acta* **952**, 115–130.
Surewicz, W. K., and Mantsch, H. H. (1990) In *Protein Engineering: Approaches to the Manipulation of Protein Folding* (S. Narang, ed), pp. 131–157, Butterworths, Boston.
Surewicz, W. K., Mantsch, H. H., and Chapman, D. (1993) *Biochemistry* **32**, 389–394.
Susi, H., and Byler, M. (1986) *Methods Enzymol.* **130**, 290–311.
Susi, H., and Byler, M. (1987) *Arch. Biochem. Biophys.* **258**, 465–469.

Tadesse, L., Nazarbaghi, R., and Walters, L. (1991) *J. Amer. Chem. Soc.* 113, 7036–7037.

Takahashi, K., and Sturtevant, J. M. (1981) *Biochemistry* 21, 6185–6190.

Takesada, H., Nakanishi, M., Hirakawa, A. Y., and Tsuboi, M. (1976) *Biopolymers* 15, 1929–1938.

Tonge, P. J., Pusztai, M, White, A. J., Wharton, C. W., and Carey, P. R. (1991) *Biochemistry* 30, 4790–4795.

Tonge, P. J., and Carey, P. R. (1992) *Biochemistry* 31, 9122–9125.

Torii, H., and Tasumi, M. (1992) *J. Chem. Phys.* 96, 3379–3387.

van der Goot, F. G., Gonzalez-Manas, J. M., Lakey, J. H., and Pattus, F. (1991) *Nature (London)* 354, 408–410.

van Stokkum, I. H. M., Spoelder, H. J. W., Bloemendal, M., van Grondelle, R., and Groen, F. C. A. (1990) *Anal. Biochem.* 191, 110–118.

Venyaminov, S. Y., and Kalnin, N. N. (1990) *Biopolymers* 30, 1243–1257.

Wilder, C. L., Friedrich, A. D., Potts, R. O., Daumy, G. O., and Francoeur, M. L. (1992) *Biochemistry* 31, 27–31.

Woodruff, W. H., Dyer, R. B., and Einarsdóttir, O. (1993) In *Advances in Spectroscopy* (R. J. H. Clark and R. E. Hester, Eds.), Vol. 21, Chap. 4. Wiley, Chichester, UK.

Woody, R. W. (1985) In *The Peptides* (E. R. Blout, F. A. Bovey, M. Goodman, and N. Lotan, Eds.), pp. 15–114. Academic Press, New York.

Yang, J. T., Wu, C.-S. A., and Martinez, M. (1986) *Methods Enzymol.* 130, 208–269.

Zhang, M., Fabian, H., Mantsch, H. H., and Vogel, H. J. (1994) *Biochemistry* 33, 10883–10888.

Part IV

APPLICATIONS

9

The Design of Polymeric Biomaterials from Natural α-L-Amino Acids

ARUNA NATHAN

JOACHIM KOHN

Department of Chemistry,
Rutgers University,
New Brunswick, New Jersey 08903

I. INTRODUCTION

Over the past 20 years significant efforts have been devoted to the development of polymeric biomaterials (Hench and Ethridge, 1982; Lyman, 1983). Prior to about 1966, these efforts were focused on the identification of biologically inert materials that were stable under physiological conditions. Such materials are suitable for long-term applications such as artificial organ parts, permanent bone and joint replacements, dental devices, and cosmetic implants (Table I). However, in recent years, the emphasis has shifted toward the development of bioresorbable polymers for short-term applications such as drug delivery, sutures, temporary vascular grafts, or temporary bone fixation devices (Table II). The major advantage of using degradable implants in short-term applications is that the problems associated with the long-term safety of permanently implanted devices (e.g., chronic inflammation, contact carcinogenesis) can be circumvented. Also, unlike a biostable prosthesis that either remains in the body of a patient lifelong or has to be explanted, a bioresorbable implant may be replaced by fully functional tissue as part of the natural healing process.

The toxicity of the polymer degradation products represents a major concern associated with the use of bioresorbable polymers. Attempts have

Protein Engineering and Design
Copyright © 1996 by Academic Press, Inc. All rights of reproduction in any form reserved.

TABLE I   Some Long-Term Applications of Nondegradable Biomaterials

Application	Specific examples
Ocular devices	Artifical vitreous humor, corneal prostheses, intraocular lens, artificial tear duct
Electrical devices	Percutaneous leads, auditory and visual prostheses, electrical analgesia, bladder control
Cardiac and vascular devices	Heart pacer, chronic shunts, heart valves, arterial and vascular prostheses, heart assist devices
Orthopedic devices	Bone plates, screws, and wires, intramedullary nails, Harrington rods, artificial hip, knee, shoulder, elbow, wrist replacement
Dental devices	Alveolar bone replacement, mandibular reconstruction, endosseous tooth replacement, subperiosteal tooth replacement, orthodontic anchors
Soft-tissue prostheses	Cosmetic implants for contouring or filling of nose, ear, or neck, implants for augmentation or reconstruction of breast, maxillofacial reconstruction, artificial ureter, bladder, or intestinal wall, artificial skin

been made to address these concerns by using polymers that degrade to naturally occurring metabolites during the process of bioresorption. The lactic/glycolic acid copolymers obtained from naturally occurring hydroxy acids represent an excellent example of this approach (Yolles and Sartori, 1980). These polymers were the first synthetic degradable polymers to be used widely in humans (Sanders, 1985). Originally developed as suture materials, the lactic/glycolic acid copolymers are now intensively investigated in a variety of biomedical applications (Lewis, 1990).

Since poly(amino acids) are structurally related to natural proteins, these polymers were recognized as potential biomaterials early on. The degradation *in vivo* of poly(amino acids) to naturally occurring α-amino acids gave rise to the expectation that these polymers will be highly biocompatible. Consequently, starting from about 1970, many investigators studied the use of both homo- and copolymers of amino acids for a variety of biomedical applications (Anderson *et al.*, 1985; Marchant *et al.*, 1985; Lescure *et al.*, 1989). The structural diversity of these materials represents another potential advantage: A wide range of surface and bulk properties can, in principle, be obtained by copolymerizing two or more of the natural α-L-amino acids. Also, the side chains of some of the amino acids offer attachment points for drugs, crosslinkers, or other pendent groups. On the other hand, the unavoidable enzymatic degradation of the recurring amide bonds in the polymer backbone represents an important limitation, since poly(amino acids) cannot usually be considered for the long-term applications listed in Table I. Within the context of implant materials, poly(amino

TABLE II Applications of Conventional Poly(amino acids)
as Implant Materials[a]

Application	Specific Poly(amino acid) explored	Illustrative reference
Drug administration	Derivatives of poly(glutamic acid), polyglutamine, poly(aspartic acid), polyasparagine, polylysine, polyleucine.	(Seno and Kuroyanagi, 1986); (Bennett et al., 1988)
Degradable sutures	Filaments drawn from γ-alkyl esters of poly(glutamate). Degradation time could be varied from 2 to 60 days.	(Miyamae et al., 1968)
Artificial skin substitutes	Polypeptide laminates with nylon–velour. Clinical studies in humans were reported.	(Spira et al., 1969); (Hall et al., 1970)
Membranes for artificial kidney	The unusual permeability properties of poly(ᴅ,ʟ-methionine-co-leucine) stimulated this research effort.	(Martin et al., 1971)
Wound dressings	Several laboratories used polyglutamate and/or its derivatives (often in combination with other polymers) in the design of wound dressings or adhesion barriers. A poly-ʟ-leucine sponge is in clinical trials in Japan.	(Shiotani et al., 1987); (Shioya et al., 1988); (Brack, 1989); (Kuroyanagi et al., 1992); (Shiotani et al., 1988)
Hemostatic agents	A carrier containing thrombin and blood coagulation factor XIII made of a derivative of poly(glutamic acid).	(Sakamoto et al, 1986)
Hard-tissue prostheses	Composite material consisting of calcium phosphate, poly(γ-benzyl ʟ-glutamate) and poly(2-ethyl-2-oxazoline).	Dorman and Meyers, 1986)

[a] From Nathan and Kohn (1994). Reprinted with permission.

acids) are thus most suitable for use in the short-term applications listed in Table II.

II. POLY(AMINO ACIDS) AS BIOMATERIALS

One of the earliest efforts in the use of poly(amino acids) as biomaterials was that of Miyamae et al., who used filaments drawn from γ-alkyl esters of poly(glutamic acid) as suture material (Miyamae et al., 1968). The degradation time of these polymers could be varied from 2 to about 60 days. This pioneering work of Miyamae *et al.* led to the investigation of poly(amino acids) as medical implant materials by several other researchers.

Polypeptide films have been suggested for use as artificial skin substitutes in burn wound therapy (Spira *et al.*, 1969; Hall *et al.*, 1970). A polypeptide–nylon velour laminate was found to adhere to a surgically created wound and function as a satisfactory temporary biological dressing. Application of the polymer laminate dressing material was followed by the rapid development of a mature fibroblastic granulation response as evidenced by growth of neovasculature and collagen bundles. Laminates composed of polypeptides and elastomers have also been suggested as burn wound dressings.

Martin *et al.* have studied the permeability properties of copolymers of L-leucine and D,L-methionine (Martin *et al.*, 1971). These copolymers formed clear flexible films having low water vapor transmission rates. The hydrophilicity and thus the water vapor transmission rates of these membranes could be enhanced by oxidation of the pendent methyl thioethyl groups to methyl sulfinoethyl groups. Also, the oxygen and carbon dioxide permeability of these membranes could be varied by varying the ratio of leucine to methionine. Potential applications of these membranes would be in hemodialysis membranes and artificial kidneys.

Glutamic acid and its derivatives have been extensively investigated as matrices for drug delivery (Sidman *et al.*, 1980). It has been shown that the rate of degradation of a partially esterified poly(glutamic acid) and the rate of drug release from this copolymer could be manipulated by varying the ratio of glutamic acid to alkyl glutamate. Such copolymers have also been used to prepare hydrogel membranes for the controlled delivery of proteins (Sidman *et al.*, 1983). Drugs such as 5-fluorouracil, adriamycin, and pepliomycin have been released from membranes obtained by copolymerizing poly(γ-benzyl L-glutamate) and poly(N^5-hydroxyethylaminopropyl-L-glutamine) (Seno and Kuroyanagi, 1986).

In addition, release of drugs from poly(γ-benzyl L-glutamate) in the presence of an electric or magnetic field has been studied by Sparer *et al.*, (Sparer and Bhaskar, 1985). The authors have exploited the liquid crystalline properties of the polymer that produce a phase transition and a simultaneous change in the permeability of the membrane. The permeability of synthetic membranes can also be varied by pH-induced reversible conformational changes (Minoura *et al.*, 1986). The reversible conformational change of poly(glutamic acid) from an α-helix to a random coil was found to control solute permeation through copolymer membranes.

In addition to the above-mentioned applications, poly(glutamic acid) and its derivatives have been suggested for use as hemostatic agents and as hard tissue prosthesis (Dorman and Meyers, 1986; Sakamoto *et al.*, 1986). For a detailed review on the biomedical applications of poly(amino acids) that describes work done up to 1985, the reader is referred to a review article by Anderson *et al.* (1985).

The *in vivo* degradation of poly(amino acids) has been evaluated. It was found that polymers with uncharged side chains degraded faster than those that had a high percentage of negatively charged side chains (Hayashi *et al.*, 1985; Hayashi and Iwatsuki, 1990). The degradation process was found to follow Michaelis–Menten kinetics. The enzymatic degradation of poly(N^5-hydroxyalkyl-L-glutamine) using a variety of different proteases showed that all of these enzymes degraded the polymer by an endopeptidase mechanism, although at different rates (Pytela *et al.*, 1989). Also, surface modification was found to influence the biodegradation of poly(γ-benzyl L-glutamate) and other polymers (Chandy and Sharma, 1991). Other important aspects of biomaterials research such as tissue compatibility and absorption of proteins at the polymer surface have also been investigated (Cho *et al.*, 1990). The interaction of implanted poly(amino acids) with living tissue was studied extensively by Walton *et al.* (Walton, 1980).

III. DISADVANTAGES OF POLY(AMINO ACIDS) AS BIOMATERIALS

Although an intense research effort in many laboratories has been directed toward the development of poly(amino acids) as implant materials, these polymers have so far not found significant clinical applications. One of the major problems lies in the synthesis of poly(amino acids), which typically requires the formation of the corresponding N-carboxyanhydride, which then undergoes a ring-opening polymerization to give a homopolymer, a random copolymer, or an A–B- or A–B–A-type block copolymer. The N-carboxyanhydrides used in the synthesis of poly(amino acids) are not only expensive to prepare but also difficult to handle due to their high reactivity and moisture sensitivity (Block, 1983).

Synthetic poly(amino acids) have also been shown to have unfavorable engineering properties. Most homopolymers are nonmelting solids with poor solubility in water or common organic solvents. This makes processing of these polymers by conventional fabrication techniques (solvent casting, injection molding, extrusion) very difficult. Thus with the exception of γ-alkyl esters of poly(glutamic acids), poly(N-alkyl glutamines) and copolymers of glutamic acid with a few other amino acids, poly(amino acids) rarely have desirable properties from a material engineering point of view.

Poly(amino acids) show a pronounced tendency to swell in aqueous media (Sidman *et al.*, 1980). This is a particularly serious disadvantage when poly(amino acids) are used in the formulation of drug delivery devices, since swelling can lead to unpredictable drug release rates. In addition, since degradation of the amide bond takes place predominantly through enzymatic hydrolysis, the variability in the level of enzymatic activity from person to person makes it difficult to reproduce and control the rate of degradation *in vivo* (Kohn and Langer, 1984).

The immunogenicity of poly(amino acids) is another important concern. Structures and conformation have been found to influence strongly the immunogenicity of an amino acid-derived polymer (Sela *et al.*, 1956). The immunogenicity has been shown to increase with increase in molecular complexity, for instance, branching, and with the number of different amino acid residues present in the copolymer (Sela, 1974). A branched copolymer of lysine, alanine, tyrosine, and aspartic acid was found to be a powerful synthetic immunogen (Sela *et al.*, 1962), whereas an unbranched copolymer of L-tyrosine and L-aspartic acid was found to be nonimmunogenic (Sela, 1974). Thus in spite of the structural diversity of poly(amino acids), the choice of a poly(amino acid) for biomaterial applications based on immunological considerations is limited to unbranched homopolymers and copolymers containing not more than two different amino acid residues in random sequence.

One of the reasons for the selection of poly(amino acids) as implant materials was the perception that these polymers would resemble natural tissue (Anderson *et al.*, 1985). This, however, is not necessarily true. Although the polymers themselves are relatively nontoxic and tissue compatible, their interaction with living cells bears very little resemblance to natural tissue. Simple homopolymers or copolymers of two amino acids do not selectively support the growth of specific cell types, stimulate growth of healthy tissue at an injured site, or prevent attachment and activation of platelets. Thus one of the challenges in biomaterials research today is to identify biologically functional tissue analogs that would exhibit some of the above properties.

In summary, the extensive research effort of many laboratories since about 1970 resulted in the exploration of numerous applications for poly(amino acids) as medical implant materials. In view of this extensive research effort, it is somewhat disappointing that conventional poly(amino acids) have not yet found significant medical applications. It appears that the slow and often incomplete biodegradation of many poly(amino acids) *in vivo*, their marginal engineering properties, and their high cost were the most significant obstacles for the practical application of these polymers as medical implant materials.

IV. STRATEGIES FOR THE MODIFICATION OF CONVENTIONAL POLY(AMINO ACIDS)

Attempts have been made to overcome at least some of the limitations of conventional poly(amino acids) as biomaterials. In principle, three different approaches (copolymerization, side-chain modification, and backbone modification) have so far been adopted (Fig. 1).

The basic rationale for the synthesis of copolymers of amino acids and

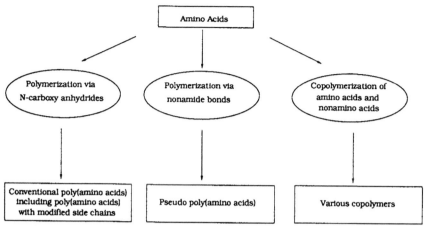

FIGURE 1   Schematic representation of the three approaches used for the design of polymeric biomaterials from natural α-L-amino acids.

non-amino acids was the expectation that copolymers would exhibit new and/or improved material properties not obtainable by any combination of amino acids alone. A patent issued in 1972 represents the first specific design of a copolymer of amino acids and hydroxy acids for medical applications (Goodman and Kirshenbaum, 1973). Although these polymers were suggested as possible suture materials, no practical applications emerged. Obviously, a virtually unlimited number of structurally different copolymers are theoretically possible and indeed, a very wide range of amino acid-containing copolymers have been described in the literature. For example, A–B-type block copolymers with a poly(amino acid) as the A block and poly(L-lactide) as the B block have been explored as controlled drug delivery systems by Kim *et al.* (Kim *et al.*, 1990), and water-soluble drug carriers were obtained by the use of poly(ethylene oxide) as the B block by Yokoyama *et al.* (Yokoyama *et al.*, 1989a,b, 1990a). A–B–A-type block copolymers with a poly(amino acid) as the A block and polyoxypropylene (Kang *et al.*, 1987), polydimethylsiloxane (Kumaki *et al.*, 1985), polyoxyethylene (Cho *et al.*, 1991), or polybutadiene (Nakajima *et al.*, 1979a,b; Kugo *et al.*, 1982, 1983) as the B block are also known. In addition, segmented multiblock copolymers of poly(ethylene oxide) and poly(β-benzyl L-aspartate) (Yokoyama *et al.*, 1990b), graft copolymers of poly(butyl methacrylate) and poly(glutamic acid) (Chung *et al.*, 1986), and alternating copolymers of glutamic acid with various diols have been suggested as drug carriers (Pramanick and Ray, 1987).

Modification of the amino acid side chains either prior to polymerization or after polymerization via conventional N-carboxyanhydrides is another intensively investigated approach (Anderson *et al.*, 1985; Lescure *et al.*, 1989). For example, by attaching hydrophobic residues to the γ-carboxylic acid side chain of poly(glutamic acid), the polymer was rendered water-insoluble. Consequently, γ-alkyl-substituted derivatives of poly(glutamic acid) were suggested for a wide range of medical applications (Table II).

The approach of side-chain modification, however, is not generally applicable to all naturally occurring amino acids. Besides glutamic and aspartic acids, only L-lysine has a reactive side chain (the ε-amino group) that can be easily modified. Modifications of the hydroxyl groups of serine, threonine, or tyrosine have not been explored within the context of biomaterials research and due to the known toxicity of poly(L-lysine), derivatization of this polymer as a means to improve its material properties has only rarely been suggested.

The third and most recently explored approach to improve the material properties of poly(amino acids) is backbone modification. The backbone amide bonds of poly(amino acids) are known to promote interchain hydrogen bonding that leads to the poor processibility, poor solubility, and slow biodegradation of conventional poly(amino acids). Replacement of the amide bonds by a variety of other linkages such as ester, carbonate, iminocarbonate, or urethane linkages represents therefore the most direct means to eliminate the underlying cause for some of the poor engineering properties of conventional poly(amino acids). The backbone modification of poly(amino acids) has its analogy in "pseudopeptide" chemistry where peptide bond surrogates are used in the design of specific structural probes or the synthesis of pharmacologically active peptide drugs (Spatola, 1983). Backbone-modified, "pseudo"-poly(amino acids) were prepared and investigated as potential biomaterials for the first time by Kohn and Langer in 1984 (Kohn and Langer, 1984).

The literature on the backbone modification of poly(amino acids) is surprisingly limited. The apparently first (albeit unsuccessful) attempt to prepare a backbone-modified poly(amino acid) was made as early as 1937 by Greenstein who used cyclo(L-cysteinyl-L-cysteine) as monomer (Greenstein, 1937). Later the synthesis of some low-molecular-weight polymers containing amino acids linked by nonamide bonds was reported (without detailed description of the polymer properties) by Katchalski in one of the earliest reviews on poly(amino acids) (Katchalski and Sela, 1958). Noteworthy are the attempts by Fasman (Fasman, 1960) and by Jarm and Fles (Jarm and Fles, 1977) to synthesize poly(serine ester), a polymer whose backbone consisted of serine units linked by ester bonds derived from the side-chain hydroxyl groups and the C-termini. Fasman's attempt to use the facile N → O shift of polyserine resulted only in a random amide-ester copolymer. Jarm and Fles polymerized N-protected serine β-lactones and obtained poly(N-benzenesulfonamido serine ester). However, since these authors used the bioincompatible benzenesulfonamido group for protection of the serine N-terminus, their polymer was not suitable for use as a biomaterial.

Since the introduction of pseudo-poly(amino acids) as biomaterials by Kohn and Langer in 1984, a series of backbone-modified poly(amino acids) have been synthesized and characterized. The objective of this work has been to develop a new class of biomaterials derived from amino acids that would retain the nontoxicity and biocompatibility of poly(amino acids) and at the same time exhibit improved physicomechanical properties (Pulapura *et al.*, 1990; Pulapura and Kohn, 1992b; Zhou and Kohn, 1990; Kohn, 1991).

V. PSEUDO-POLY(AMINO ACIDS)

Since it is not possible by simple synthetic methods to replace the backbone amide bonds of conventional poly(amino acids) by nonamide linkages, the synthesis of pseudo-poly(amino acids) can only be accomplished by carefully designed polymerization reactions using suitably functionalized amino acid derivatives as monomers. Whereas peptide bond surrogates are often prepared by introducing new functional groups at the N and C terminal positions of the amino acids, few attempts have been made to explore reaction schemes where amino acids are linked through their side chains. Depsipeptides, which contain an ester linkage involving the hydroxyl group of serine and the carboxylic acid group of the adjacent amino acid and unnatural peptides containing amide bonds derived from the ε-amino group of lysine and the β or γ carboxylic acid groups of aspartic or glutamic acid are some of the few examples where the amino acid side chains are part of the peptide backbone.

One possible approach to the synthesis of pseudo-poly(amino acids) involves the use of side-chain functionalities of trifunctional amino acids. As is illustrated in Fig. 2, a trifunctional amino acid can, in principle, be polymerized in three different ways. In Fig. 2a, the polymer backbone consists of linkages formed between the C-terminus and the side-chain functional group R; in Fig. 2b, the polymer backbone consists of linkages formed between the N-terminus and the side-chain functional group; and in Fig. 2c, the polymer backbone consists of linkages formed between the C-terminus and the N-terminus. Obviously, if the linking bond in Fig. 2c is an amide bond, a schematic representation of a conventional poly(amino acid) is obtained.

This approach is applicable, among others, to serine, hydroxyproline, threonine, tyrosine, cysteine, glutamic acid, and lysine, and is limited only by the requirement that the nonamide backbone linkages give rise to polymers with desirable material properties. Recently, several new pseudo-poly(amino acids) have been prepared by variations of this basic synthetic approach (Kohn and Langer, 1984, 1987; Kohn, 1990, 1991; Pulapura *et al.*, 1990, Zhou and Kohn, 1990; Pulapura and Kohn 1992a,b). Some of these polymers are now commercially available.[1] Several representative examples of this new class of polymers will be described below.

[1]Several pseudo-poly(amino acids) have recently become commercially available through Sigma Chemical Company.

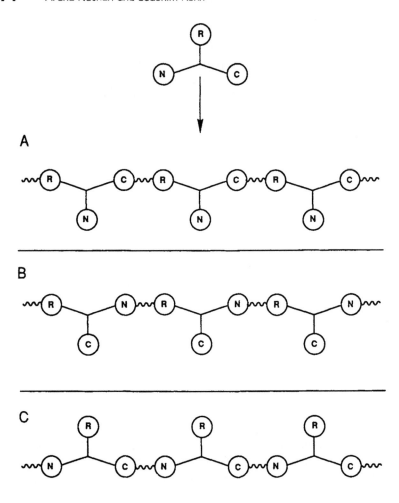

FIGURE 2 The use of trifunctional amino acids for the preparation of pseudo-poly(amino acids). The zigzag line represents a nonamide linkage between individual amino acids. In conventional poly(amino acids), amino acids are linked via amide bonds derived from the N- and C-termini. However, there is no reason to limit polymerization reactions to the termini and, at least in principle, polymers can also be prepared by reactions that involve the side-chain functional groups. Reproduced, by permission, from Nathan and Kohn (1994).

A. Polyesters Derived from *trans*-4-Hydroxy-ʟ-Proline

One of the first examples of a pseudo-poly(amino acid) developed by Kohn and Langer was a polyester derived from *trans*-4-hydroxy-ʟ-proline (Kohn and Langer 1984, 1987). This polymer was synthesized by a transesterification reaction commonly used for the preparation of polyester fibers such as Dacron (Fig. 3). Prior to polymerization, protection of the secondary amine function of the amino acid is mandatory. Since the N-protecting group was found to influence the polymer properties, it is possible to design materials with specific properties by careful selection of the N-terminus-protecting group. For the permanent protection of the N-terminus a range of biocompatible carboxylic acids varying only in the number of carbon atoms was utilized (Yu, 1988; Yu Kwon and Langer, 1989).

The reaction conditions for the synthesis of the polymer were optimized using N-palmitoyl-*trans*-4-hydroxy-ʟ-proline as a model compound. A Lewis acid catalyst such as titanium isopropoxide was most effective in catalyzing this melt transesterification reaction. Highest molecular weights were obtained at a reaction temperature of 180°C using 1 mol% of the catalyst. Under these conditions, a molecular weight of 40,000 (weight average, relative to polystyrene standards) could be obtained.

When palmitic acid, a natural constituent of body fat, was used for N-protection, poly(N-palmitoyl *trans*-4-hydroxy-ʟ-proline ester) [poly(Pal-Hyp ester)] was obtained. This is a highly hydrophobic, wax-like, white material that is insoluble in water, but readily soluble in most nonpolar organic solvents, including hexane. The polymer is apparently amorphous and is thermally stable up to at least 300°C in air with a glass transition temperature of 97°C. It is readily processible by commonly used fabrication techniques and can be prepared at a lower cost than the isometric, conventional poly(amino acid) derived from O-palmitoyl *trans*-4-hydroxy-ʟ-proline.

The polymer has a degradable ester backbone but the hydrophobicity of the palmitoyl side chain and the rigidity of the polymer backbone cause this degradation to be slow (Yu, 1988). As expected, the rate of degradation increased when smaller pendent groups were used. Release studies using model drugs suggest that the release time depends on the loading and the hydrophobicity of the released drug. A potential application of this polymer is in long-term drug delivery, for instance, as a multiyear contraceptive formulation. Biocompatibility studies in an animal model showed that

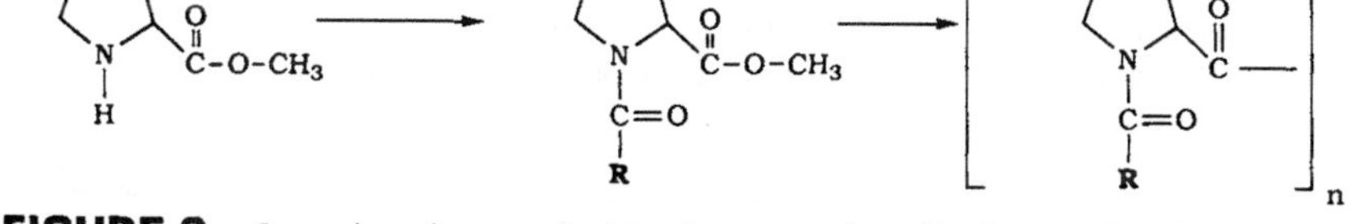

FIGURE 3   General synthetic method for the preparation of hydroxyproline derived polyesters. In the first step, the N-terminus is chemically protected. A wide range of carboxylic acids can be used for this purpose. A polymer particularly useful for drug delivery applications is obtained when R = palmitoyl. Reproduced, by permission, from Nathan and Kohn (in press).

poly(Pal-Hyp ester) caused only a very mild local tissue response, comparable to that of medical grade stainless-steel or poly(lactic acid)/poly(glycolic acid) implants.

B. Polyesters Derived from L-Serine

In order to find a generally applicable method for the preparation of optically pure poly(L-serine ester), a number of different polyesterification reactions involving N-protected serine derivatives were explored (Fig. 4).

The most straightforward approach is the direct esterification of N-protected serine in the presence of a suitable catalyst (Fig. 4A). However, due to the facile β-elimination of the serine side chain under the dehydrating conditions of the polyesterification reaction this approach failed (Zhou and Kohn, 1990).

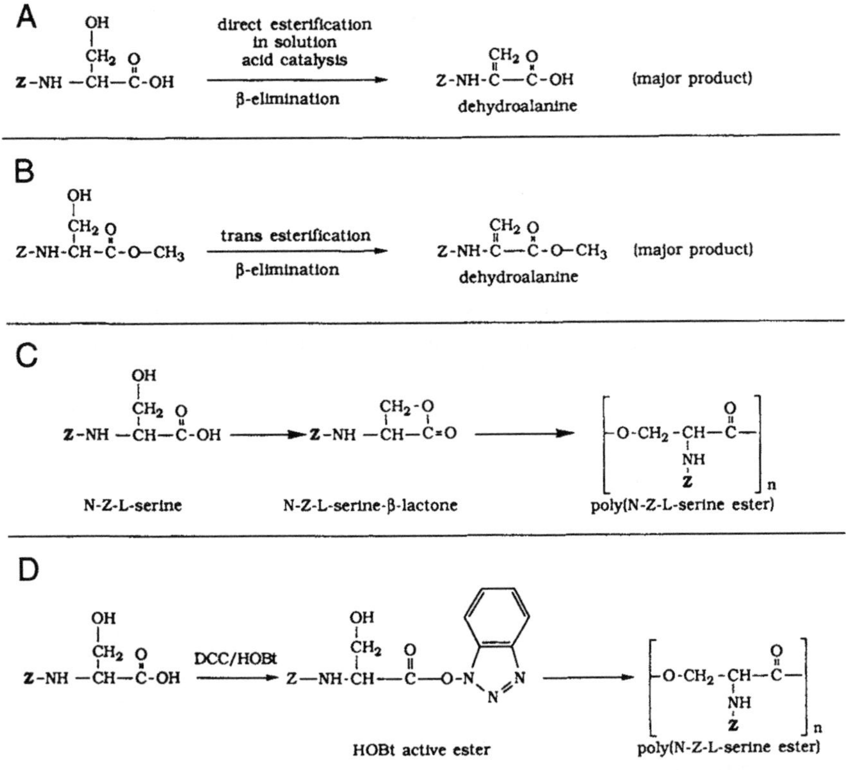

FIGURE 4 For the preparation of poly(N–Z serine ester) (Z = benzyloxycarbonyl) four different synthetic methods were explored. (A) Direct esterification; (B) melt transesterification; (C) ring-opening polymerization of serine-β-lactones; (D) bulk polymerization of hydroxybenzotriazole active esters of serine. Reproduced, by permission, from Nathan and Kohn.

An alternative approach is the transesterification of N-benzyloxycarbonyl-ʟ-serine methyl ester (Fig. 4B). Although polymers of low molecular weight ($M_n \sim 600$) were obtained, dehydroalanine formation was observed even under optimized conditions (Zhou and Kohn, 1990).

The ring-opening polymerization of lactones is known to be the method of choice for the preparation of several important poly(hydroxy acids) such as poly(lactic acid), poly(glycolic acid), or polycaprolactone. In analogy, the ring-opening polymerization of N-protected serine-β-lactones proceeded smoothly and yielded poly(N-protected serine esters) of reasonable molecular weight (Fig. 4C). With tetraethylammonium benzoate (TEAB), an anionic initiator, polymers of molecular weight up to 40,000 (relative to polystyrene standards) and high optical purity (less than 0.5% of racemization) were obtained under mild conditions (37°C, solution in THF). The disadvantages of this approach are the high cost of preparing the labile serine-β-lactone and the occurrence of a chain transfer reaction during the polymerization (Zhou and Kohn, 1990).

Very recently, a convenient and cost-efficient synthesis of poly(N-benzyloxycarbonyl serine ester) via the bulk polymerization of the hydroxybenzotriazole (HOBt) active ester of N-benzyloxycarbonyl ʟ-serine (Fig. 4D) was reported (Gelbin and Kohn, 1991, 1992). Since the need to isolate and purify the reactive serine-β-lactone is eliminated, this approach is the first practical method for the multigram synthesis of serine-derived polyesters of reasonably high molecular weight.

As a representative example, poly(N-benzyloxycarbonyl-serine ester) was synthesized. This polymer is a white solid that is soluble in polar, aprotic solvents such as dimethylformamide. A detailed evaluation of its material properties has not yet been published. The major advantage of poly(serine esters) over the currently available polyesters is the presence of pendent amino groups that can, in principle, be used for the attachment of drugs and crosslinkers. Thus, poly(N-benzyloxycarbonyl-serine ester) could, after removal of the benzyloxycarbonyl group, be further modified by the attachment of suitable biocompatible moieties.

C. Polyiminocarbonates Derived from Tyrosine

Tyrosine is the only major, natural nutrient containing an aromatic hydroxyl group. Derivatives of tyrosine dipeptide can be regarded as diphenols and were employed as replacements for industrially used diphenols such as Bisphenol A in the design of medical implant materials (Fig. 5A, 5B). Tyrosine-derived polyiminocarbonates are examples of pseudo-poly(amino acids) in which a dipeptide (and not a single amino acid) is the basic polymer repeat unit (Pulapura et al., 1990; Pulapura and Kohn, 1992a). The observation that aromatic backbone structures can significantly increase the stiffness and mechanical strength of polymers provided the rationale for the use of tyrosine dipeptides as monomers.

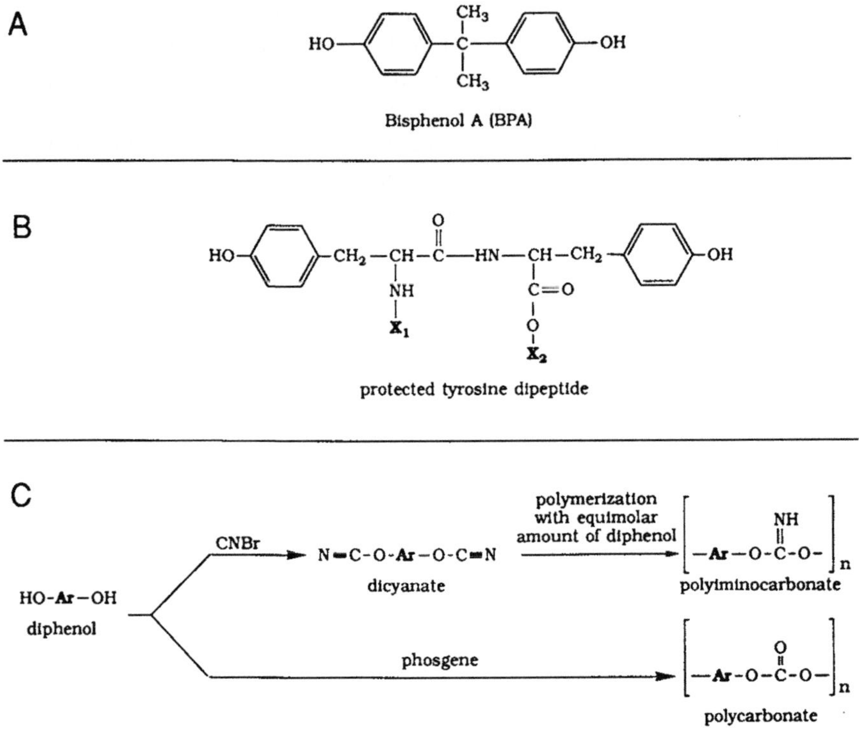

A

Bisphenol A (BPA)

B

protected tyrosine dipeptide

C

FIGURE 5 Structures of (A) Bisphenol A, a widely used diphenol in the manufacture of commercial polycarbonate resins; (B) tyrosine dipeptide with chemical protecting groups X_1 and X_2 attached to the N and C termini, respectively; (C) general synthetic methods for the preparation of polyiminocarbonates and polycarbonates from either industrial diphenols such as Bisphenol A or tyrosine dipeptide derivatives. The only structural difference between polycarbonates and polyiminocarbonates is the replacement of the carbonyl oxygen in the carbonate linkage by an imino group. Reproduced, by permission, from Nathan and Kohn (in press).

Polyiminocarbonates are the "imine analogs" of the industrially used polycarbonates obtained by the replacement of the carbonyl oxygen by an imino group (Fig. 5C). In the late 1960s, the first synthesis of low-molecular-weight polyiminocarbonates by reaction of aqueous solutions of various chlorinated diphenolate sodium salts with cyanogen bromide was reported (Hedayatullah, 1967). Later, the reaction of a diphenol and a dicyanate to give a polyiminocarbonate (Fig. 5C), using both solution and bulk polymerization techniques, was suggested by Schminke et al. in a now abandoned patent (Schminke et al., 1970). The first comprehensive and systematic study of polyiminocarbonate synthesis in general was published in 1989 (Li and Kohn, 1989), followed in 1990 by a detailed exploration of the synthesis and characterization of tyrosine-derived polyiminocarbonates (Pulapura et al. 1990). A comprehensive review of interfacial and solution polymerization procedures for the preparation of polyiminocarbonates of high molecular weight has recently been published (Kohn, 1990).

In order to explore the correlations between the structure of the N- and C-terminus protecting groups and the properties of the resulting polymers, several series of tyrosine-derived polyiminocarbonates were prepared and their physicomechanical properties were evaluated. In one such series, the ethyl, hexyl, and palmitoyl esters of N-benzyloxycarbonyl-ʟ-tyrosyl-ʟ-tyrosine were synthesized and the corresponding polymers, poly(N-Z-Tyr-Tyr-OEt iminocarbonate), poly(N-Z-Tyr-Tyr-OHex iminocarbonate), and poly(N-Z-Tyr-Tyr-OPal iminocarbonate), were used for a detailed study of the effect of the length of the alkyl group attached to the C-terminus (Pulapura and Kohn, 1992a).

The main conclusion of this study was that a relatively small increase in the length of the C-terminus alkyl chain has a significant effect on the polymer properties. Thus, the differences between poly(N-Z-Tyr-Tyr-OEt iminocarbonate) and poly(N-Z-Tyr-Tyr-OHex iminocarbonate) were pronounced: Whereas poly(N-Z-Tyr-Tyr-OEt iminocarbonate) was a high melting and virtually insoluble polymer that could not be processed into shaped objects by any one of the conventional polymer processing techniques, poly(N-Z-Tyr-Tyr-OHex iminocarbonate) was readily soluble in many common organic solvents including methylene chloride, chloroform, and tetrahydrofuran. In addition, the melting range and glass transition temperature were significantly lowered by the replacement of the ethyl ester by a hexyl ester.

Any further increase in the length of the C-terminus alkyl ester chain, however, did not seem to alter significantly the melting range or solubility of the corresponding polymer. Thus, the solubility and melting range of poly(N-Z-Tyr-Tyr-OPal iminocarbonate) were not drastically changed in spite of a large increase in the length of the C-terminus-protecting group. This seemed to indicate that the C-terminus-protecting group must have a certain minimum length in order for the tyrosine polyiminocarbonate to become soluble in organic solvents, but that very long pendent chains are not necessary.

In a related series of experiments the amino group and/or the carboxylic acid group of tyrosine was replaced by hydrogen atoms. The corresponding tyrosine derivatives are 3-(4'-hydroxyphenyl)propionic acid, commonly known as desaminotyrosine (Dat), and tyramine (Tym) (Fig. 6). Based on these monomeric building blocks a series of four structurally related polyiminocarbonates were synthesized carrying either no pendent chains at all, an N-benzyloxycarbonyl group as pendent chain, a hexyl ester group as pendent chain, or both types of pendent chains simultaneously (Fig. 6) (Pulapura et al., 1990).

This series of polymers made it possible to investigate the contribution of each type of pendent chain separately. Here the most important conclu-

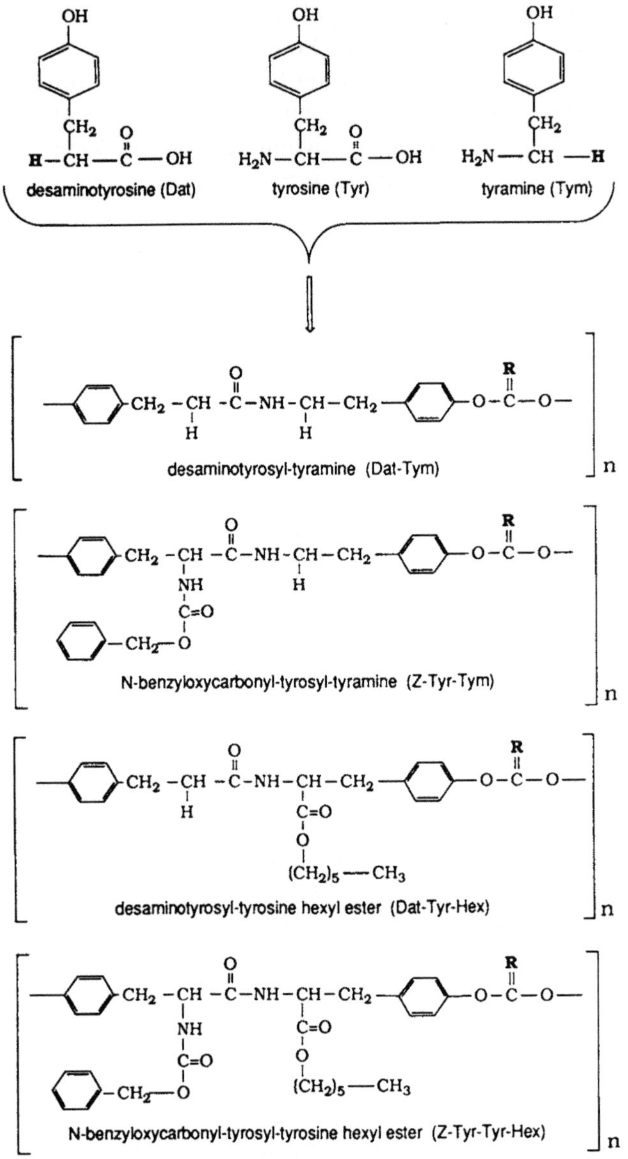

FIGURE 6   Using 3-(*p*-hydroxyphenyl)propionic acid (desaminotyrosine), L-tyrosine, and tyramine as starting materials, four structurally related peptides were prepared that carry various combinations of protecting groups. Next, the corresponding polyiminocarbonates (R = NH) and polycarbonates (R = O) were prepared and their physicomechanical properties were evaluated. In this way, a library of 8 related polymers was used to identify correlations between the structure of the pendent chains and the properties of the polymers. See text for further details. Reproduced, by permission, from Nathan and Kohn (1994).

sions were (i) that the presence of C-terminus pendent chains is required and (ii) that the potential for interchain hydrogen bonding has to be minimized to obtain readily processible polymers. Thus, there were significant differences between the effect of the *urethane-linked* N-terminus pendent chains (which are strong promoters of interchain hydrogen bonding) and the *ester-linked* C-terminus pendent chains (which are devoid of hydrogen bond donating moieties): All polymers carrying N-terminus pendent chains were inferior in ductility, processibility, and solubility to the polymers carrying only C-terminus pendent chains. Consequently, of the four test polymers shown in Fig. 6, poly(Dat-Tyr-Hex iminocarbonate) was identified as the most promising polymer for biomedical applications.

In view of the nonprocessibility of conventional polytyrosine, which cannot be used as an engineering plastic, the above results are highly significant: The development of poly(Dat-Tyr-Hex iminocarbonate) represents the first time a tyrosine-derived polymer with favorable engineering properties has been identified.

D. Polycarbonates Derived from Tyrosine

To confirm the general applicability of the structure–property correlations obtained for tyrosine-derived polyiminocarbonates, the corresponding tyrosine-derived polycarbonates were investigated as well. As shown in Fig. 5C, polycarbonates are obtained from the reaction of diphenols with phosgene or the less hazardous triphosgene (Pulapura and Kohn, 1992b).

Using desaminotyrosyl-tyrosine hexyl ester (Dat-Tyr-Hex or DTH) as a model monomer, the molecular weight of the resulting polymer was found to be strongly dependent on the reaction conditions and the amount of phosgenating agent used. With an exactly equivalent amount of phosgenating agent, only low-molecular-weight oligomers were obtained. Up to a certain maximum, the molecular weight increased with increasing molar excess of phosgenating agent. The optimum molar ratio of Dat-Tyr-Hex to phosgenating agent was 1:1.6 for triphosgene and 1:2 for phosgene. As expected, the polymerization was somewhat slower for triphosgene (maximum molecular weight after 60 min) than with phosgene (maximum molecular weight after 30 min). Longer reaction times or using a higher molar excess of phosgenating agent resulted in a gradual decrease in the polymer molecular weight. When using highly purified monomers (chemical purity > 99.5% by melting point depression) and under optimized reaction conditions, polymers with molecular weights of up to 400,000 (by GPC, relative to polystyrene standards) were obtained (Pulapura and Kohn, 1992b).

The four model structures shown in Fig. 6 were used also for the preparation of a series of polycarbonates. Structure–property correlations were obtained that were similar to those observed for the corresponding polyiminocarbonates. Among the four polycarbonates tested, poly(Dat-Tyr-Hex carbonate) was again the most promising material: This polymer was freely soluble in a variety of organic solvents, was readily processible by thermal processing techniques (extrusion, injection molding), and formed strong, highly ductile, transparent films. Unoriented samples had a tensile strength of about 340 kg cm^{-2}, a tensile modulus of about 13,000 kg cm^{-2}, and an elongation at break of close to 100%. X-ray diffraction did not show any ordered domains. Since the DSC thermograms showed only the glass transition and decomposition exotherm but no melting endotherm, poly(Dat-Tyr-Hex carbonate) appears to be completely amorphous.

The effect of the polymer backbone linkages on the polymer properties was explored by a detailed comparison of poly(Dat-Tyr-Hex iminocarbonate) and poly(Dat-Tyr-Hex carbonate). The replacement of the carbonyl oxygen by an NH group presents the only molecular difference between those two polymers. Poly(Dat-Tyr-Hex iminocarbonate) and poly(Dat-Tyr-Hex carbonate) are completely amorphous materials and solvent cast films are virtually indistinguishable in appearance. Both polymers exhibit high tensile strength of between 300 and 450 kg cm^{-2} with the polyiminocarbonate being slightly stronger than the polycarbonate. For comparison, for wet films of poly(γ-methyl-D-glutamate) a much lower tensile strength of only 50 to 100 kg cm^{-2} has been reported (Mohadger and Wilkes, 1976).

The most striking difference between poly(Dat-Tyr-Hex iminocarbonate) and poly(Dat-Tyr-Hex carbonate) was in their ductility; whereas poly(Dat-Tyr-Hex iminocarbonate) is a brittle material that failed at about 7% of elongation, poly(Dat-Tyr-Hex carbonate) is a tough material that failed only at about 100% elongation. Since those polymers are derived from identical monomers, this difference in ductility must be attributed to the relatively small change in backbone structure (Kohn, 1991).

In general, polycarbonates are more stable than polyiminocarbonates. This general trend is also evident in the tyrosine-derived polymers. When solvent cast films of poly(Dat-Tyr-Hex iminocarbonate) were exposed to aqueous buffer solutions, the films became turbid within a matter of days and swelled noticeably due to the absorption of water. The molecular weight decreased rapidly to about 1000 to 6000 (weight average) with a concomitant loss of mechanical strength. On the other hand, films of poly(Dat-Tyr-Hex carbonate) absorbed little water and remained intact when immersed in phosphate buffer solution at pH 7.4 (37°C). Overall, a 50% reduction of the initial molecular weight of poly(Dat-Tyr-Hex carbonate) was observed over 6 months. Blending poly(Dat-Tyr-Hex carbonate) with increasing amounts of poly(Dat-Tyr-Hex iminocarbonate) accelerated the degradation of the blends (Pulapura *et al.*, 1990).

An evaluation of the tissue compatibility of Dat-Tyr-Hex-derived polyiminocarbonate and polycarbonate in rats has recently been concluded (Silver *er al.* 1992). Those studies included medical grade poly(D,L-lactic acid) and medical grade polyethylene as controls. At times ranging from 7 days to 4 months postimplantation, the subcutaneous implantation sites were histologically evaluated. The general conclusion was that the tissue response elicited by the two tyrosine-derived polymers was not significantly different from the mild responses seen for medical grade polyethylene or medical grade poly(D,L-lactic acid).

Currently, poly(Dat-Tyr-Hex iminocarbonate) and poly(Dat-Tyr-Hex carbonate) are the only tyrosine-derived polymers known whose engineering properties appear to be suitable for implant materials. Since those polymers are also tissue compatible, their use is currently being explored in the formulation of drug delivery devices and in the design of orthopedic implants (Lin *et al.*, 1991).

VI. CONCLUSIONS

This chapter reviews the use of polymers derived from the naturally occurring α-L-amino acids within the context of biomaterials and medical applications. The authors have assumed that the reader is familiar with the basic chemical, conformational, and physicomechanical properties of conventional poly(amino acids). Due to the availability of several comprehensive prior reviews (Katchalski *et al.*, 1958; Lotan *et al.*, 1972; Katchalski, 1974; Anderson *et al.*, 1985; Fasman, 1987), developments that occurred after 1985 were emphasized.

Amino acid-derived polymers can be conveniently classified into three groups: conventional poly(amino acids), pseudo-poly(amino acids), and copolymers of amino acids and non-amino acids. Here poly(amino acids) are defined as synthetic polymers composed of α-amino acids linked by *peptide* bonds, whereas pseudo-poly(amino acids), are synthetic polymers composed of α-amino acids linked by *nonpeptide* bonds such as ester, carbonate, or urethane linkages. The third category comprises a large number of structurally diverse copolymers that contain both amino acid and non-amino acid units within the polymer backbone (Nathan and Kohn, 1994).

Poly(amino acids) are intensely investigated polymers that have captured the imagination of biologists, biochemists, polymer chemists, theoreticians, and material scientists for over 40 years. The initial and most significant application of those polymers was in biochemistry where they served as readily accessible models for natural polypeptides and proteins. Early attempts to identify large-scale commercial applications for poly(amino acids) were uniformly unsuccessful. Usually, the high cost of the *N*-carboxyanhydrides required as monomeric starting materials made poly(amino acids) economically unattractive. Thus, Katchalski and Sela wrote as early as 1958 that "the primary importance of poly(amino acids) has been, and will probably remain, as simple synthetic protein models" (Katchalski and Sela, 1958). It is therefore not surprising that the initial excitement and interest waned during the late 1970s when large-scale practical applications appeared unrealistic and poly(amino acids) were no longer needed as model compounds in studies relating to structure and function of proteins (Nathan and Kohn, in press).

More recently, the emphasis has been shifting from the conventional poly(amino acids) to *amino acid-derived polymers* in which amino acids are either linked by nonamide bonds, e.g., pseudo-poly(amino acids), or in which amino acids are copolymerized with a wide variety of non-amino acid components. These polymers are currently investigated in a large number of medical applications ranging from drug delivery systems, to wound dressings, to degradable bone fixation devices. These new, amino acid-derived materials become available at a time when the successful development of a large number of "high tech" applications requires a range of specialty polymers with carefully designed properties. Thus, one may predict that some of the amino acid-derived polymers described in this chapter will find important commercial applications in the future.

ACKNOWLEDGMENTS

J. Kohn acknowledges the support of NIH Research Career Development Award GM 00550. Some of the work reviewed in this chapter was supported by NIH grant GM 39455, by a research contract from Zimmer, Inc., and by a Focused Giving Award from Johnson & Johnson.

REFERENCES

Anderson, J. M., Spilizewski, K. L., and Hiltner, A. (1985) *Biocompatibility of Tissue Analogs* (D. F. Williams, Ed.), pp. 67–88. CRC Press, Boca Raton, FL.

Bennett, D. B., Adams, N. W., Li, X., Feijen, J., and Kim, S. W. (1988) *J. Bioact. Compat. Polym.* **3**, 44–52.

Block, H. (1983) *Poly(γ-benzyl L-glutamate) and Other Glutamic Acid Containing Polymers.* Gordon and Breach, New York.

Brack, A. (1989) *Preparation of α-Amino Acid Copolymer for Wound Dressings,* French Patent 2625507, assigned to S. A. Delalande, France.

Chandy, T., and Sharma, C. P. (1991) *Biomaterials* **12**, 677–682.

Cho, C. S., Kim, H. Y., and Akaike, T. (1990) *Pollimo* **14**(5), 570–573.

Cho, C. S., Park, J. W., Kwon, J. K., Jo, B. W., Lee, K. C., Kim, K. Y., and Sung, Y. K. (1991) *Pollimo* **15**(1), 27–33.

Chung, D., Higuchi, S., Maeda, M., and Inoue, S. (1986) *J. Amer. Chem. Soc.* **108**, 5823–5826.

Dorman, L. C., and Meyers, P. A. (1986) *Hard Tissue Prosthetics,* European Patent Application 192068, filed August 27, 1986, assigned to Dow Chemical Co., U.S.A.

Fasman, G. D. (1960) *Science* **131**, 420–421.

Fasman, G. D. (1987) *Biopolymers* **26**, S59–S79.

Gelbin, M. E., and Kohn, J. (1991) *Polym. Prepr.* **32**(2), 241–242.

Gelbin, M. E., and Kohn, J. (1992) *J. Amer. Chem. Soc.* **114**, 3962–3965.

Goodman, M., and Kirshenbaum, G. S. (1973) *Hydrolyzable Polymers of Amino Acids and Hydroxy Acids,* US Patent 3,773,737, assigned to Sutures, Inc.

Greenstein, J. P. (1937) *J. Biol. Chem.* 118(2), 321–329.

Hall, C. W., Spira, M., Gerow, F., Adams, L., Martin, E., and Hardy, S. B. (1970). *Trans. Amer. Soc. Artif. Intern. Org.* 16, 12–16.

Hayashi, T., and Iwatsuki, M. (1990). *Biopolymers* 29(3), 549–557.

Hayashi, T., Tabata, Y., and Nakajima, A. (1985) *Polym. J. (Tokyo)* 17(3), 463–471.

Hedayatullah, M. (1967) *Bull. Soc. Chim. (France),* 416–421.

Hench, L. L., and Ethridge, E. C. (1982) *Biomaterials—An Interfacial Approach: Biophysics and Bioengineering Series* (A. Noordergraaf, Ed.). Academic Press, New York.

Jarm, V., and Fles, D. (1977) *J. Polym. Sci. Polym. Chem. Ed.* 15, 1061–1071.

Kang, I., Ito, Y., Sisido, M., and Imanishi, Y. (1987) *Polym. J.* 19(12), 1329–1339.

Katchalski, E. (1974). In *Peptides, Polypeptides, and Proteins—Proceedings of the Rehovot Symposium on Poly(Amino Acids), Polypeptides, and Proteins and Their Biological Implications* (E. R. Blout, F. A. Bovey, M. Goodman, and N. Lotan, Eds.), pp. 1–13. Wiley, New York.

Katchalski, E., and Sela, M. (1958) In *Advances in Protein Chemistry* (C. B. Anfinsen, M. L. Anson, K. Bailey, and J. T. Edsall, Eds.), pp. 243–492. Academic Press, New York.

Kim, H., Sung, Y. K., Jung, J., Baik, H., Min, T. J., and Kim, Y. S. (1990) *Taehan Hwahakhoe Chi* 34(2), 203–210.

Kohn, J. (1990) In *Biodegradable Polymers in Drug Delivery. Systems* (M. Chasin and R. Langer, Ed.), pp. 195–229. Dekker, New York.

Kohn, J. (1991) In *Polymeric Drugs and Drug Delivery Systems* (R. L. Dunn and R. M. Ottenbrite, Eds.), pp. 155–169. American Chemical Society, Washington, DC.

Kohn, J., and Langer, R. (1984) *Polym. Mater. Sci. Eng.* 51, 119–121.

Kohn, J., and Langer, R. (1987) *J. Amer. Chem. Soc.* 109, 817–820.

Kugo, K., Hata, H., Hayashi, T., and Nakajima, A. (1982) *Polym. J.* 14(5), 401–410.

Kugo, K., Murashima, M., Hayashi, T., and Nakajima, A. (1983) *Polym. J.* 15(4), 267–277.

Kumaki, T., Sisido, M., and Imanishi, Y. (1985) *J. Biomed. Mater. Res.* 19, 785–811.

Kuroyanagi, Y., Kim, E., Kenmochi, M., Ui, K., Kageyama, H., Nakamura, M., Takeda, A., and Shioya, N. (1992) *J. Appl. Biomater.* 3, 153–161.

Lescure, F., Gurny, R., Doelker, E., Pelaprat, M. L., Bichon, D., and Anderson, J. M. (1989) *J. Biomed. Mater. Res.* 23, 1299–1313.

Lewis, D. H. (1990) In *Biodegradable Polymers as Drug Delivery Systems* (M. Chasin and R. Langer, Eds.), pp. 1–41. Dekker, New York.

Li, C., and Kohn, J. (1989) *Macromolecules* 22(5), 2029–2036.

Lin, S., Krebs, S., and Kohn, J. (1991) *Proceedings, 17th Annual Meeting of the Society of Biomaterials, Scottsdale Arizona,* pp. 187. Society for Biomaterials, Algonquin, IL.

Lotan, N., Berger, A., and Katchalski, E. (1972) In *Annual Review of Biochemistry* (E. E. Snell, P. D. Boyer, A. Meister, and R. L. Sinsheimer, Eds.), pp. 869–901. Annual Reviews, Palo Alto, CA.

Lyman, D. J. (1983) In *Polymers in Medicine. Biomedical and Pharmacological Applications* (E. Chiellini and P. Giusti, Eds.), pp. 215–218. Plenum, New York.

Marchant, R. E., Sugie, T., Hiltner, A., and Anderson, J. M. (1985) *ASTM Spec. Tech. Publ. (Corros. Degrad. Implant Mater.)* 859, 251–266.

Martin, E. C., May, P. D., and McMahon, W. A. (1971) *J. Biomed. Mater. Res.* 5, 53–62.

Minoura, N., Aiba, S., and Fujiwara, Y. (1986) *J. Appl. Polym. Sci.* 31, 1935–1942.

Miyamae, T., Mori, S., and Takeda, Y. *Poly-L-Glutamic Acid Surgical Sutures.* U.S. Patent 3,371,069, assigned to Ajinomoto Co., Inc.

Mohadger, Y., and Wilkes, G. L. (1976) *J. Polym. Sci. Polym. Phys. Ed.* 14, 963–980.

Nakajima, A., Hayashi, T., Kugo, K., and Shinoda, K. (1979a) *Macromolecules* 12(5), 840–843.

Nakajima, A., Kugo, K., and Hayashi, T. (1979b) *Polym. J.* 11(12), 995–1001.

Nathan, A., and Kohn, J. (1994) In *Designed -to-Degrade Biomedical Polymers* (S. Shalaby, Ed.), pp. 117–151. Hanser, New York.

Pramanick, D., and Ray, T. T. (1987) *Poym. Bull.* 18(4), 311–315.

Pulapura, S., and Kohn, J. (1990) *Proceedings, 17th International Symposium for the Controlled Release of Bioactive Materials, Reno, Nevada* (V. H. Lee, Ed.), pp. 154–155. Controlled Release Society, Lincolnshire, IL.

Pulapura, S. and Kohn, J. (1992a). In *Peptides—Chemistry and Biology: Proceedings of the 12th American Peptide Symposium* (J. A. Smith and J. E. Rivier, Eds.), pp. 539–541. Escom Science Publishers, Leiden, The Netherlands.

Pulapura, S., and Kohn, J. (1992b) *Biopolymers* 32, 411–417.

Pulapura, S., Li, C., and Kohn, J. (1990) *Biomaterials* 11, 666–678.

Pytela, J., Saudek, V., Drobnik, J., and Rypacek, F. (1989) *J. Control. Rel.* 10(1), 17–25.

Sakamoto, I., Unigame, T., and Takagi, K. (1986) *Hemostatic agent*, European Patent Application 172710, filed February 26, 1986, assigned to Unitika Ltd., Japan.

Sanders, H. J. *Chem. Eng. News* (1985) (April 1), 31–48.

Schminke, H. D., Grigat, E., and Putter, R. *Polyimidocarbonic Esters and Their Preparation.* U.S. Patent 3,491,060.

Sela, M. (1974) In *Peptides, Polypeptides, and Proteins—Proceedings of the Rehovot Symposium on Poly(Amino Acids), Polypeptides, and Proteins and Their Biological Implications* (E. R. Blout, F. A. Bovey, M. Goodman, and N. Lotan, Ed.), pp. 495–509. Wiley, New York.

Sela, M., Fuchs, S., and Arnon, R. (1962) *Biochem. J.* 85, 223–235.

Sela, M., Katchalski, E., and Olitzki, A. L. (1956) *Science* 123, 1129.

Seno, M., and Kuroyanagi, Y. (1986) *J. Membr. Sci.* 27, 241–252.

Shiotani, N., Kuroyanagi, T., Koganei, Y., and Miyata, T. (1987) *Bandage Containing a Mixture of Poly(amino acid) and Antimicrobial Agents*, Japanese Patent 62246370, assigned to Koken Co., Ltd., Japan.

Shiotani, N., Kuroyanagi, T., Koganei, Y., and Yoda, R. (1988) *Bandages Containing Amino Acid Polymer Films Impregnated with Biocompatible Materials*, Japanese Patent 63115564, assigned to Nippon Zeon Co., Ltd., Japan.

Shioya, N., Kuroyanagi, Y., Koganeo, Y., and Yoda, R. (1988) *Porous Layer Wound Dressing with Good Tissue Affinity*, European Patent Application 265906, filed 4 May 1988, assigned to Nippon Zeon Co., Ltd. Japan.

Sidman, K. R., Schwope, A. D., Steber, W. D., Rudolph, S. E., and Poulin, S. B. (1980) *J. Membr. Sci.* 7, 277–291.

Sidman, K. R., Steber, W. D., Schwope, A. D., and Schnaper, G. R. (1983) *Biopolymers* 22, 547–556.

Silver, F. H., Marks, M., Kato, Y. P., Li, C., Pulapura, S., and Kohn, J. (1992) *J. Long-Term Effects Med. Implants* 1(4), 329–346.

Sparer, R. V., and Bhaskar, R. K. (1985) *Variable Permeability Liquid Crystalline Membranes*, U.S. Patent 4,513,034, assigned to Merck and Co.

Spatola, A. F. (1983) In *Chemistry and Biochemistry of Amino Acids, Peptides, and Proteins* (B. Weinstein, Ed.), pp. 267–357. Dekker, New York.

Spira, M., Fissette, J., Hall, C. W., Hardy, S. B., and Gerow, F. J. (1969) *J. Biomed. Mater. Res.* 3, 213–234.

Walton, A. G. (1980) In *Biomedical Polymers. Polymeric Materials and Pharmaceuticals for Biomedical Use* (E. P. Goldberg and A. Nakajima, Eds.), pp. 53–83. Academic Press, New York.

Yokoyama, M., Anazawa, H., Takahashi, A., and Inoue, S. (1990a) *Makromol Chem.* 191, 301–311.

Yokoyama, M., Miyauchi, M., Yamada, N., Okano, T., Sakurai, Y., Kataoka, K., and Inoue, S. (1990b) *Cancer Res.* 50, 1693–1700.

Yokoyama, M., Inoue, S., Kataoka, K., Yui, N., Okano, T., and Sakurai, Y. (1989a) *Makromol. Chem.* **190**, 2041–2054.

Yokoyama, M., Okano, T., Sakurai, Y., Kataoka, K., and Inoue, S. (1989b) *Biochem. Biophys. Res. Commun.* **164**(3), 1234–1239.

Yolles, S., and Sartori, M. F. (1980) In *Drug Delivery Systems* (R. L. Juliano, Ed.), pp. 84–111. Oxford Univ. Press, New York.

Yu, H. (1988) *Pseudopoly(amino Acids): A study of the Synthesis and Characterization of Polyesters Made from α-L-Amino Acids.* Massachusetts Institute of Technology, Cambridge, MA.

Yu Kwon, H., and Langer, R. (1989) *Macromolecules* **22**, 3250–3255.

Zhou, Q.-X., and Kohn, J. (1990) *Macromolecules* **23**, 3399–3406.

10

The Nicotinic Acetylcholine Receptor as a Model for a Superfamily of Ligand-Gated Ion Channel Proteins

K. E. MCLANE*

S. J. M. DUNN†

A. A. MANFREDI*

B. M. CONTI-TRONCONI*

M. A. RAFTERY*

*Department of Biochemistry, University of Minnesota, St. Paul, Minnesota 55108, and Department of Pharmacology, University of Minnesota School of Medicine, Minneapolis, Minnesota 55455; and †Department of Pharmacology, Faculty of Medicine, University of Alberta, Edmonton, Alberta, Canada T6G 2HF

The acetylcholine receptors (AChRs) are the best characterized members of a large superfamily of ligand-gated ion channel proteins, which, in spite of their different ligand specificity and ion gating preferences, are structurally related and are likely to have evolved from a common ancestor. We will discuss how structural and functional information of the AChRs relates, by analogy, to the structure and function of other related ligand-gated ion channel proteins.

I. THE SIMILAR SEQUENCE AND MULTISUBUNIT STRUCTURE OF THE NICOTINIC ACETYLCHOLINE RECEPTOR AND OTHER LIGAND-GATED ION CHANNELS DEFINE A SUPERFAMILY OF PROTEINS

Several common features of receptors for neurotransmitters have emerged with the cloning of their genes and their sequencing: (i) an unexpected degree of homology between different subunits of a single neurotransmitter receptor, suggesting that they originated through duplications of a common ancestral gene; (ii) a degree of structural and functional heterogeneity of receptors for a given neurotransmitter, due to multiple subunit subtypes, greater than anticipated from the pharmacological profiles; (iii) receptors for different transmitters share common structural elements: this defines receptor superfamilies, comprising homologous classes of receptors. In the present review we discuss how these general features apply to the superfamily of ligand-gated channel proteins (ionotropic receptors), of which the AChR is the prototype. The ionotropic receptor family includes the subfamily of the type A γ-aminobutyric acid receptors (GABA$_A$), the serotonin (5-hydroxytryptamine, 5HT) subtype-3 receptor (5HT-3), and the glycine receptor (Betz, 1990a,b: Maricq, 1991). Distant members of the AChR superfamily are the excitatory glutamate/aspartate receptors (Sommer and Seeburg, 1992; Meguro *et al.*, 1992; Kutsuwada *et al.*, 1992)—the NMDA N-methyl-D-asparate (NMDA) receptors (Moriyoshi *et al.*, 1991; Monyer *et al.*, 1992), the non-NMDA receptors for kainate and α-amino-3-hydroxy-5-methyl-isoxazole-4-propionate (AMPA) (Gasic and Heinemann, 1991; Dingledine, 1991), and the insect muscle glutamate receptor (Schuster *et al.*, 1991).

Ionotropic receptors are ligand-gated ion channels that elicit rapid (submillisecond) conductance changes. The *Torpedo* electric organ AChR, which is the best characterized ion channel protein (Maelicke, 1988; Claudio, 1989; Stroud *et al.*, 1990; Betz, 1990a; Galzi *et al.*, 1991), provides a structural model for the other ligand-gated channels. By analogy with the known pentameric structure of the *Torpedo* AChR (discussed in sections I,D and I,E), the members of this superfamily are believed to be multimeric proteins composed of one to four different but homologous subunits. Thus, basic structural features identify ionotropic neurotransmitter receptors.

Excellent reviews have been published on the structure and function of the AChR and of other members of the ionotropic receptor superfamily (e.g., Maelicke, 1988; Claudio, 1989; Stroud *et al.*, 1990; Betz, 1990a,b; Maricq *et al.*, 1991). We will discuss more recent findings elucidating (i) how heterogeneity generated by multiple subtypes and subfamilies within the AChR family has originated functional diversity and (ii) how structural features of the AChR complex confer special functional attributes, some of which are conserved in other members of the AChR superfamily.

A. Primary Structure of the AChRs in Peripheral and Neuronal Tissues

1. AchRs from Peripheral Tissues (Skeletal Muscle, Electric Organ, Thymus)

Aminoterminal sequencing of *Torpedo california* (Hunkapiller *et al.*, 1979) and *Torpedo marmorata* (Devillers-Thiery *et al.*, 1979) AChR α subunits paved the way for elucidation of AChR structure and recognition of the highly homologous nature of the AChR subunits. Aminoterminal sequencing and subsequent cloning and sequencing of the genes encoding the four *Torpedo* AChR subunits (α, β, γ, and δ) demonstrated a sequence identity of 40–50% between them (Raftery *et al.*, 1980, Noda *et al.*, 1982, 1983b,c; Maelicke, 1988; Claudio, 1989; Stroud *et al.*, 1990). *Torpedo* and *Electrophorus* are distant species, which diverged ~400 million years ago and whose similar electric organs originated by convergent evolution. The AChRs from the electric organs of both these fish were found to be pentamers of homologous subunits sharing ~60% amino acid identity, in a stoichiometry $\alpha_2\beta\gamma\delta$ (Raftery *et al.*, 1980; Conti-Tronconi *et al.*, 1982a), first demonstrating that AChRs from different species are members of the same protein family, and that the subunit structure and primary sequence of peripheral AChRs is highly conserved through evolution. This was verified by isolation and sequencing of the α, β, γ, δ, and ϵ subunits of calf muscle AChRs (Conti-Tronconi *et al.*, 1982b; Noda *et al.*, 1983a, Tanabe *et al.*, 1984; Takai *et al.*, 1984; Kubo *et al.*, 1985; Takai *et al.*, 1985). AChR subunit sequences from human (Noda *et al.*, 1983a), calf (Noda *et al.*, 1983a), mouse (Boulter *et al.*, 1985), chicken (Barnard *et al.*, 1986, Nef *et al.*, 1986,1988), *Xenopus* (Baldwin *et al.*, 1988), and cobra (Neumann *et al.*, 1989) muscle are also conserved.

Mammalian muscle AChR exists in two developmentally regulated isoforms (Schuetze, 1986). Embryonic muscle expresses AChRs composed of α, β, γ, and δ subunits. Upon innervation, the γ subunit is substituted by a homologous ϵ subunit, to yield adult AChR, an $\alpha_2\beta\epsilon\delta$ oligomer (Mishina *et al.*, 1986; Gu and Hall, 1988). This change in subunit composition alters the pharmacological and metabolic properties of the AChR and the conductance characteristics of the channel (Trautman, 1982, Hall *et al.*, 1985; Schuetze, 1986, Mishina *et al.*, 1986; Gu and Hall, 1988; Mishina *et al.*, 1986; Sakmann, 1992).

The thymus contains a component(s) immunologically cross-reactive with muscle AChR (Aharonov *et al.*, 1975; Ueno *et al.*, 1980; Schleup *et al.*, 1987; Kirchner *et al.*, 1988) and binding sites for α-bungarotoxin (α-BGT), which specifically recognize AChR from peripheral tissues (Engel *et al.*, 1977; Kao and Drachman 1977; Kawanami *et al.*, 1988). The thymus α-BGT binding component has the subunit structure and physico-chemical properties expected for a true AChR, and cross-reacts with antisera raised against *Torpedo* AChR (Kawanami *et al.*, 1988). Subunit-specific antibodies demonstrated that the thymus AChR-like protein contains subunits immunologically related or identical to all the subunits forming the embryonic muscle AChR, i.e., α, β, γ, and δ subunits (Nelson and Conti-Tronconi, 1990).

2. AChRs of Autonomic Ganglia and Central Nervous System: Multiple Subtypes of α and β Subunits

Neuronal AChR subunits have been identified from mammalian (rodent) and nonmammalian (chicken, goldfish, *Drosophila*, locust) tissues

(Table I). Like the muscle-type α1 subunits, neuronal AChR α subunits contain a vicinal pair of cysteine residues, marking a potential cholinergic site within the N-terminal segment (see Section II). The amino acid sequences of different rodent AChR α and β subunits are 40–70% identical (Boulter *et al.*, 1990a). It is not surprising that these subunits are homologous to the muscle AChR sequences and to each other, since they were selected on the basis of homology. The number of different AChR subunits expressed in the brain, however, was not anticipated.

The subunits of neuronal AChRs are likely to share common structural features with the electric organ and muscle AChR subunits, due to their similar primary sequences. These include (i) a large extracellular N-terminal

TABLE I Neuronal AChR Subunits Identified by Low Stringency Hybridization

AChR subunit	Probe	Source	Reference
α3	Mouse muscle α1 subunit	PC12 cell line (rat pheochromocytoma)	Boulter *et al.*, 1986, 1987
α2, α4, α5, β2, β3	Rat α3 subunit	PC12 and rodent brain	Wada *et al.*, 1988; Goldman *et al.*, 1987; Boulter *et al.*, 1990a; Deneris *et al.*, 1988, 1989
β4, α6, β5[a]	Rat neuronal α and β subunits	Rodent brain	Duvoisin *et al.*, 1989; Deneris *et al.*, 1991
α2, α3, α4	Chicken muscle α1 subunit	Chicken brain and autonomic nervous system	Nef *et al.*, 1988; Barnard *et al.*, 1986; Ballivet *et al.*, 1988; Schoepfer *et al.*, 1988
α5, nα3[b]	Chick α3 gene cluster	Chicken brain	Courturier *et al.*, 1990a
nα1, nα2[b]	Chicken muscle α1 subunit	Chicken brain and autonomic nervous system	Ballivet *et al.*, 1988; Schoepfer *et al.*, 1988; Courturier *et al.*, 1990a; Deneris *et al.*, 1991
α7 (or αBGTBPα1) α8 (or αBGTBPα2)	N-terminus of an α-BGT binding, 48 kDa subunit	Chicken brain	Conti-Tronconi *et al.*, 1985; Schoepfer *et al.*, 1990; Courturier *et al.*, 1990b
GFα-3 GFnα-2, GFnα-3	*Torpedo* and rat muscle α1 and rodent α4 subunits	Common goldfish (*Carassius auratus*) brain	Cauley *et al.*, 1989, 1990
ARD (β homologue) ALS or Dα1 (α3 homologue)	Vertebrate AChR subunits	*Drosophila*	Schlosse *et al.*, 1988; Bossy *et al.*, 1988
SAD or Dα2 Dα3	ALS conserved M4 oligonucleotides	*Drosophila*	Sawruk *et al.*, 1990b; Gundelfinger, 1992
αL1 or ARL2 αRL1	Chick β2 subunit	*Schistocerca gregaria*	Marshall *et al.*, 1990; Hermsen *et al.*, 1991; Gundelfinger, 1992

[a] α6, β5 sequences and characterization have not been reported to date.
[b] nα means non α; also called structural αβ subunits.

domain containing two cysteine residues separated by ~15 amino acids, known as the Cys-Cys loop; (ii) four putative transmembrane regions, designated M1 to M4; (iii) conservation of a proline in the M1 segment; (iv) an abundance of serine, threonine, and small aliphatic amino acids in the M2 segment; and (v) a long non-conserved region between M3 and M4 that is at least partly cytoplasmic (see also Section I,C).

3. Functional Heterogeneity in AChR Subtypes Induced by Different Subunit Combinations

Functional diversity conferred by different combinations of subunits was first described for the AChRs of embryonic and adult mammalian muscle (Mishina *et al.*, 1986) (see Section IA,1). Neuronal AChRs differ from the muscle-type AChRs in that functional receptors may contain only one or two subunits, α and β (Goldman *et al.*, 1987; Wada *et al.*, 1988; Duvoisin *et al.*, 1989; Couturier *et al.*, 1990b; Bertrand *et al.*, 1992). Coexpression of rodent $\alpha2$, $\alpha3$, or $\alpha4$ subunits with either the $\beta2$ or the $\beta4$ subunit in *Xenopus* oocytes results in acetylcholine-gated cation channels with different properties: the different α subunits expressed in the central and autonomic nervous systems endow neurons of cholinergic pathways with multiple response states, potentially of functional significance (Papke *et al.*, 1989).

Different subunits of rodent neuronal AChRs confer different pharmacological characteristics to AChRs expressed in *Xenopus* oocytes. The α subunit subtype is important in determining the differential sensitivity of the resulting AChR complex to neurotoxins from invertebrates and snake venoms. Two classes of snake neurotoxins from *Bungarus multicinctus* and *Bungarus flavus* venum distinguish AChR subtypes, i.e., α-neurotoxins, such as α-BGT, and the κ-neurotoxins, κ-bungarotoxin (κ-BGT) and κ-flavitoxin (κ-FTX) (Chiappinelli, 1985; Grant *et al.*, 1988). κ-Bungarotoxin [also referred to as Toxin F (Loring *et al.*, 1984), bungarotoxin 3.1 (Ravdin and Berg, 1979), and neuronal bungarotoxin (Lindstrom *et al.*, 1987)] and α-BGT were initially regarded as specific antagonists of ganglionic AChR and muscle AChRs, respectively. Molecular genetic approaches, however, have revealed that this simple dichotomy does not hold.

α-Bungarotoxin is a potent inhibitor of the AChRs formed by the coexpression of the subunit combinations $\alpha1\beta1\gamma\delta$ and $\alpha1\beta2\gamma\delta$ (Deneris *et al.*, 1988). Many avian and rodent neuronal AChRs expressed as α/β subunit combinations—$\alpha2\beta2$, $\alpha3\beta2$, $\alpha4\beta2$, and $\alpha3\beta4$—are insensitive to α-BGT (Deneris *et al.*, 1988; Wada *et al.*, 1988; Duvoisin *et al.*, 1989). In contrast, neuronal AChRs formed by $\alpha3\beta2$ and $\alpha4\beta2$ are sensitive to κ-BGT (Deneris *et al.*, 1988). The sensitivity of the $\alpha3\beta2$ AChR to κ-BGT is 10-fold greater than the $\alpha4\beta2$ complex (Luetje *et al.*, 1990a,b). Interestingly, the $\alpha3\beta4$ complex is insensitive to κ-BGT (Duvoisin *et al.*, 1989), indicating that the β subunit is also able to affect the ligand binding characteristics. The $\alpha2\beta2$ AChR is insensitive to both α-BGT and κ-BGT (Wada *et al.*, 1988). The chicken α7 subunit can form a functional homomeric AChR sensitive to α-BGT (Couturier *et al.*, 1990b; Bertrand *et al.*, 1992). Neuronal α subunits from *Drosophila* (Sawruk *et al.*, 1990a,b) and locust (Marshall *et al.*, 1990) can also form homomeric AChRs. It remains to be determined whether other subunits contribute to physiologically relevant AChR complexes comprising α subunits able to form functional homomeric AChRs. The pharmacology of a homomeric locust AChR is worth noting, as it exemplifies the heterologous ligand-binding properties of the AChRs from species representing different levels of evolution: locust AChRs composed of only αL1 subunits are blocked by α-BGT, κ-BGT, bicuculline (a $GABA_A$ receptor ligand), and strychnine (a glycine receptor antagonist).

Luetje *et al.* (1990a,b) studied the effects on AChRs expressed in *Xenopus* oocytes of other neurotoxins known to interact with AChRs. Neosugarotoxin (NeSuTx), isolated from the Japanese ivory shell, *Babylonia japonica*, blocks receptors formed by the β2 subunit in combination with the α2, α3, or α4 subunit, whereas the α1β1γδ AChR is relatively insensitive. In contrast, α-conotoxin (α-CnTx) isolated from the venom of marine snails (Gray *et al.*, 1988) blocks only the α1β1γδ AChR. Lophotoxin, a cyclic diterpene from gorgonian corals (Culver *et al.*, 1985), covalently labels Tyr_{190} of the *Torpedo* α subunit (Abramson *et al.*, 1989), which is conserved in all rodent AChR subunits with the notable exception of the α5 subunit (Boulter *et al.*, 1990a). As predicted from the presence of this tyrosine, AChRs formed by combinations of α1β1δγ, α2β2, α3β3, and α4β2 subunits are sensitive to lophotoxin, although the α2β2 AChR is less sensitive for reasons that remain unclear. A functional AChR has not been successfully expressed in *Xenopus* oocytes using the α5 subunit, which is likely to form a lophotoxin-insensitive complex. Table II summarizes the contribution of different neuronal subunits to the toxin sensitivity of the resulting AChRs.

Sumikawa and Miledi (1989) initially demonstrated that the *Torpedo* β subunit affects the time course and extent of desensitization in a species-specific manner. Functional expression of neuronal AChR subunits in *Xenopus* oocytes indicated that both the α and the β (also called nα, non α) subunits influence agonist sensitivity and desensitization (Cachelin and Jaggi, 1991; Gross *et al.*, 1991; Luetje and Patrick, 1991; Papke and Heinemann, 1991). Coexpression of the chick subunit pairs α3/nα1 and α4/nα1 demonstrated that the α3 subunit lowers the sensitivity for acetylcholine and enhances desensitization. Symmetric hybrid subunits containing the N-terminal sequence of either the α3 or the α4 subunit indicated that the N-terminal region of the α subunit determines acetylcholine sensitivity. The β subunit affects both agonist sensitivity and AChR activation kinetics. The β2 subunit confers sensitivity to cytisine of the AChR formed with the α2, α3, or α4 subunits, and the β4 subunit forms cytisine-insensitive AChRs (Luetje and Patrick, 1991). The β subunit also confers different relative sensitivities

TABLE II Contribution of Different Subunits
to the Toxin Sensitivity of Neuronal AChRs

AChR	Neurotoxin sensitivity	Reference
$\alpha1\beta1\gamma\delta$; $\alpha1\beta2\gamma\delta$; $\alpha7$	α-BGT sensitive	Deneris et al., 1988; Couturier et al., 1990b; Bertrand et al., 1992
$\alpha2\beta2$; $\alpha3\beta4$	α-BGT insensitive; κ-BGT insensitive	Deneris et al., 1988; Wada et al., 1988; Duvoisin et al., 1989
$\alpha3\beta2$; $\alpha4\beta2$	α-BGT low sensitivity; κ-BGT sensitive	Deneris et al., 1988; Luetje et al., 1990a,b
$\alpha2\beta2$; $\alpha3\beta2$; $\alpha4\beta2$	NeSuTx sensitive; α-CnTx insensitive	Luetje et al., 1990a,b
$\alpha1\beta1\gamma\delta$	NeSuTx insensitive; α-CnTx sensitive	Luetje et al., 1990a,b
$\alpha1\beta1\gamma\delta$; $\alpha2\beta2$; $\alpha3\beta3$; $\alpha4\beta2$	Lophotoxin sensitive	Luetje et al., 1990a,b

to acetylcholine and 1,1-dimethyl-4-phenylpiperazinium (DMPP). Papke and Heinemann (1991) have shown that AChR complexes expressed in *Xenopus* oocytes composed of $\alpha3\beta2$ or $\alpha3\beta4$ subunits have different conductance, open times, and burst kinetics.

Chick sympathetic neurons form four different classes of functional AChRs and express six subunits ($\alpha3$, $\alpha4$, $\alpha5$, $\alpha7$, $\beta2$, and $\beta4$). Listerud et al. (1991) used antisense oligonucleotides to delete selectively individual different subunits and determine their functional contribution to cholinergic function. Antisense oligonucleotides against the $\alpha3$ subunit decreased the number of channel openings of all four classes of AChRs and resulted in the predominance of a new class of channels composed of the $\alpha7$ subunit that were sensitive to α-BGT. The α-BGT insensitivity of untreated neurons suggests that the $\alpha7$ subunit forms AChR complexes involving other α subunit subtypes, which alter the α-BGT binding properties of the $\alpha7$ subunit.

Expression of mRNAs for the $\alpha2$, $\alpha3$, $\alpha4$, $\alpha5$, and $\beta2$ subunits has been mapped in rat brain by *in situ* hybridization (Wada et al., 1988, 1989; Boulter et al., 1986; Goldman et al., 1986, 1987, Deneris et al., 1988). The $\alpha4$ and $\beta2$ subunits are the most highly and extensively expressed in mammalian brain (Goldman et al., 1987; Deneris et al., 1988; Wada et al., 1988, 1989). Other neuronal AChR subunits exhibit less diffuse expression, and may therefore be related to specific functions, although these remain to be determined (Wada et al., 1988, 1989; Duvoisin et al., 1989; Boulter et al., 1990a). The most prominent regions of hybridization of the different α and β probes are listed in Table III.

In summary, functional diversity of neuronal AChRs results from different combinations of subunits. This diversity is reflected in differences in channel conductance and open time, sensitivity to neurotoxins, and binding

TABLE III *In Situ* Hibridization of AChR Subunit Probes in the Nervous System[a]

CNS region	Probes
Substantia nigra	$\alpha3$, $\alpha4$, $\alpha5$, $\beta2$, and $\beta3$
Medial habenula	$\alpha3$, $\alpha4$, $\beta3$, and $\beta4$
thalamus	$\alpha3$, $\alpha4$, $\alpha5$, and $\beta2$
Trigeminal ganglia	$\alpha3$, $\alpha5$, $\beta2$, and $\beta3$
Interpeduncolar nucleus	$\alpha2$, $\alpha3$, and $\beta2$
Hippocampus	$\alpha2$, $\alpha3$, $\alpha4$, $\alpha5$, and $\beta2$
Cerebral cortex	$\alpha3$, $\alpha4$, $\alpha5$, and $\beta2$
Hypothalamus	$\alpha3$, $\alpha4$, and $\beta2$
Adrenal medulla	$\alpha3$, $\alpha4$, $\alpha5$, $\alpha7$, $\beta2$, and $\beta4$

[a] From: Boulter *et al.*, 1986; Deneris *et al.*, 1988; Duvoisin *et al.*, 1989; Goldman *et al.*, 1986, 1987; Wada *et al.*, 1988, 1989.

properties for agonists and antagonists. The regional expression of different neuronal AChR subtypes indicates that these different functional properties may be of physiological importance.

B. Sequence Homology between the Other Members of the Ligand-Gated Ion Channel Superfamily

Comparison of the deduced amino acid sequences of cDNA clones for subunits of the $GABA_A$, glycine, and 5HT-3 receptors indicates that they belong to the same protein superfamily as the AChR. Based on more limited sequence homology and predicted transmembrane topology, certain structural features identify other more distant members of this protein superfamily, including the insect neuromuscular glutamate receptor and several subunits of the mammalian glutamate receptor family. Conservation of sequence and structure of the proteins forming this heterogeneous class of neurotransmitter receptors, as described below, suggests that they evolved from a common ancestor gene.

Functional similarities between the receptor complexes formed by these related proteins suggest conservation of important structural elements: They are all ionotropic receptors, containing ion channels that undergo rapid conductance changes upon agonist binding. The $GABA_A$ and glycine receptors in the brain and spinal cord mediate a rapid increase in inhibitory chloride conductance (Stevens, 1987; Betz, 1990a,b, 1991). The 5HT-3 receptor and the kainate-AMPA-sensitive glutamate receptors mediate excitatory monovalent cation conductance, similar to the AChR, in the peripheral and central nervous system (Yakel *et al.*, 1990; Monaghan *et al.*, 1989).

The NMDA receptor ion channel is permeable to Ca^{2+}, Na^+, and K^+ and is blocked by Mg^{2+} in a voltage-dependent manner (Monaghan *et al.*, 1989).

Analysis of conserved and divergent structural features between different receptors of the AChR superfamily suggests important structure–function relationships. Regions of homology might identify common structural requirements for ligand-gating and ion channel formation, whereas diversity might reflect sites mediating specific ligand and ion selection interactions. We will review here the progress made in determining the primary sequence of the related members of the AChR superfamily, and identifying conserved structural features. We also discuss how the heterogeneity of these related receptors confers functional diversity to the receptor subclasses of the AChR superfamily.

1. Conserved Structural Features of the AChR, GABAA, Glycine, and 5HT-3 Receptors

Several structural features are conserved by the $GABA_A$, glycine, and 5HT-3 receptor subunits (Stevens, 1987; Stroud *et al.*, 1990; Betz, 1990a,b) (see also Fig. 1): (i) a long N-terminal putative extracellular segment; (ii) four putative transmembrane domains (M1 to M4), which are the most highly conserved sequence regions between subfamilies: the region of highest homology is the M2 segment, which is believed to line the ion channel (discussed in Section V), and is rich in serine, threonine, and small aliphatic amino acid residues: (iii) a long sequence segment between the M3 and the M4 domains, which is the region of highest diversity between the members of the AChR superfamily, and may be partially or completely involved in formation of a cytoplasmic domain; (iv) a Cys-Cys loop in their N-terminal, putative extracellular domain; and (v) a conserved proline residue in the M1 segment, which may be involved in ion channel function.

Conspicuously lacking from $GABA_A$ and glycine receptor subunit sequences is a vicinal cysteine pair, the hallmark of the AChR α subunit. In addition, positively charged residues are clustered around the putative transmembrane domains in the $GABA_A$ and glycine subunit sequences, whereas these residues are negatively charged in the AChR subunits. This structural difference may account for the difference in their ion selectivity.

2. Heterogeneity of the GABAA, Glycine, and 5HT-3 Receptors

a. The GABA$_A$ receptors. The $GABA_A$ receptor was initially purified from the bovine brain, and appeared to be composed of two types of subunits—α (53 kDa) and β (57 kDa) (Mamalaki *et al.*, 1987). The β subunit contains the binding site for GABA, and the α subunit the binding site for benzodiazepines, which potentiate the action of GABA (Casalotti *et al.*, 1986; Mamalaki *et al.*, 1987). Later, cDNAs for several other $GABA_A$ receptor subunits were isolated, indicating that the $GABA_A$ receptor complex

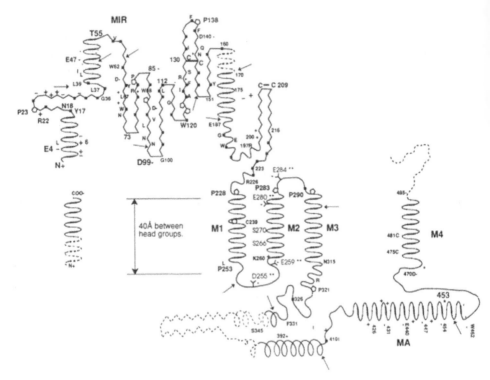

FIGURE 1   Consensus model of a possible topology of a peripheral AChR subunit and secondary structure predicted by amphipathic analysis. The four hydrophobic, putative membrane-spanning α helices are labeled M1–M4. Letters identify the usual residue of highly conserved amino acids. Square blocks identify conservation of hydrophobic on non-transmembrane domains. Residues in M2 whose mutation causes change in conductance are indicated in narrow letters with (**) on those that are charged. Positive and negative signs identify regions that generally carry charged side chains. The conserved cysteines, known to form a disulfide-linked loop in all four neuromuscular AChR subunits (indicated as C130–C144, according to the numbering of the consensus alignement, corresponding to Cys$_{128}$ and Cys$_{142}$ of the *Torpedo* α subunit), enclose the conserved site of N-linked glycosylation site, found in all neuromuscular AChR subunits and the GABA$_A$ receptor β subunits. The adjacent, disulfide-linked cysteines at the agonist binding site, found only in AChR α subunits, are labeled C208 and 209, following the numbering of the consensus alignement (corresponding to Cys$_{192}$ and Cys$_{193}$ of the *Torpedo* α subunit). The main immunogenic region (MIR) is located on the N-terminal putative extracellular domain. Dashed lines indicate sequence regions nonconserved between species, and arrows indicate common intron boundaries. Reproduced, by permission, from Stroud *et al.* (1990).

includes subunits not identified by protein purification, and that extensive subtype heterogeneity exists for this receptor. These GABA$_A$ subunits have been designated α1 to α6, β1 to β3, γ1 to γ3, δ and ρ (Schofield *et al.*, 1987, Levitan *et al.*, 1988a, Ymer *et al.*, 1989; Lolait *et al.*, 1989; Pritchett *et al.*, 1989a; Shivers *et al.*, 1989; Pritchett and Seeburg, 1990; Luddens *et al.*,

1990; Herb *et al.*, 1992a). The α1 and β1 subunits, coexpressed in *Xenopus* oocytes, form GABA-sensitive chloride channels, inhibited by bicuculline and picrotoxin and potentiated by barbiturates—all responses characteristic of the native GABA$_A$ receptor (Schofield *et al.*, 1987; Levitan *et al.*, 1988 a,b). Later, it was shown that the γ subunit is required for a robust response and modulation by benzodiazepines (Pritchett *et al.*, 1989b; Sigel *et al.*, 1990). Thus, the minimal GABA$_A$ receptor complex is α$_x$β$_x$γ$_x$, where x could be any subunit subtype, in an unknown stoichiometry. Expression studies demonstrated that the different GABA$_A$ α subunits confer different GABA sensitivity (Levitan *et al.*, 1988a,b) and distinct pharmacological profiles—i.e., the α1 subunit yields a receptor complex characteristic of a type I benzodiazepine receptor with high affinity for triazolopyridines and β-carbolines, whereas the α2, α3, and α5 subunits form a receptor complex that behaves as a type II receptor, with lesser affinity for these compounds (Pritchett *et al.*, 1989b; Luddens and Wisden, 1991). Novel classes of GABA$_A$ receptor complexes composed of the α4 and α6 subunits lack benzodiazepine sensitivity and/or have lower affinity for benzodiazepines (Luddens *et al.*, 1990).

Given the number of subunit variants that are expressed in the mammalian brain, more than 500 different GABA$_A$ receptor subtypes could be formed by different combinations of subunits (Luddens and Wisden, 1991). Further, more than one α subunit type can be immunoprecipitated from a single, homogeneous population of GABA$_A$ receptors (Luddens *et al.*, 1991), and multiple α, β, and γ subunit variants can be coexpressed in the same cell population (Wisden *et al.*, 1992, Laurie *et al.*, 1992). This level of heterogeneity far exceeds that previously anticipated from pharmacological profiles.

b. The glycine receptors.   The glycine receptor purified from porcine spinal cord appeared to be composed of three subunits (M_r ~48 kDa, ~50 kDa, and ~93 kDa, respectively). The 48-kDa subunit is photoaffinity-labeled with [^{3}H]-strychnine (Graham *et al.*, 1985). Several α subunits have been sequenced from mammalian cDNA and genomic libraries, designated α1, α2, α2*, α3, and α4 (Grenningloh *et al.*, 1987, 1990; Kuhse *et al.*, 1990a,b, 1991; Betz, 1991). The overall homology with the AChR subunits is relatively low (~20%). Expression in *Xenopus* oocytes of the α1, α2, or α3 subunits yields functional homomeric receptors sensitive to strychnine at nanomolar concentrations (Schmieden *et al.*, 1989; Sontheimer *et al.*, 1989; Grenningloh *et al.*, 1990; Kuhse *et al.*, 1990b, 1991). The α2* variant forms a homomeric channel with a lower affinity for strychnine (Kuhse *et al.*, 1990a).

A second subunit of the glycine receptor, designated β, has been identified (Grenningloh *et al.*, 1990). Coexpression of α and β subunits in *Xenopus* oocytes elicits a large glycine-gated response (Grenningloh *et al.*, 1990).

The β subunit mRNA is expressed in many regions of the brain, as demonstrated by *in situ* hybridization, including regions where none of the identified glycine receptor α subunits are expressed, and sites where glycine receptors have not been previously identified by ligand binding studies (Malosio *et al.*, 1991). Therefore, it remains unclear if the β subunit represents a subunit unique to the glycine receptor, or if it also participates in formation of other receptor complexes (Malosio *et al.*, 1991).

c. The 5HT-3 receptor.   The 5HT-3 receptor is the only ionotropic serotonin receptor. The other identified 5HT receptors (5HT-1, 5HT-2, and 5HT-4) are coupled to G-proteins and mediate slow conductance changes via second-messenger pathways (Peroutka, 1988; Julius, 1991). One pharmacological property of the 5HT-3 receptor that suggested that it might belong the the AChR superfamily was its sensitivity to the nicotinic antagonist *d*-tubocurare (Neijt *et al.*, 1988). The deduced amino acid sequence for the 5HT-3 receptor α subunit exhibits ~20% identity with the *Torpedo* AChR α, the GABA$_A$ β1, and the glycine α1 subunits (Maricq *et al.*, 1991). The 5HT-3 α subunit has been expressed in *Xenopus* oocytes as a homomeric complex that exhibits a divalent cation-mediated conductance, similar to the NMDA-sensitive glutamate receptor (Maricq *et al.*, 1991). However, the ion selectivity of the native 5HT-3 receptor is more similar to the AChR receptor, i.e., for monovalent cations (Yakel *et al.*, 1990). This discrepancy may indicate that other unidentified subunits participate in the formation of the native 5HT-3 receptor complex.

3. The Glutamate Receptors

a. The AMPA/kainate subfamily.   Another structurally related ligand-gated receptor family is formed by the glutamate receptors (Gasic and Heinemann, 1991; Dingeldine *et al.*, 1991). Several different kainate/AMPA receptor subunits have been identified and expressed in *Xenopus* oocytes. Based on sequence homology and pharmacological profiles, the kainate/AMPA receptor subunits can be classified into α, β, and γ subfamilies (Sakimura *et al.*, 1992), as summarized in Table IV. Expression in *Xenopus* oocytes of single subunits of the α subfamily, GluR-1 to GluR-4, yields functional homomeric complexes with agonist selectivity in the rank order quisqualate>AMPA>glutamate>kainate (Hollmann *et al.*, 1989; Keinanen *et al.*, 1990; Boulter *et al.*, 1990b; Nakanishi *et al.*, 1990; Dawson *et al.*, 1990; Sakimura *et al.*, 1992). Homomeric channels are also formed by a member of the β subfamily, GluR-6 (β2) with rank order of ligand selectivity kainate>quisqualate>glutamate (Egebjerg *et al.*, 1991). The subunits of the kainate-selective γ subfamily can bind agonists as homomeric complexes, but require other subunits for channel formation (Sommer *et al.*, 1991; Herb *et al.*, 1992b; Sakimura *et al.*, 1990), and at least two distinct subunits are required to form complexes with current–voltage relationships similar to kainate receptors found on neurons (Nakanishi *et al.*, 1990).

TABLE IV   Kainate/AMPA Receptor Subunit Subfamilies

Subunit	Subfamily	Pharmacological properties	Reference
GluR-1 (α1) (or GluR-A); GluR-2 (α2) (or GluR-B); GluR-3 (or GluR-C); GluR-4 (or GluR-D)	α	High affinity for AMPA	Hollmann et al., 1989; Keinanen et al., 1990; Boulter et al., 1990b; Nakanishi et al., 1990; Sakimura et al., 1990
GluR-5; GluR-6 (β2)	β	High affinity for domoate and moderate affinity for kainate	Bettler et al., 1990; Egebjerg et al., 1991; Werner et al., 1991; Sommer et al., 1992
KA-1; KA-2; γ2	γ	High affinity for kainate	Werner et al., 1991; Herb et al., 1992b; Sakimura et al., 1992

Although the deduced amino acid sequences of the kainate/AMPA-sensitive glutamate receptor subunits share little homology with AChR, glycine, and GABA$_A$ subunits, a number of structural characteristics are shared among these ionotropic receptors. Hydropathy analysis of the kainate/AMPA-sensitive glutamate receptor subunits indicates a large N-terminal putative extracellular domain, at least four potential transmembrane domains (M1 to M4), a large putative cytoplasmic region between M3 and M4, and conservation of a proline residue in the M1 segment. Although the kainate/AMPA glutamate receptor subunits lack the Cys-Cys loop conserved between the AChR, 5HT-3, glucine, and GABA$_A$ receptor subunits, sequence comparisons of this region indicate conservation of several residues (Hollmann et al., 1989). The N-terminal region of the glutamate receptor subunits is highly variable, whereas the cytoplasmic loop between M3 and M4 is relatively conserved. This is the reverse of the trend observed for the AChRs, 5HT-3 glycine, and GABA$_A$ receptor subunits.

b. The NMDA receptor subfamily.   The NMDA receptors play a key role in long-term potentiation, an activity-dependent enhancement of synaptic efficacy that may be related to memory and learning (Monaghan et al., 1989). The NMDA receptors expressed in neurons contain several distinct binding sites for glutamate, Zn^+, and phencyclidine, and a voltage-dependent site for Mg^{2+}. Several subunits for NMDA-sensitive receptors have been identified (Moriyoshi et al., 1991; Monyer et al., 1992; Meguro et al., 1992; Kutsuwada et al., 1992). The NMDA receptor subunits have been assigned to the ζ and ε subfamilies (Meguro et al., 1992; Kutsuwada et al., 1992). The ζ subfamily includes the NM1 (ζ1) subunit (Moriyoshi et al., 1991). The present members of the ε subfamily are ε1 (NM2a), ε2 (NM2b), and ε3 (NM2c) subunits (Meguro *et al.*, 1992; Monyer *et al.*, 1992; Kutsuwada *et al.*, 1992). The NMDA-type subunits share only 11–21% amino acid sequence identity with the AMPA/kainate receptor subunits (the α, β, and γ glutamate receptor subfamilies), but contain several structural motifs in common with other members of the AChR superfamily: (i) a large N-terminal extracellular domain, (ii) four predicted transmembrane-spanning segments, and (iii) a cluster of negatively charged residues flanking the M2 segment. Within the M2 segment a conserved Asn residue, common to all NMDA subunits and to the AMPA/kainate receptor α2 subunit, is believed to be involved in the control of Ca^{2+} permeability (Burnashev *et al.*, 1992; Kutsuwada *et al.*, 1992). A stretch of glutamic acid residues preceding M2 is peculiar to the NM1 (ζ1) subunit, and may be involved in Mg^{2+} voltage-dependent gating. Although homomeric complexes of the NM1 (ζ1) subunit expressed in *Xenopus* oocytes exhibit NMDA-activated currents that are enhanced by glycine and inhibited by Zn^+, coexpression of the NM1 (ζ1) and one of the ε subunits results in receptor complexes that are capable of larger and more physiologically comparable NMDA-induced currents (Moriyoshi *et al.*, 1991; Monyer *et al.*, 1992). The different NMDA receptor of ε subunits do not form glutamate-sensitive ion channels when expressed alone in *Xenopus* oocytes (Meguro *et al.*, 1992; Kutsuwada *et al.*, 1992). This suggests that the NM1 (ζ1) subunit forms the ligand-binding sites for agonists, and that the ε subunits are "structural" subunits. The ε2/ζ1 and ε3/ζ1 receptor complexes exhibit higher affinities for glutamate and glycine than the ε1/ζ1 subunit, indicating that different pharmacological properties can be conferred by the ε subunits.

c. The insect muscle glutamate receptors.   Insects and other invertebrates use glutamate as neurotransmitter at the neuromuscular junction. The deduced amino acid sequences of the *Drosophila* neuromuscular glutamate receptor subunits, designated DGluR-I and DGluR-II, share only ~27% identity with the identified rat glutamate receptor subunits (Schuster *et al.*, 1991). Conservation of structural elements, however, indicated by hydropathy analysis predicts at least four potential transmembrane domains. The DGluR-II subunit expressed as a functional homomeric complex is sensitive to glutamate and aspartate, and less sensitive to quisqualate, AMPA, and kainate.

C. Transmembrane Topology of the AChR Subunits

All *Torpedo* AChR subunits form both extracellular and cytoplasmic domains (Strader *et al.*, 1979). Due to their strong sequence similarity, all AChR subunits should have similar transmembrane folding. Hydropathy analysis of a "typical" AChR subunit (Claudio, 1989; Stroud *et al.*, 1990) identifies a long N-terminal region of ~200 amino acids rich in hydrophilic residues that could form an extracellular domain. This is followed by four hydrophobic potentially α-helical segments ~20 amino acid long (see Fig. 1), referred to as M1 to M4, which could form membrane-spanning regions (Finer-Moore and Stroud, 1984; Guy, 1983). Between M3 and M4 there is a long sequence region, most diverged in the different AChR subunits, containing a segment, called MA, that has the periodicity of an amphipathic α helix (Finer-Moore and Stroud, 1984). The M4 segment is followed by a short carboxyl terminal region.

Different models of the transmembrane folding of the AChR subunits have been proposed, with four (M1 to M4) or five (M1 to M4 and MA) transmembrane segments. In both models the sequence region preceding M1 is extracellular, whereas the COOH terminus is extracellular in the four transmembrane domain model and cytoplasmic in the five transmembrane domain model. We will summarize here the experimental evidence supporting each model. For more detailed reviews on these matters, see Maelicke, 1988; Claudio, 1989; Stroud *et al.*, 1990; Betz, 1990a,b; Galzi *et al.*, 1991.

1. Region between the Aminoterminus and M1

The *Torpedo* AChR δ subunit contains a processed signal peptide, indicating that the mature amino terminus is extracellular (Anderson *et al.*, 1982). The aminotermini of AChR subunits expressed *in vitro* are translocated into the lumen of microsomal vesicles (topologically equivalent to the extracellular space), as are the amino termini of native AChR subunits (Anderson *et al.*, 1983, Chavez and Hall, 1991).

At least part of the sequence region between the amino terminus and the putative transmembrane segment M1 is extracellular, because it contains (i) an N-glycosylation site(s) (Asn_{141} of the muscle-like AChR subunits, Asn_{30}, and sometimes Asn_{141} of neuronal AChR subunits) (Claudio, 1989), (ii) residues involved in formation of a cholinergic ligand binding site (see Section II), and (iii) a sequence loop (within residues 67–76) that is an important constituent element of the main immunogenic region (MIR) (Bellone *et al.*, 1989)—an extracellular area of the AChR that dominates the autoantibody response in the human disease Myasthenia Gravis (Lindstrom *et al.*, 1988).

The transmembrane topology of the N-terminus of the α subunit was investigated by introducing novel glycosylation sites and expressing fragments of the α subunit sequence. A fragment terminating at position α_{207} (just before the M1 segment) was a nonintegral membrane protein, and glycosylation sites introduced at position α_{154} and α_{200} were found on the lumenal side of microsomal vesicles (equivalent topologically to the extracellular space). This suggests that the entire N-terminal domain preceding M1 is extracellular (Chavez and Hall, 1991), a conclusion supported by studies on the tridimensional structure of *Torpedo* AChR employing low-dose electron microscopy and X-ray diffraction, which concluded that the volume of protein protruding toward the extracellular space is 215,000 cubic Å (157 kDa)—the predicted molecular mass formed by the N-terminal region, up to M1, of all five glycosylated subunits (Noda *et al.*, 1983c; Finer-Moore and Stroud, 1984) plus the oligosaccharide moieties (Poulter *et al.*, 1989).

2. Potential Transmembrane Segments (M1 to M4 and MA)

The hydrophobic domain M2 is rich in uncharged hydrophilic residues and may contribute to the lining of the ion channel (Miller, 1989; Dani, 1989; discussed in detail in Section V) because in *Torpedo* AChR mutations of this segment subunits alter the ion conductance properties of AChRs expressed in *Xenopus* oocytes (Imoto *et al.*, 1988; Leonard *et al.*, 1988) and channel blockers label residues within this segment (Hucho, 1986; Giraudet *et al.*, 1989). Synthetic peptides corresponding to the M2 sequence form cation channels in lipid bilayers (Oiki *et al.*, 1988). The M1 segment may also be involved in formation of the channel, as suggested from labeling experiments with noncompetitive blockers (Karlin *et al.*, 1986). A highly conserved proline residue in the middle of the M1 segment of all AChR subunits might be important in conferring structural flexibility and facilitate ion channel gating (Dani, 1989). A transmembrane disposition of M4 is suggested by the selective labeling of Lys residues of AChR in sealed vesicles, in the presence and in the absence of saponin (Dwyer, 1991): $Lys_{\alpha 380}$, which is N-terminal to M4, has cytoplasmic location, whereas and $Lys_{\gamma 486}$, which is C-terminal to M4, is extracellular.

The potential amphipathic α-helical MA segment (Finer-Moore and Stroud, 1984; Guy, 1983; Stroud *et al.*, 1990) was proposed to contribute to the lining of the ion channel. A synthetic peptide corresponding to the MA sequence of the *Torpedo* AChR β subunit forms ion channels in artificial phospholipid bilayers (Ghosh and Stroud, 1991). However, this segment can be deleted from the *Torpedo* sequence without affecting the formation of the ion channel by subunits expressed in *Xenopus* oocytes (Mishina *et al.*, 1985), and antibodies to epitopes within the MA sequence regions bind to the cytoplasmic surface of membrane bound AChR (Ratnam *et al.*, 1986b, Maelicke *et al.*, 1989). That MA may not have a transmembrane disposition is also suggested by the disappearance of antibody epitopes in this region upon trypsin treatment of native AChR (Roth *et al.*, 1987), and by the results of experiments where the transmembrane disposition of the different putative transmembrane regions of the AChR, including MA, was deduced using proteolysis protection assays of fusion proteins containing a reporter group fused after the nucleic acid sequence encoding each putative transmembrane domain (Chavez and Hall, 1992).

3. Putative Cytoplasmic Domain Between M3 and M4

The region between M3 and M4 is highly divergent between different AChR subunits (Claudio, 1989) and is particularly long in the α4 subunit

(Goldman *et al.*, 1987). This segment may be involved in the differential regulation of AChR complexes. The carboxyl terminal part of this sequence region corresponds to the proposed MA amphipathic helix (see Section I,C,2). Several studies using sequence-specific antibodies indicate that the N-terminal part of the region between M3 and M4 is cytoplasmic (Ratnam *et al.*, 1986a,b; Young *et al.*, 1985; Kordossi and Tzartos, 1987; LaRochelle *et al.*, 1985; Lei *et al.*, 1993). Expression studies using fusion proteins containing a reporter group after each putative transmembrane segment M1–M4 and MA, also indicated that the region between M3 and M4 is cytoplasmic (Chavez and Hall, 1992).

In conflict with those results, a study on the sequence of AChR fragments released upon brief proteolytic treatment of sealed AChR rich membrane vesicles concluded that the sequence regions α341–380, β351–385, γ353–414, and δ328–341, which were quickly released by trypsin treatment, are exposed on the extracellular surface (Moore *et al.*, 1989).

4. Carboxyl Terminus

The transmembrane topology of the COOH terminus of the AChR subunits has been investigated by immunological, biochemical, and genetic approaches. Studies using monoclonal antibodies to the carboxyl terminus of different AChR subunits consistently suggested a cytoplasmic location of the carboxyl terminus (Ratnam and Lindstrom, 1984; Lindstrom *et al.*, 1984; Young *et al.*, 1985). However, two important observations indicate that the carboxyl terminus is extracellular: (i) *Torpedo* AChR exist as dimers held together by a disulfide bridge occurring between the second last residue of the δ subunit of each monomer (DiPaola *et al.*, 1988), which can be reduced from the extracellular surface (Dunn *et al.*, 1986; McCrea *et al.*, 1987; DiPaola *et al.*, 1988, 1989); (ii) residue Lys_{486} of the γ subunit, which is on the carboxyl terminal side of the transmembrane domain M4, can be labeled in closed vesicles (Dwyer, 1991); (iii) proteolysis protection studies of fusion proteins containing the carboxyl terminus of the mammalian muscle α and δ subunits concluded that the carboxyl termini of these subunits are extracellular when expressed in mammalian cells (Chavez and Hall, 1992).

D. Pentameric Quaternary Structure of the AChRs and Other Members of the Ionotropic Receptor Superfamily

The muscle-type AChR of *Torpedo* electric organ is composed of four subunits—α, β, γ, and δ. The subunit stoichiometry of *Torpedo* AChR $[(\alpha)_2\beta\gamma\delta]$ was determined by simultaneous quantitative N-terminal microsequencing of the polypeptides obtained from purified AChR (Raftery *et al.*, 1980). The same subunit stoichiometry was obtained for the *Electrophorus* electric organ AChR (Conti-Tronconi *et al.*, 1982a), and for the AChRs from fish, calf, and chicken muscle (Conti-Tronconi *et al.* 1982b, 1984; B. M.

Conti-Tronconi, S. M. J. Dunn, and M. A. Raftery, unpublished observations).

Neuronal AChRs exist either as homomeric complexes, as in the case of the chick brain α7 subunit (Couturier et al., 1990b) and the α-BGT-sensitive AChRs of Drosophila (Sawruk et al., 1990a,b) and of locust (Marshall et al., 1990), or as heteromeric complexes composed of α and β subunits (Deneris et al., 1991). A pentameric structure for neuronal AChRs is consistent with their molecular weight (Conti-Tronconi et al., 1985; Smith et al., 1985; Whiting and Lindstrom, 1986, 1987, 1988; Whiting et al., 1987, 1991) and was verified for the α4/β2 AChR subtype using two different approaches. Quantitative [^{35}S]-methionine incorporation in α4 and β2 subunits expressed in fibroblasts (Whiting et al., 1991) and Xenopus oocytes (Anand et al., 1991) indicated a relative subunit ratio of 2:3 for α4 and β2 subunits, respectively. Site-directed mutagenesis of the α4 and β2 subunits was used to obtain hybrid AChR complexes with distinguishable electrophysiological characteristics (Cooper et al., 1991). Quantitation of the hybrid channels formed suggested a pentameric complex of $(\alpha 4)_2(\beta 2)_3$.

A pentameric structure for other members of the AChR superfamily has been inferred, given their sequence and functional similarities. Cross-linking studies of the glycine receptor suggest that it contains three ligand binding subunits (α) and two structural subunits (β) (Langosch et al., 1988). The subunit stoichiometry of different GABA$_A$ receptor complexes has not been reported, but given the multimeric structure of the functional complexes it is likely that these receptors have a quaternary structure similar to the muscle-type AChRs.

E. Tridimensional Structure of the AChR from *Torpedo* Electric Organ: Electron Microscopy, Scanning Tunneling Imaging, and Low Resolution X-Ray Studies

The tridimensional shape of the AChR has been studied by different approaches, including electron microscopy, low-dose electron microscopy and scanning tunneling microscopy (STM). Detailed reviews of those structural studies are reported in Mitra et al. (1989), Unwin et al. (1988), Stroud et al. (1990), and Bertazzon et al. (1992). We summarize here the overall conclusions from those studies, which all yielded consistent pictures of the AChR at different resolutions [between 22 and 11.25 Å (Mitra et al., 1989; Unwin et al., 1988; Bertazzon et al., 1992)].

The AChR is almost cylindrical, having essentially constant lateral dimensions (mean diameter ~65 Å) in the extracellular and transmembrane part, and narrower on the cytoplasmic side (Brisson and Unwin, 1985). The total length of the AChR was originally estimated to be about 140 Å (Brisson and Unwin, 1985). In improved density maps obtained by low-density electron microscopy and X-ray diffraction (Mitra et al., 1989) the estimated total length was 115 Å in native AChR and 130 Å after alkali treatment, due to disordering of the protein domains.

Native, nondesensitized AChR molecules are almost perfectly symmetrical: all subunits are rod-shaped structures, approximately perpendicular to the plane of the membrane, arranged symmetrically around the central pore and extending radially toward the surrounding lipids by the same distance from the center (Unwin *et al.*, 1988; Mitra *et al.*, 1989). The lateral dimensions of the AChR in its transmembrane portion is compatible with the close packing of four or five α helices extending through the membrane with cross-sectional area of 550–540 Å² (Brisson and Unwin, 1985; Stroud *et al.*, 1990). They delineate a water-filled opening, presumed to be the ion channel, along the axis of the AChR molecule (Kistler *et al.*, 1982; Brisson and Unwin, 1985). The pentagonal symmetry of the AChR is almost perfect in the transmembrane region and over the contiguous regions of the extracellular side (Brisson and Unwin, 1985; Unwin *et al.*, 1988). Exposure to carbamylcholine (Carb) and consequent AChR desensitization reduces this symmetry, by rearrangement and protrusion from the center of two subunits, which were proposed to be the γ and δ subunits, assuming that the β subunit lies between the two α subunits (Unwin *et al.*, 1988). This assumption may or may not be true (see Section IV). The almost perfect pentameric symmetry of the AChRs is supported by the angle separating the two α subunits (144° ± 4°), which correspond precisely to two sectors of a pentagon (Fairclough *et al.*, 1983). It has been proposed (Brisson and Unwin, 1985) that the symmetric organization of the AChR subunits may be important for the coordinated movement of the subunits around the channel, involved in a switch to different functional states as a result of ligand binding.

The AChR subunits are asymmetrically placed in relation to the bilayer, since they have two to three times more of their combined mass on the synaptic than on the cytoplasmic side (Kistler *et al.*, 1982; Brisson and Unwin, 1985). A relatively detailed tridimensional structure of the extracellular domain of the AChR has been obtained by electron microscopy and STM approaches. It forms a cylindrical "vestibule," which extends 54 Å above the plane of the membrane (Mitra *et al.*, 1989). The individual height of the subunits above the plane of the synaptic membrane is 50–60 Å; the total outer diameter is 74–81 Å. The part of the channel contained in this vestibule is relatively uniform in diameter (25.5 Å in Mitra *et al.*, 1989; 30 Å in Brisson and Unwin, 1985), but it becomes more narrow and difficult to follow as it approaches the cytoplasmic domain (Kistler *et al.*, 1982; Brisson and Unwin, 1985). The protein walls surrounding the cylindrical channel of the vestibule are 24.5 ± 1.5 Å thick (Stroud *et al.*, 1990), which would exactly accommodate the dimensions of an antiparallel β-barrel structure predicted on the basis of amphipathic secondary structure analysis for the extracellular domain of *Torpedo* AChR (Finer-Moore and Stroud, 1984).

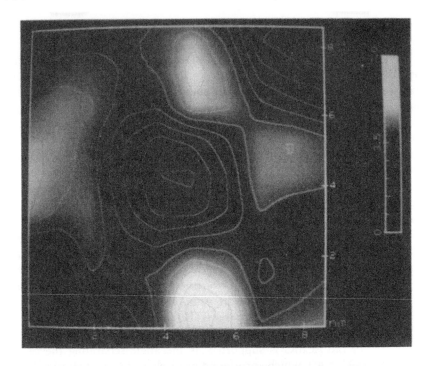

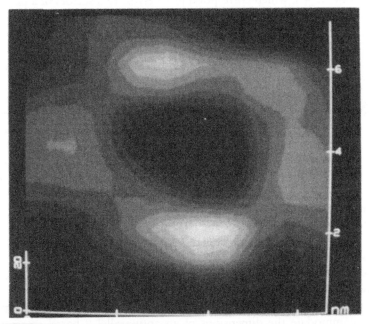

FIGURE 2   Subunit localization on unfiltered STM images of single AChR molecules. The general pentameric geometry observed in filtered images is generated by five separate peaks on

The structure summarized above has been recently been confirmed and detailed by STM imaging of *Torpedo* post-synaptic membrane fragments (Bertazzon *et al.*, 1992). The average outer diameter is 69 ± 10 Å, and the central cavity, taken on contour maps of filtered images at the largest delimiting line, is 26 ± 7 Å. Contour maps of STM images of single molecules yielded further structural details (Fig. 2). The total height above the background is ~50 Å, which is close to the dimensions reported earlier for the extracellular part of the AChR that protrudes from the plane of the membrane (Brisson and Unwin, 1985; Mitra *et al.*, 1989; Stroud *et al.*, 1990). Five peaks can be observed, two of which are not well resolved. The major peak protrudes ~15 Å from the largest contour line delimiting the central pit. On the opposite side of the central opening, a second peak protrudes ~9 Å. On the left side of the AChR molecule (as defined by the two above peaks) there are two poorly resolved peaks: the lower protrudes ~3 Å, the upper protrudes ~9 Å above this same plane. On the right side, a peak protruding to ~6 Å can be observed. These five peaks may be related to the five subunit domains. The average width of the walls of the pseudosymmetric rosette surrounding the central pit is 25 Å. This contour image closely matches those previously reported from hybrid maps (Mitra *et al.*, 1989).

II. STRUCTURE OF CHOLINERGIC LIGAND BINDING SITES ON THE ACHR α SUBUNIT

A. Studies with Affinity Labels and Expression of Mutant Subunits Identify Amino Acid Residues and Peptide Loops in Close Proximity to Ligand Binding Sites on the AChR α Subunit

The AChR has two distinct binding sites for agonists and competitive antagonists that are believed to reside primarily on the two α subunits. Other AChR subunits may contribute to the formation of these and perhaps other cholinergic sites, as discussed in Sections IV,D and IV,F.

1. Potential Sulfydryl/Disulfide Groups Involved in Ligand Binding

A sulfhydryl group within 1 nm of the binding site for acetylcholine on the AChR is specifically labeled with cholinergic affinity labels like 4-(*N*-

the surface. The upper image is enlarged from a 193 × 193-nm field, obtained with a bias voltage of 18 mV, a tunneling current 1.9 nA, a scanning rate of 1.3 cycles/s (650 nm/s) and sampling of 18 μs. The contour lines are spaced 0.3 nm in the vertical dimension. The color scale is in nanometers. In the lower panel, five peaks can be distinguished. The image was obtained from a 230 × 230-nm field with instrumental settings as above and a scanning rate of 2.7 nm/s. Reproduced, by permission, from Bertazzon *et al.* (1992).

maleimido)benzyltri-[³H]methylammonium (³H-MBTA) and bromoacetyl-choline (BAC) (Maelicke, 1988; Claudio, 1989). Alkylation with these lig-ands blocks α-BGT binding and AChR function. They label the AChR α subunit (Maelicke, 1988; Claudio, 1989).

Three potential disulfide bonds exist in the *Torpedo* AChR α subunit, which can be labeled with disulfide reagents and can be distinguished on the basis of their sensitivity to reduction: they involve the Cys pairs Cys_{128}/Cys_{142}, Cys_{412}/Cys_{418}, Cys_{192}/Cys_{193}, respectively (Kao *et al.*, 1984; Kao and Karlin, 1986; Mosckovitz and Gershoni, 1988). Cys_{412}/Cys_{418} are within M4, Cys_{128}/Cys_{142} and Cys_{192}/Cys_{193} are within the N-terminal putative extracellular domain (Noda *et al.*, 1982). A free cysteine at position 222, which then alkylated blocks the ion channel, has been associated with a hydrophobic pocket (Huganir and Racker, 1982; Yee *et al.*, 1986; Clarke and Martinez-Carrion, 1986). Models have been proposed in which either Cys_{128}/Cys_{142} or Cys_{192}/Cys_{193} were contained within the cholinergic bind-ing site (Noda *et al.*, 1982; Smart *et al.*, 1984; Criado *et al.*, 1985; Kao *et al.*, 1984; Kao and Karlin, 1986).

That Cys_{128}/Cys_{142} were within the cholinergic site was first proposed based on theoretical modeling of the AChR sequence (Noda *et al.*, 1982; Smart *et al.*, 1984). Mutations of either Cys_{128} or Cys_{142} of the *Torpedo* α subunit resulted in undetectable levels of α-BGT binding and no detectable response to acetylcholine (Mishina *et al.*, 1985). However, Sumikawa and Gehle (1992) demonstrated that mutation of either Cys_{128} or Cys_{142} causes a lower efficiency of complex assembly and membrane incorporation, but the few complexes that do form with either mutant α subunit are responsive to acetylcholine, although insensitive to α-BGT. Therefore in the native AChR an intact disulfide bond of the Cys loop is important for formation or accessibility of the α-BGT binding site.

Residues Cys_{192} and Cys_{193} of the *Torpedo* α subunit were identified by amino acid sequencing as the disulfide pair specifically labeled by the cholinergic affinity labels ³H-MBTA and *p*-dimethylamino)-[³H]ben-zenediazonium fluoroborate (DDF) following reduction (Kao and Karlin, 1986; Dennis *et al.*, 1988). The ability of α-BGT and MBTA to inhibit labeling reciprocally (Kao *et al.*, 1984; Gershoni *et al.*, 1983) indicated that they may bind to a common site. When either Cys_{192} or Cys_{193} was mutated to serine and coexpressed with the other *Torpedo* AChR subunits in *Xenopus* oocytes, α-BGT binding was reduced by 60–70% (Mishina *et al.*, 1985). However, the role of a disulfide bond between Cys_{192} and Cys_{193} in formation of this ligand binding site remains elusive. In the intact *Torpedo* AChR, reduction, or reduction and alkylation, does not change the number of a α-BGT binding sites (Moore and Raftery, 1979; Walker *et al.*, 1981). Sim-ilarly, selective alkylation of the vicinal cysteine residues with a large adduct does not interfere with the binding of α-BGT to the intact *Torpedo* α subunit (Mosckovitz and Gershoni, 1988). Expression studies in *Xenopus* oocytes of *Torpedo* α subunits carrying mutations of Cys_{192} or Cys_{193} also indicated that α-BGT binding can occur in the absence of an intact disulfide (Mishina *et al.*, 1985). In addition, the consensus of the results from binding studies of α-BGT to synthetic peptides (discussed in Sections II,B and II,C) indicates that although the sequence segment surrounding the vicinal Cys_{192} and Cys_{193} has an important role in forming a binding site for α-BGT, a vicinal disulfide bond between Cys_{192} and Cys_{193} is not critical for α-BGT binding.

2. Other Amino Acid Residues in Proximity to Cholinergic Ligand Binding Sites Identified by Affinity Labeling

Studies employing affinity labeling identified other amino residues within the N-terminal extracellular domain of the *Torpedo* α subunit involved in formation of cholinergic binding sites. Lophotoxin competes directly with the α-BGT binding and covalently labels Tyr-190 of the AChR α subunit, via a reactive epoxide linkage (Abramson *et al.*, 1989). The photoaffinity cholinergic reagent DDF labels Tyr_{190} as efficiently as Cys_{192} and Cys_{193}, and to a lesser extent Trp_{149} and Tyr_{93} (Dennis *et al.*, 1988; Galzi *et al.*, 1990).

The agonist binding site(s) has been mapped by affinity labeling of the intact *Torpedo* AChR with [³H]acetylcholine mustard (Middleton and Cohen, 1991) and a [³H]nicotine photoaffinity analogue (Cohen *et al.*, 1991). [³H]Acetylcholine (Middleton and Cohen, 1991) and [³H]nicotine (Cohen *et al.*, 1991) specifically labeled only Tyr_{93} and Tyr_{198} of the α subunit, respectively. Therefore a binding site for acetylcholine may be distinct from that for nicotine and other agonists, and formation of these (sub)sites may involve several peptide loops of the α subunit. This conclusion is also substantiated by studies using monoclonal antibodies recognizing different epitopes within the cholinergic binding site. Competition studies using such antibodies indicate that different subregions within the cholinergic site form the ligand binding interfaces for (Carb) α-BGT and *d*-tubocurare (Watters and Maelicke, 1983; Mihovilovic and Richman, 1987). For example, four antibodies that blocked both α-BGT and Carb binding to *Torpedo* AChR did not affect *d*-tubocurare binding, thus distinguishing between agonist and antagonist subsites. Another antibody did not affect agonist binding, but inhibited ~50% of α-BGT binding and high affinity *d*-tubocurare binding. Thus, it is likely that the binding sites for competitive agonists and antagonists are both distinct and overlapping.

B. Identification of Sequence Segments Contributing to Cholinergic Binding Sites on the AChR α Subunit by the Use of Proteolytic Fragments and Synthetic and Biosynthetic Peptides

1. Studies on Torpedo AChR and Muscle AChRs from Different Species

a. The torpedo AChR α subunit. The α subunit of *Torpedo* AChR was identified as containing an α-BGT binding site by ¹²⁵I-α-BGT labeling of protein blots after SDS–PAGE (Haggerty and Froehner, 1981; Tzartos and Changeux, 1983; Gershoni *et al.*, 1983). Also, *Torpedo* AChR expressed in *Xenopus* oocytes does not bind α-BGT unless the α subunit is expressed (Mishina *et al.*, 1984). Several laboratories have used proteolytic fragments (Wilson *et al.*, 1984, 1985; Pederson *et al.*, 1986; Oblas *et al.*, 1986; Neumann *et al.*, 1986a), synthetic peptides (Neumann *et al.*, 1986b; Ralston *et al.*, 1987; Wilson *et al.*, 1988; Wilson and Lentz, 1988; Conti-Tronconi *et al.*, 1988, 1989, 1990a, 1991; McLane *et al.*, 1990a,b, 1991a,b,c,d, 1993; Gotti *et al.*, 1987, 1988; Griesmann *et al.*, 1990), and biosynthetic peptides (Barkas *et al.*, 1987; Aronheim *et al.*, 1988; Ohana and Gershoni, 1990; Ohana *et al.*, 1991) to locate sequence regions of the *Torpedo* AChR α subunit able to bind α-BGT, and perhaps other cholinergic ligands. Such continuous peptide sequences, able to form independent ligand-binding sites in the absence of surrounding structural elements, have been called "prototopes" (Wilson *et al.*, 1988).

Proteolytic mapping of *Torpedo* AChR α subunit was an initial approach used to define the cholinergic and α-BGT binding site(s) (Wilson *et al.*, 1984, 1985; Oblas *et al.*, 1986; Pederson *et al.*, 1986; Neumann *et al.*, 1985, 1986a). The results of these studies are summarized in Table V.

Synthetic peptides have been widely used to investigate the sequence requirements of the cholinergic site that binds α-BGT. Atassi and co-workers identified several sequence segments of the *Torpedo* AChR α subunit that bound α-BGT—α125–148, α182–198, and α388–408 (McCormick and Atassi, 1984; Mulac-Jericevic and Atassi, 1987; Atassi *et al.*, 1988). Although α-BGT binding to peptides containing the Cys-Cys loop (Cys_{128}/Cys_{142}) and the sequence α388–408 has not been reproducible, several studies confirmed that the sequence region of the *Torpedo* α subunit flanking the vicinal Cys residues at position 192 and 193 forms a prototope for α-BGT (Neumann *et al.*, 1986b; Ralston *et al.*, 1987; Wilson *et al.*, 1988; Wilson and Lentz, 1988; Conti-Tronconi *et al.*, 1988, 1989, 1990a, 1991;

TABLE V   Proteolytic Mapping of the Cholinergic Binding Site on the *Torpedo* α Subunit Using V8 Protease

Fragment binding α-BGT (kDa)	Residues identified within the peptide fragment	Other binding ligands	Reference
18	Asn141, Cys128, Cys142	None	Pederson *et al.*, 1986
20	Cys192, Cys193	d-TC	Pederson *et al.*, 1986
17	Asn141, Cys128, Cys142	None	Oblas *et al.*, 1986
19	Cys192, Cys193	MBTA	Oblas *et al.*, 1986
18	Cys192, Cys193	MBTA	Wilson *et al.*, 1984, 1985
18	Residues 169–181	None	Neumann *et al.*, 1985, 1986a

McLane *et al.*, 1991a; Gotti *et al.*, 1987, 1988; Griesmann *et al.*, 1990). The results of these studies are summarized in Table VI.

Conti-Tronconi *et al.* (1989, 1990a) further defined the structural elements forming an α-BGT binding site on *Torpedo* AChR by testing a panel of overlapping synthetic peptides corresponding to the complete α subunit. Two sequence segments corresponding to α55–74 and α181–200 directly bound ^{125}I-α-BGT. Several monoclonal antibodies that compete for α-BGT binding (Watters and Maelicke, 1983; Fels *et al.*, 1986) recognized peptides α55–74 and α181–200, confirming a role of both sequence segments in formation of a cholinergic binding site. Therefore α-BGT binding involves multipoint attachments to the α subunit in the formation of a high affinity complex.

Circular dichroism (CD) and fluorescence spectroscopy have been used to study the structural characteristics of the *Torpedo* peptides α55–74 and α181–200 (Conti-Tronconi *et al.*, 1991; M. A. Raftery, A. Bertazzon and B. M. Conti-Tronconi, unpublished observations). Both peptides have a high content of β-sheet and β-turn (Table VII). Differential CD spectroscopy, in the presence and absence of α-BGT, indicates that peptides α55–74 and α181–200 undergo structural changes upon α-BGT binding, with a net increase in the β-structure component (Conti-Tronconi *et al.*, 1991). These structural changes may reflect a mechanistic basis for the essentially irreversible inactivation of the AChR by α-BGT.

Noncompetitive inhibitors (discussed in Section V) bind to at least three sites on the *Torpedo* AChR: (i) a binding site for acetylcholine, (ii) a high affinity site within the ion channel, and (iii) several low affinity binding sites. Sequence regions contributing to binding sites for the noncompetitive inhibitor phencyclidine (PCP) have been identified using synthetic peptides corresponding to amino acid residues α72–227 on the *Torpedo* AChR α subunit (Donnelly-Roberts and Lentz, 1991). Phencyclidine bound a 56-mer synthetic peptide (α172–227) and two smaller peptide segments (α173–204 and α205–227). Two distinguishable sites that bind PCP within this region were identified—a low affinity site within amino acid residues α173–204, competitive with α-BGT, and a high affinity site within residues α205–227. The latter segment contains the transmembrane segment M1, which may be exposed to the inner lining of the ion channel (as discussed in Section V). Both the high and the low affinity binding of PCP were inhibited by other noncompetitive inhibitors, chloropromazine, tetracaine, and dibucaine, indicating that PCP specifically binds to common binding sites for noncompetitive inhibitors. This study clearly indicates that the use of synthetic peptide sequences is a powerful approach for studying the binding sites of multiple ligands.

b. Muscle AChR α subunits.

The ^{125}I-labeled-α-BGT binds to the sequence regions of human and calf muscle AChR α subunits flanking

TABLE VI Mapping of the Cholinergic Binding Site on the α Subunit of AChRs from *Torpedo* Electric Organ and Vertebrate Muscle Using Synthetic Peptides and Fusion Proteins

Sequence (synthetic peptides)	Species	Apparent K_d or IC_{50}[a]	Type of assay	Reference
Isolated α subunit[b]	*Torpedo*	100–200 nM	Membrane blot assay	Gershoni *et al.*, 1983; Haggerty and Froehner, 1981; Oblas *et al.*, 1986
α173–204	*Torpedo*	500 nM IC_{50} for d-TC: 2 mM	Membrane blot assay	Wilson *et al.*, 1985
α185–196	*Torpedo*	35 μM inhibited by 10 mM d-TC	Peptide conjugated Sepharose	Neumann *et al.*, 1986b
α125–148	*Torpedo*, human	150 nM	Peptide conjugated Sepharose	McCormick and Atassi 1984
α182–198	*Torpedo*	IC_{50} 70 nM	Peptide conjugated Sepharose	Mulac-Jericevic and Atassi, 1987
α388–408	*Torpedo*	IC_{50} 100 nM	Peptide conjugated Sepharose	Atassi *et al.*, 1988
α[Lys]388–408	*Torpedo*	1 μM IC_{50} for d-TC:500 mM	Peptide conjugated Sepharose	Gotti *et al.*, 1987
α172–205	*Torpedo*	200 nM	Membrane blot assay	Ralston *et al.*, 1987
α185–199	*Torpedo*	20 μM	Membrane blot assay	Ralston *et al.*, 1987
α173–204	*Torpedo*	500 nM	Membrane blot assay	Wilson *et al.*, 1985
AChR	*Torpedo*	0.4 nM	Plate assay	Wilson *et al.*, 1988; Wilson and Lentz, 1988
Isolated α subunit	*Torpedo*	46 nM	Plate assay	Wilson *et al.*, 1988; Wilson and Lentz, 1988
α173–204	*Torpedo*	42 nM (10 nM with 0.01% SDS) IC50: 86 μM for d-TC, 8 mM for nicotine NaCl 16 mM α-cobratoxin 44 nM	Plate assay	Wilson *et al.*, 1988; Wilson and Lentz, 1988
α181–198	*Torpedo*	20 μM	Plate assay	Wilson *et al.*, 1988; Wilson and Lentz, 1988
α185–196		24 μM		
α193–204		24 μM		
α173–180		>100 μM		
α194–204		>100 μM		
α179–192		>100 μM		
AChR[c]	*Torpedo*	3 nM	Competition assay with *Torpedo* AChR	Wilson *et al.*, 1988

(*continues*)

TABLE VI *(continued)*

(Sequence synthetic peptides)	Species	Apparent K_d or IC_{50}[a]	Type of assay	Reference
Isolated α subunit	Torpedo	10 nM	Competition assay with Torpedo AChR	Wilson et al., 1988
α173–204	Torpedo	100 nM	Competition assay with Torpedo AChR	Wilson et al., 1988
α181–198	Torpedo	IC_{50} 9 μM	Competition assay with Torpedo AChR	Wilson et al., 1988
α179–192		IC_{50} 17 μM		
α185–196		IC_{50} 13 μM		
α186–196		IC_{50} 22 μM		
α193–204	Torpedo	87 μM	Competiton assay with Torpedo AChR	Wilson et al., 1988
α173–180		>400 μM		
α194–204		>500 μM		
α181–200	Torpedo	1–3 μM	Membrane blot assay	Conti-Tronconi et al., 1990a, 1991; Maelicke et al, 1989
α181–199	Human	25 μM	Membrane blot assay	Griessman et al., 1990
α185–199		80 μM		
α193–208		Inactive		
α177–192		40% binding of α185–199		
α173–204	Human	1 μM (861 nM with 0.01% SDS) IC_{50} d-TC 48 μM	Plate assay	Wilson and Lentz, 1988
α173–204	Bovine	226 nM (48 nM with 0.01% SDS) IC_{50} d-TC 250 μM	Plate assay	Wilson and Lentz, 1988
α179–191[d]	Human	αCTX 100 nM ErabuT a inactive	Quencing of intrinsic Trp fluorescence	Radding et al., 1988
α179–191	Bovine	αCTX 50 nM ErabuT a 50 nM	Quencing of intrinsic Trp fluorescence	Radding et al., 1988
α181–200	Frog	IC50 1–2 μM	Competition assay with Torpedo AChR	McLane et al., 1991b
α181–200	Chicken	IC50 1–2 μM		
α181–200	Mouse	IC_{50} ~ 150 μM		
α181–200	Calf	IC_{50} ~ 150 μM		
α181–200	Human	IC_{50} ~ 150 μM		
α181–200	Cobra	Inactive		

(Sequence fusion proteins)	Source	Apparent K_d or IC_{50}[a]	Type of assay[b]	Reference
α160–216	Mouse	75 nM	Filtration assay	Barkas et al., 1987
Isolated α subunit[c]	Torpedo	~20 nM	Filtration assay	Aronheim et al., 1988

(continues)

TABLE VI   (*continued*)

Sequence (synthetic peptides)	Species	Apparent K_d or $IC_{50}{}^a$	Type of assay	Reference
α166–315		~20 nM		
α166–200		~20 nM		
α184–196		~400 nM		
α184–200	*Torpedo*	~20 nM	Membrane blot assay	Aronheim *et al.*, 1988
		IC_{50} 600 μM DTC		
		IC_{50} 20 mM Carb		
		IC_{50} 3 μM αCTX		
α183–204	*Torpedo*	63 nM	Membrane blot assay	Ohana and Gershoni, 1990
	Xenopus	536 nM		
	Chick	150 nM		
	Mouse	3200 nM		
	Calf	6200 nM		
	Human	6470 nM		
	Cobra	Inactive		Ohana *et al.*, 1991

[a] K_d values were obtained when direct binding of α-BGT (or α cobratoxin, or erabutoxin a, see footnote *c*) to peptides immobilized on a solid support was measured, IC_{50}s were obtained when competition assays were used, measuring the ability of the synthetic or biosynthetic sequence to compete with native AChR for cholinergic ligand binding. Unless specified (see footnote *b*), α-BGT binding was measured.

[b] The affinity of α-BGT for the isolated α subunit of *Torpedo* AChR, measured in the same type of assay as the peptides listed in the box below, is reported for the sake of comparison.

[c] The affinity of α-BGT for the intact *Torpedo* AChR, measured in the same type of assay as the peptides listed in the box below, is reported for the sake of comparison.

[d] In these studies the binding of α-cobratoxin and erabutoxin a, not of α-BGT, was assessed.

Cys$_{192}$/Cys$_{193}$ (Gotti *et al.*, 1988; Griesmann *et al.*, 1990; Wilson and Lentz, 1988). The α-BGT binding properties of synthetic peptides corresponding to the sequence segment 181–200 from muscle AChR α subunits

TABLE VII   Secondary Structure of Synthetic Sequences of *Torpedo* AChR α Subunit that Bind α-BGT, in Absence and in Presence of Denaturants

Synthetic sequence	α-helix	β-sheet	β-turn	Random coil
α55–74[a]	0	64.1	10.9	25
α55–74 (GuCl 7 M)	0	20.8	30.9	48.3
α181–200[b]	0	68.1	14.8	17.1
α181–200[c]	11.5	47.7	28.3	12.5
α181–200 (DTT 10 mM)[c]	8.5	58.3	27.1	16.1
α181–200 (GuCl 7 M)	0	76.5	12.3	11.2

[a] 1 mg/ml in 10 mM Tris–HCl buffer, pH 8.

[b] at pH 4.2.

[c] at pH 8.2.

of different species have been studied (McLane *et al.*, 1991a) (Table VI). The *Torpedo*, frog, and chicken muscle synthetic sequences bound ^{125}I-α-BGT with relatively high affinity (apparent IC_{50} 1–2 μM), whereas the mammalian muscle peptides (human, murine, and bovine) had lower affinity (apparent IC_{50} ~15 μM) (McLane *et al.*, 1991a). The use of a homologous peptide corresponding to cobra muscle AChR α subunit demonstrated that the snake insensitivity to its own toxin is likely due to the inability of this sequence segment to form a prototope for α-BGT (McLane *et al.*, 1991a).

A disulfide between Cys_{192} and Cys_{193} has been demonstrated in the native *Torpedo* α subunit (Kellaris *et al.*, 1989). We tested for a role of such disulfide in α-BGT binding to the synthetic sequences α181–200 of muscle AChRs from different species by cysteine/cystine modifications (McLane *et al.*, 1991a). Reduction and alkylation reduced α-BGT binding, whereas oxidation to a disulfide-containing peptide monomer had no effect. Therefore a vicinal disulfide is a possible structure formed by the synthetic peptides, but is not critical for α-BGT binding.

Nuclear magnetic resonance (NMR) spectroscopy has been used to study the low affinity interaction of acetylcholine with the peptide segments *Torpedo* α184–200 and human α183–206 expressed in a trpE fusion protein (Fraenkel *et al.*, 1991). These studies indicate that the quaternary ammonium methyl groups of acetylcholine interact the Trp_{184} of both peptides, indicating that the α-BGT binding site formed by this sequence region is also a competitive binding site for acetylcholine.

2. Studies on Different Neuronal AChR Subtypes

a. κ-Bungarotoxin sensitive neuronal AChRs.   As discussed in Section I,A, a number of deduced amino acid sequences for α and β subunits of avian and rodent neuronal AChRs have been reported. To date seven different neuronal AChR α subunits (α2, α3, α4, α5, α6, α7, and α8) and four β subunits (β2, β3, β4, β5) have been identified (Deneris *et al.*, 1991; see Table I). Different α/β subunit pairs affect the pharmacological profile of the resulting AChR complex (Luetje *et al.*, 1990b; Papke *et al.*, 1989; see Table II).

Using overlapping peptides corresponding to the complete α3 subunit, we mapped a potentially important constituent segment for the κ-BGT and κ-FTX binding sites to the sequence region α3(51–70) (McLane *et al.*, 1990a, 1993). κ-Bungarotoxin binds to this sequence with a K_d of ~300 nM, whereas α-BGT does not bind to any of the α3 peptides. The sequence segment α3(51–70) (shown in Fig. 3) contains a number of negatively charged residues that may interact with the Lys and Arg residues present in the disulfide-stabilized sequence loops of κ-BGT, as well as several aromatic amino acids, which are a consistent structural feature of the α-BGT binding sequences. We will discuss further the effect of amino acid substitutions on κ-BGT and κ-FTX binding in Section II,C,2,c.

Two other largely overlapping peptide sequences that may contribute to the κ-BGT binding site, α3(180–199) and α3(183–201), were identified

$$\cdot \quad \pi \quad + \quad \pi \quad - \pi + \ + \ \pi +$$

$$\textbf{ETNLWLKQIWNDYKLKWKPS}$$

FIGURE 3   The deduced amino acid sequence of the sequence segment 51–70 of the rat α3 AChR subunit. The sequence and numbering of the deduced amino acid sequence of the rat α3 subunit are taken from Boulter *et al.* (1986). Positive and negative charges over residues indicate charged side chains, and π indicates hydrophobic residues capable of π-electron charge interactions with κ-bungarotoxin and κ-flavitoxin.

using a competition assay with native neuronal AChR on PC-12 cells (most likely the α3/β2 subtype) (McLane *et al.*, 1990a). Both peptides contain the vicinal Cys pair and are homologous to, although relatively divergent from, the muscle-type α-BGT binding sequence α181–200 of different species. An involvement of this region of the α3 subunit in κ-BGT binding has also been suggested by *Xenopus* oocyte expression studies using α2/α3 chimeras (Luetje and Patrick, 1991; Luetje *et al.*, 1992). Thus, it appears that, like α-BGT, κ-BGT makes multipoint attachments to the α3 subunit, and that the segments of the α3 subunit contributing to κ-BGT binding are homologous to those contributing to the α-BGT site in the *Torpedo* α subunits.

b. α-Bungarotoxin sensitive neuronal AChRs.   Our laboratory has used synthetic peptides to determine the neurotoxin sensitivity of other α subunits, not successfully expressed in *Xenopus* oocytes as functional AChR complexes. The α5 subunit is one such neuronal AChR component. It is unclear if failure to demonstrate α-BGT binding to *Xenopus* oocytes injected with α5 transcripts is the result of low levels of expression (Sumikawa and Gehle, 1992), failure of assembly of complexes, and/or membrane insertion, or if other unidentified subunits are required for functional expression (Boulter *et al.*, 1990a; Couturier *et al.*, 1990a). However, α5 mRNA expression correlates directly with the expression of neuronal α-BGT binding AChRs in a number of cell lines (Chini *et al.*, 1992).

Overlapping peptides corresponding to the sequence region between residues 171 and 205 of the α5 subunit, and of the mouse muscle α1 and rat neuronal α2, α3, and α4 subunits, which all contain the vicinal Cys_{192} and Cys_{193}, were compared for their ability to bind α-BGT. Peptides corresponding to this region of the *Torpedo* α subunit (Conti-Tronconi *et al.*, 1990a) and the muscle α subunits of a number of species (McLane *et al.*, 1991a) form prototopes for α-BGT (see Sections II,B,1,a and II,B,1,b). In a solid phase assay used to test for direct binding of ^{125}I-α-BGT, and two different competition assays, in which peptides were tested for their ability to sequester ^{125}I-α-BGT from binding to native AChRs in *Torpedo* or PC-12 membranes, only peptides corresponding to the mouse muscle α1 and rat neuronal α5 subunits were found to bind α-BGT (McLane *et al.*, 1990b). This is consistent with the known pharmacology of the α1, α2, α3, and α4 AChR subtypes, and suggested that the α5 subunit is able to bind α-BGT. The most prominent α-BGT binding peptide of the α5 subunit was α5(180–199).

We have also performed synthetic peptide mapping studies of α-BGT binding regions on the complete sequence of the chick brain α7 subunit, and selected sequence regions of the highly homologous α8 subunit (McLane *et al.*, 1991c). The cDNAs for these neuronal α subunits were isolated using oligonucleotides corresponding to the N-terminal sequence of an α subunit from AChR proteins isolated from chick brain using α-BGT affinity chromatography (Conti-Tronconi *et al.*, 1985; Schoepfer *et al.*, 1990). In both solid phase and competition assays, peptides corresponding to the sequence segments α7(181–200) and α8(181–200) bound α-BGT with affinities similar to those reported for the muscle AChR α subunits (McLane *et al.*, 1991a). The ability of the α7 and α8 subunit peptides to bind α-BGT was surprising given their highly divergent sequences relative to the muscle α subunits. These results have been confirmed by expression studies, which indicate that the homomeric complex formed by the α7 subunit is blocked by α-BGT (Couturier *et al.*, 1990b; Bertrand *et al.*, 1992). Our studies of the α5, α7, and α8 α-BGT binding sites exemplifies how the synthetic peptide approach can be used to predict the potential pharmacology of a subtype prior to its functional reconstitution.

C. Identification of Individual Residues Contributing to Cholinergic Binding Sites on the AChR α Subunit

1. Comparison of Homologous Sequences

a. Muscle AChR α (181–200) sequences from different species. Naturally occurring species-specific amino acid substitutions provided a first step toward identification individual residues interacting with α-BGT. The muscle α subunits are highly conserved between different species, and all contain an α-BGT binding sequence within amino acid residues α181 and 200, with the notable exception of the cobra α subunit (Ohana and Gershoni, 1990; McLane *et al.*, 1991a; Ohana *et al.*, 1991) (Fig. 4). Inspection of these sequences reveals that six amino acid residues of the cobra sequence differ from the other α subunits, and may represent important residues for α-BGT binding. Notable nonconservative substitutions in the cobra α181–200 sequence include replacements of Lys_{185}, Trp_{187}, Tyr_{189}, and Pro_{194} by Trp, Ser, Asn, and Leu, respectively. Single residue mutations of the *Torpedo* α sequence to each of the six substitutions of the cobra α sequence demonstrated that conversion of Tyr_{189} to Asn or Pro_{194} to Leu in the *Torpedo* sequence suffices to eliminate α-BGT binding (Ohana *et al.*, 1991; Chaturvedi *et al.*, 1992).

Comparison of the vertebrate muscle α sequences (Fig. 4) indicates that the sequence VVY at position 188–190 is common to peptides that bind

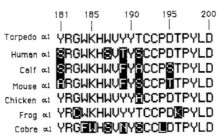

FIGURE 4   Comparison of aligned sequence segments 181–200 of the α subunits of AChRs from *Torpedo* electroplax and muscle of different species. The sequence regions *Torpedo* α1(181–200), Human α1(181–200), Calf α1(181–200), Mouse α1(182–201), Chicken α1(179–198), Frog α1(181–200), and Cobra α1(181–200) are aligned and numbered according to the *Torpedo* AChR α1 subunit sequence. The highlighted amino acids represent *nonconserved* residues. Reproduced, by permission, from McLane *et al.* (1991a).

α-BGT with high affinity (the *Torpedo*, frog, chick sequences), and that substitution of amino acids at position 189 to Phe (calf and mouse sequences) or Thr (human sequence) reduces the affinity of the sequence for α-BGT (McLane *et al.*, 1990a). The convergence of these results indicates that Tyr_{189} is a critical residue in the interaction of α subunits from different muscle AChRs with α-BGT.

b. Neuronal AChR α (181–200) sequences from different subtypes/ species.   The sequences of the neuronal α-BGT binding α subunits are highly diverged with respect to the muscle α1 subunits (Fig. 5). This indicates a low predictive value of sequence homology to infer neurotoxin sensitivity, which can best be appreciated by comparison of the sequences of AChR α subunits that bind α-BGT (i.e., the *Torpedo* α, chick muscle α1, chick brain α7, and α8, and *Drosophila* ALS and SAD subunits), with those that do not bind α-BGT (i.e., the cobra muscle α1, and the rat and avian neuronal α2, α3, and α4 subunits). Seven amino acid residues are characteristic of all α subunits regardless of α-BGT binding activity: i.e., Gly_{183} (or the conservative substitution Ala), Tyr_{190}, Cys_{192}, Cys_{193}, Asp_{195} (or the conservative substitution Glu), Tyr_{198}, and Asp_{200}. All of the α subunits in Fig. 5 that bind α-BGT have Tyr_{189} (or the conservative substitution Phe) and Pro_{197}, whereas Lys_{189} and Ile_{197} are characteristic of α subunits that do not bind α-BGT.

This general rule, however, does not hold for α5 subunit sequence, whose sequence is highly divergent from other AChR α subunits and offers a unique opportunity to study the structural requirements for α-BGT binding. The sequences of the AChR α subunits within the α-BGT binding region α181–200 (relative to the *Torpedo* sequence) are aligned in Fig. 6. This sequence comparison contrasts the highly conserved muscle α1 subunit sequences with the highly diverged α-BGT binding neuronal α subunits.

AChR α subunit sequences that bind αBGT

```
            181   185   190   195   200
             |     |     |     |     |
Torpedo  α1  YRGWKHWVYYTCCPDTPYLD
Chick Muscle α1  YRGWKHWVYYACCPDTPYLD
Chick αBGTBP α1  IPGKRTESYYECCKE-PYPD
Chick αBGTBP α2  VPGKRNELYYECCKE-PYPD
Drosophila ALS  VPAVRNEKYYSCCEE-PYLD
Drosophila SAD  VPAERHEKYYPCCAE-PYLD
```

AChR α subunit sequences that do not bind αBGT

```
            181   185   190   195   200
             |     |     |     |     |
Cobra muscle α1  YRGFWHSVNYSCCLDTPYLD
Chick brain α2  AIGRYNSKKYDCCTE-PYPD
Chick brain α3  APGYKHDIKYNCCEE-PYTD
Chick brain α4  AVGNYNSKKYECCTE-PYPD
Rat brain α2  ATGTYNSKKYDCCAE-PYPD
Rat brain α3  APGYKHEIKYNCCEE-PYQD
Rat brain α4  AVGTYNTRKYECCAE-PYPD
```

FIGURE 5 Comparison of the sequence regions surrounding the vicinal Cys pair at positions 192–192 from AChRs that bind αBGT, with those that cannot. Homologous sequences from the different α subunits are aligned according to the *Torpedo* nAChR α1 sequence numbering. Amino acids present in all α sequences, or at positions where there are only occasional conservative substitutions, are indicated by a black background. Amino acids characteristic only of those sequences that bind αBGT are indicated by a dotted background, whereas those characteristic of only those sequences that cannot bind αBGT are indicated by a hatched background. The sources of the sequences shown are *Torpedo* α1 (Noda *et al.*, 1982), chick α1, α2, α3, and α4 (Nef *et al.*, 1988), rat α2 (Wada *et al.*, 1988), rat α3 (Boulter *et al.*, 1986), rat α4 (Goldman *et al.*, 1987), chick αBGTBP α1 (also designated α7) and αBGTBP α2 (also designated α8) (Schoepfer *et al.*, 1990), *Drosophila* ALS (Bossy *et al.*, 1988) and *Drosophila* ARD (Sawruk *et al.*, 1990a), and cobra muscle α1 (Neumann *et al.*, 1989). Reproduced, by permission, from McLane *et al.* (1991c).

The inability to correlate critical structural features required for α-BGT binding with a particular amino acid sequence indicates in a broader sense a serious limitation to the use of sequence homology to define structurally related families of proteins. It is obvious from comparison of the α-BGT binding sequences that different primary sequences must fold into three-dimensional structures with comparable hydrophobic, hydrogen-bonding, and charge interactions. This fact is also illustrated by the lack of sequence homology between any of the nicotinic AChR α subunits and the acetylcholine binding sites of the muscarinic acetylcholine receptor and acetylcholinesterase (Schumacher *et al.*, 1986; Hulme *et al.*, 1990). The failure to find a common α-BGT binding motif is also similar to the search for targeting signals that are involved in sorting proteins into different cellular compart-

SPECIES/TISSUE   SEQUENCE   αBTX BINDING

SPECIES/TISSUE	SEQUENCE	αBTX BINDING
Torpedo electroplax	V R G W K H W W Y T C C P D T P W L D	+
Frog muscle	V R C W K H W W Y T C C P D K P W L D	+
Chick muscle	V R G W K H W W Y R C C P D T P W L D	+
Bovine muscle	S R G W K H W W F W R C C P S T P W L D	+
Human muscle	S R G W K H S W T W S C C P D T P W L D	+
Mouse muscle	A R G W K H W W F W S C C P T T P W L D	+
Cobra muscle	W R G F W H S W N W S C C L D T P W L D	–
Rat neurons: α2	A T G T Y N S K K W D C C R E – I W P D	–
Rat neurons: α3	A P G Y K H E I K W N C C E E – I W Q D	–
Rat neurons: α4	A U G T Y N T R K W E C C R E – I W P D	–
Rat neurons: α5	A M G S K G N R T D S C C W Y – F W I T	+
Chick neurons: α8	U P G K R N E L W W E C C K E – P W P D	+
Chick neurons: α7	I P G K R T E S F W E C C K E – P W P D	+
Drosophila SAD	U P R E R H E K W W P C C R E – P W L D	+

FIGURE 6 Comparison of the sequence α5(180–199) with the corresponding sequence regions of different α subunits. The sequences flanking the vicinal cysteine pair, which is the hallmark of a AChR α subunit, are taken from Conti-Tronconi *et al.* (1990a), McLane *et al.* (1990b), McLane *et al.* (1991a,b), and references therein. The chick neuronal α7 and α8 subunits are also referred to as αBGTBP α1 and αBGTBP α2 (e.g., Schoepfer *et al.*, 1990). The sequences are aligned with respect to the vicinal cysteine pair at positions 192/193, and amino acid residues that are conserved relative to the *Torpedo* α1 sequence are indicated by a black background. Reproduced, by permission, from McLane *et al.* 1991d).

ments. In these cases, instead of primary sequence conservation, compositional motifs are found in which certain amino acids or residues with similar physical characteristics are common between proteins destined to the same cellular organelle or membrane compartment (e.g., Dice, 1990). Compensatory, multiple nonconservative substitutions that occurred during the evolution of α-BGT binding proteins has likely obscured a "universal" α-BGT binding motif.

Given the lack of sequence homology between α-BGT binding proteins, mutational analysis has been used as a second approach to determine the requirements for α-BGT binding.

2. Studies with Single Residue Substituted Peptide Analogues

a. Torpedo and mouse muscle α subunit sequences. To further assess the structural determinants for α-BGT binding, peptides corresponding to the *Torpedo* α181–200 sequence were synthesized with single amino acid substitutions of glycine for each native residue (Conti-Tronconi *et al.*, 1991). The substituted analogues had comparable structures, as indicated by CD spectral analysis, but differed in their ^{125}I-α-BGT binding activity. These studies identified distinct clusters of amino acid residues, discontinuously positioned along the sequence 181–200, which serve as attachment points for α-BGT [residues 188–190 (VYY) and 192–194 (CCP)].

Tzartos and Remoundos (1990) used overlapping 4–9 residue peptides to define the minimum α-BGT binding sequence as *Torpedo* α188–197. Single amino acid substitutions (to Gly or Ala) of this sequence segment indicated that Tyr_{189}, Tyr_{190}, and Asp_{195} are important for α-BGT binding.

Chaturvedi *et al.* (1992) tested the effect of several multiple amino acid substitutions of the *Torpedo* α166–211 sequence expressed as a bacterial fusion protein. Several substitutions were administered, based on nonconserved residues of the rat neuronal α3 and cobra muscle α1 subunits, neither of which bind α-BGT: (i) substitution of Phe_{184}, Lys_{185}, and Trp_{187}, to Phe, Trp, and Ser (as found in the cobra α1 subunit) had no effect on α-BGT binding, introduction of two more mutations, Thr_{191} to Ser and Pro_{194} to Leu, abolished α-BGT binding: (ii) single amino acid residue mutations of the cobra α1 residues, Pro_{194} to Leu or Tyr_{189} to Asn, abolished α-BGT binding; (iii) mutation of the *Torpedo* sequence with three substitutions found in the α3 subunit (Trp_{187}, Tyr_{189}, and Thr_{191} to Asp, Lys, and Asn, respectively) eliminated α-BGT binding activity. These results confirmed that Tyr_{189} and Pro_{194} are critical for α-BGT binding.

Tomaselli *et al.* (1991) tested several mouse muscle α subunit mutants expressed as native AChR complexes in *Xenopus* oocytes. None of the following mutants were found to affect α-BGT binding affinity: His_{186} to Phe, Tyr_{190} to Phe, Pro_{194} to Ser, and Tyr_{198} to Phe. In contrast, the affinity for acetylcholine was markedly reduced when Tyr_{190} was substituted to Phe. The differences observed between the synthetic and the biosynthetic mutant peptides for the *Torpedo* and mouse α subunits could be due to species differences, or to differences in the amino acid substitutions made, or—most likely—to the fact that binding of α-BGT to native AChR occurs *via* large interacting surfaces, and mutation of one of the several residue involved in the interaction may not suffice to change the binding affinity detectably. An interesting conclusion of those studies is that α-BGT and acetylcholine have different structural requirements for binding.

b. Neuronal AChR α-BGT binding sequences. The α-BGT binding sequences 180–200 of the rat α5 subunit is relatively divergent compared with the homologous sequence regions of *Torpedo* and muscle AChRs. We identified amino acid residues critical for α-BGT binding by testing the effects of single amino acid substitutions of Gly or Ala for each residue of the rat α5(180–199) sequence (McLane *et al.*, 1991d) on binding of α-BGT to the substituted peptide analogues. Substitutions of four residues (Lys_{184}, Arg_{187}, Cys_{191}, and Pro_{195}) abolished α-BGT binding; other substitutions (Gly_{185}, Asn_{186}, Asp_{189}, Trp_{193}, Tyr_{194}, and Tyr_{196}) lowered its affinity. The importance of several aromatic amino acids for α-BGT binding to the α5 peptide is analogous to the findings reported above for the *Torpedo* α180–200 sequence. Thus, despite the apparent divergence of the α5 sequence from other α-BGT binding α subunits, certain common structural features, such as an abundance of aromatic residues and amino acids able to contribute electrostatic and/or hydrogen bond interactions, have been conserved.

c. Neuronal AChR κ-BGT binding sequences. The sequence segment $\alpha3(50–71)$ forms a prototope for κ-BGT and κ-FTX in the rat $\alpha3$ subunit (McLane *et al.*, 1990a, 1993). Synthetic peptide analogs of the sequence $\alpha3(50–71)$, in which each amino acid was sequentially replaced by Gly, were tested for their ability to bind ^{125}I-κ-BGT and ^{125}I-κ-FTX (McLane *et al.*, 1991b, 1993). No single substitution obliterated κ-BGT binding, but several substitutions lowered the affinity for κ-BGT—two negatively charged residues (Glu_{51} and Asp_{62}) and several aliphatic and aromatic residues (Leu_{54}, Leu_{56}, and Tyr_{63}). Similar to κ-BGT, aliphatic and aromatic amino acid residues were important for κ-FTX binding (Leu_{54}, Leu_{56}, and Tyr_{63}, also involved in κ-BGT binding, and additional Trp residues at positions 55, 60, and 67). In contrast to κ-BGT, however, positively charged, rather than negatively charged, amino acids appeared to mediate electrostatic interactions with κ-FTX—Lys residues at positions 57, 64, 66, and 68. These differences in amino acid specificity can be correlated with sequence differences of κ-BGT and κ-FTX, and provide important clues to the residue interactions at the toxin $\alpha3$ subunit interface (McLane *et al.*, 1993).

D. A Summary of the Results of the Studies on the Structure of Cholinergic Binding Sites by the Use of Synthetic or Biosynthetic Peptides

The studies summarized in Sections II,B and II,C allow the following conclusions:

(i) Two or three sequence segments of the α subunits contribute to form the cholinergic binding sites recognized by snake α-neurotoxins (α-BGT, κ-BGT, and κ-FTX). These sites therefore are complex surface areas, formed by clusters of amino acid residues from different sequence regions, similar to the "discontinuous epitopes" of antibody–antigen complexes (Davies *et al.*, 1988).

(ii) In all the AChRs studied the sequence segments contributing to the cholinergic site are in similar position along the α subunit sequence, suggesting that the extracellular domain of all α subunits folds in a similar manner. The sequence segments containing Cys_{192}/Cys_{193} is part of a cholinergic site in all α subunits.

(iii) In peripheral AChRs the sequence region $\alpha180–200$ is well conserved, and well-defined clusters of residues surrounding and including the vicinal Cys_{192} and Cys_{193} are involved in interaction with α-BGT. The corresponding sequence region is not well conserved in the neuronal AChRs that bind α-BGT, and the residues identified as crucial for interaction with α-BGT are at different positions than those of peripheral AChRs. Therefore there is no universal sequence motif with predictive value for α-BGT binding, and multiple, nonconservative substitutions in these sequence regions that occurred during evolution of the AChR proteins have both obscured the original ancestral sequence and reestablished, as a result of new mutual interactions, a structure compatible with α-BGT binding.

(iv) Although Cys_{192}/Cys_{193} are involved in forming the toxin/α subunit interface, a vicinal disulfide bound is not required for α-BGT binding.

(v) Within the relatively large area forming the cholinergic site, cholinergic ligands bind with multiple points of attachment and ligand-specific patterns of attachment points exist. This may be the molecular basis of the broad spectra of binding affinities, kinetic parameters and pharmacologic properties observed for the different cholinergic ligands.

(vi) The sequence regions α181–200 and α50–75 are unusually rich in aromatic residues, and substitution of aromatic residues frequently abrogates or decreases α-BGT (or κ-BGT) binding. These findings are compatible with the suggested model for the anionic cholinergic binding site of the AChR as formed not by a single negatively charged residue, but rather by interaction of the π electrons of aromatic rings (Dougherty and Stauffer, 1990), as has been demonstrated for the cholinergic site of acetylcholinesterase (Sussman et al., 1991) (discussed in Section II,E).

E. Atomic Structure of Acetylcholine Binding Site of Acetylcholinesterase: Implications for Structural Requirements of the Cholinergic Site of the AChR

The binding site for acetylcholinesterase has been well characterized through the use of different types of inhibitors. Although it has been proposed that an anionic site within the enzymatic pocket of acetylcholinesterase stabilizes binding of the quaternary ammonium group of acetylcholine, its existence is incompatible with the high affinity binding of neutral analogs of acetylcholine (Dougherty and Stauffer, 1990, and references therein). Photoaffinity labeling of acetylcholinesterases from different sources consistently identified aromatic amino acids as contributing to the active site, whereas no charged residues have been found (Dougherty and Stauffer, 1990). Based on these findings, it was suggested that a binding site containing aromatic amino acid residues could stabilize the binding of the quaternary ammonium group through cation/π-electron interactions (Dougherty and Stauffer, 1990). Using a synthetic receptor composed of primarily aromatic rings, these authors were able to demonstrate that π-electron-rich systems can stabilize the positive charge of acetylcholine.

More recently, the atomic structure of acetylcholinesterase has been determined (Sussman et al., 1991). The catalytic triad of the enzyme, Ser-His-Glu, lies at the end of a gorge lined with ~14 aromatic residues. Thus, the predictions that aromatic residues form the binding site for acetylcholine and that cation/π-electron interactions act to stabilize the binding of the quaternary ammonium group have been filled.

These results relate to studies of the effects of amino acid substitutions on the binding of acetylcholine (Tomaselli *et al.*, 1991) and the composition ·of the competitive binding site for acetylcholine and α-BGT at positions α181–200 of the α subunit of the AChR (see Section II,C). Those studies consistently demonstrated that aromatic residues are abundant in this region and are important for ligand binding. It is likely that they contribute to stabilizing positive charges of acetylcholine and other cholinergic ligands.

III. STRUCTURE OF LIGAND BINDING SITES OF OTHER MEMBERS OF THE AChR SUPERFAMILY

The agonist and antagonist binding sites on the AChR-related receptors, such as the $GABA_A$ and glycine receptors, have not been as thoroughly characterized as on the AChR. The Cys_{192}/Cys_{193} pair, which is the hallmark of a cholinergic binding site on the AChR α subunits, is absent from the subunits of the $GABA_A$, glycine, 5-HT-3, and related glutamate receptors. On the other hand, the disulfide-linked (Mosckovotz and Gershoni, 1988; Kellaris *et al.*, 1989) Cys_{128}/Cys_{142} residues of *Torpedo* AChR α subunit are conserved in the subunits of the receptors of the AChR superfamily. This region, called the Cys-Cys loop, is highly homologous between members of the AChR superfamily (Fig. 7), suggesting that it serves an important structural and/or functional role for ionotropic receptors. Typically, the two Cys residues are separated by 14–15 amino acids, which include two invariant residues (Pro and Asp) at positions 9 and 11 (with reference to the first Cys of the loop, Cys_{128}), and two alternative amino acids at positions 8 (Phe or Tyr) and 13 (His or Gln). Based on molecular modeling of this region as a potential ligand binding domain, position 6 has been proposed to be the key determinant for selective recognition of agonists for the glycine, GABA, and ACh receptors (Cockcroft *et al.*, 1990). These predictions, however, have not been tested experimentally.

The $GABA_A$ receptor binds a number of different ligands (see also Section IV,G): (i) agonists, such as GABA and muscimol, (ii) competitive antagonists, such as bicuculline and picrotoxin, and (iii) positive and negative allosteric regulators, such as barbiturates, benzodiazepines, imidazolpyridines, β-carbolines, and steroids. Affinity labeling of the intact $GABA_A$ receptor with muscimol suggested that the β subunit contained the agonist binding site (Casalotti *et al.*, 1986; Browning *et al.*, 1990). Expression of the cloned receptors in *Xenopus* oocytes, however, indicates that either the α or the β subunits can be expressed as homoligomeric complexes that exhibit GABA-induced Cl^- currents modulated by bicuculline and barbiturates (Sigel *et al.*, 1990). Either the γ2 (Pritchett *et al.*, 1989a; Sigel *et al.*, 1990) or

```
AChR      α1     C E I D U T H F P F D E Q N C
          α2     C S I D U T F F P F D Q Q N C
          α3     C K I D U T Y F P F D Y Q N C
          α4     C S I D U T F F P F D Q Q N C
          α5     C T I D U T F F P F D L Q N C
          α7/α8  C Y I D U R H F P F D U Q K C

5HT-3     α1     C S L D I Y N F P F D U Q N C

Glycine   α1/α2/α2*   C P H D L K N F P H D U Q T C

GABA-a    α1     C P H H L E D F P H D A H A C
          α2     C P H H L E D F P H D A H S C
          α3     C P H H L E D F P H D U H S C
          α4     C P H R L U D F P H D G H A C
          α5     C P H Q L E D F P H D A H A C
          α6     C P H R L U N F P H D G H A C

GABA-a    β1/β2/β3   C H H D L R R Y P L D E Q H C

GABA-a    γ1     C Y L Q L H N F P H D E H S C
          γ2     C Q L Q L H N F P H D E H S C

GABA-a    δ1     C D H D L A K Y P H D E Q E C
          ρ1     C N H D F S R F P L D T Q T C
```

FIGURE 7   The Cys–Cys loop of different members of the AChR superfamily. The deduced amino acid sequences for the highly conserved Cys–Cys loop (corresponding to the sequence segment 128–142 of the *Torpedo* α subunit) are given for different subunits of AChRs, the 5HT-3, GABA$_A$, and glycine receptors. Residues that are conserved compared with the mouse muscle AChR α1 subunit are boxed. (References for sequences are given in the text, Section I.B.)

γ3 subunits (Knoflach *et al.*, 1991) are necessary for allosteric regulation by benzodiazepines. In complexes containing α, β, and γ2 subunits, the α subunit subtype determines the affinity for different benzodiazepines—e.g., the α1 subunit confers high affinity for diazepam, whereas the α6 subunit complex is diazepam-insensitive (Luddens *et al.*, 1990; Pritchett and Seeburg, 1990; Pritchett *et al.*, 1989b). Deletion analysis and single point mutations of the GABA$_A$ α3 subunit indicate that the sequence region(s) affecting benzodiazepine affinity is(are) within the large N-terminal extracellular domain and include(s) residue Gly$_{225}$ (Pritchett and Seeburg, 1991). In addition, the apparent affinity for GABA is affected by mutation of Phe$_{64}$ of the α1 or α5 subunits to Leu (Sigel *et al.*, 1992). Thus, as in the case of the AChR, the α subunit is involved in formation of ligand binding sites, and several sequence regions on the α subunits contribute to ligand binding. Also analogous to the AChR is the demonstration that other subunits of the GABA$_A$ receptor complex participate in ligand binding.

The α subunit of the glycine receptor was identified as the binding site for strychnine by photoaffinity labeling of the intact receptor (Graham *et al.*, 1983). Proteolytic mapping of the labeled α subunit identified a strychnine binding site between amino acid residues 170 and 220 (Grenningloh *et al.*, 1987; Ruiz-Gomez *et al.*, 1990). Site-directed mutagenesis of the α2* subunit, which is resistant to strychnine, indicated that the residue at position 167 is important for a high affinity strychnine binding site. Construction of chimeric α subunits and site-directed mutagenesis of the α1 subunit demonstrated that there are two agonist binding sites on the α subunit. The high affinity agonist binding site is missing in the α2* subunit due to a mutation of Gly_{167} to Glu, and the low affinity agonist binding site common to all glycine receptor α subunits is formed by discontinuous stretches between amino acids 111 and 212 of the α1 subunit and 118 and 219 of the α2 and α2* subunits (Schmieden *et al.*, 1989). These results indicate that the region of the α subunit of the glycine receptor homologous to the α-BGT binding segment of the AChR α subunit (α181–200) contributes to a ligand binding site, and that two sites of different agonist affinity are contained within the α subunit. This is analogous to the AChR and is functionally significant, as discussed in Section IV.

IV. MULTIPLICITY OF LIGAND BINDING SITES ON THE AChR AND OTHER RELATED RECEPTORS

A. Functional Properties of AChR Related to Activation and Desensitization

Activation of AChRs in a variety of systems is characterized by low affinity for the agonist [K_{app} ~100 mM for acetylcholine (ACh)], which is consonant with the high concentrations of ACh (10^{-4} to 10^{-3} M) in the synaptic cleft immediately following neurotransmitter release (Land *et al.*, 1981; Adams, 1981). The Hill coefficient for channel activation approaches two, suggesting that the probability of channel opening is greater if more than one molecule of ACh binds. The channel opens rapidly, but the open state is transient, having a lifetime of ~1 ms (Katz and Miledi, 1971). If the agonist remains bound, the AChR undergoes a slow conformational change leading to a desensitized state in which the channel can no longer be activated. Under these conditions, in which the AChR is desensitized, two high affinity binding sites for ^3H-ACh (K_d 10 nM) have been detected in direct radiolabeled ligand binding studies. The molecular basis for the large (10^4) difference in ACh concentrations that induce channel activation and those that occupy the high affinity binding sites measured at equilibrium has been the subject of intense study. Two possible explanations are (i) that the low affinity sites originally present in the AChR have undergone sequential conformational transitions and are no longer present in the desensitized state, or (ii) that the AChR carries multiple sites for agonists, and that the low affinity sites involved in channel activation are distinct from those detected under desensitized conditions. Equilibrium radiolabeled ligand binding studies cannot discriminate between these two possibilities. In the former case, the low affinity sites are not present at equilibrium. In the latter case, technical difficulties render it impossible to measure such low affinity ^{3}H-ACh binding due to restrictions on the experimental concentrations of AChR that may be obtained and the inevitably small depletion of added ^{3}H-ACh due to binding.

B. Studies on the Binding of Small Cholinergic Ligands

1. Equilibrium Binding of Radiolabeled Agonists and Competitive Antagonists

The equilibrium binding of cholinergic agonists and antagonists to the AChR from *Torpedo* muscle and neuronal tissue has been extensively reviewed (Conti-Tronconi and Raftery, 1982; Karlin *et al.*, 1986; Claudio, 1989; Stroud *et al.*, 1990; Galzi *et al.*, 1991). Agonists bind tightly at equilibrium (K_d ~10 nM for ACh) and in *Torpedo* AChR the stoichiometry of high affinity sites is two per receptor. In most reports, this binding has been considered to be noncooperative (but see Stroud *et al.*, 1990) and has been shown to be inhibited in a competitive manner by antagonists such as *d*-tubocurarine. Affinity labeling techniques have shown that a disulfide bond formed by adjacent cysteines at positions 192 and 193 on the α subunits is located close to these binding sites since, following reduction of this bond, the cysteines generated may be covalently labeled by alkylating analogues of ACh. Whereas the two high affinity sites for ACh appear to have identical affinities, the binding of *d*-tubocurare is heterogeneous and is characterized by binding to two sites having K_ds of 33 nM and 7.7 mM (Neubig and Cohen, 1979). In recent years, these sites have been studied in great detail leading to the identification of specific amino acids that may be involved in binding (see Sections II,A,2). In the interpretation of the equilibrium binding data it is, however, important to bear in mind that, under these conditions, the AChR is desensitized. The equilibrium binding results do not, therefore, provide information on the agonist binding to the resting state of the AChR, and their ability to open the ion channel.

2. Spectroscopic Approaches Reveal Four Binding Sites for Cholinergic Agonists

Fluorescence methods have been used to probe the properties of cholinergic binding sites and to monitor conformational changes of the AChR that may be important for receptor function (Conti-Tronconi and Raftery, 1982; Ochoa *et al.*, 1989; Galzi *et al.*, 1991). Since the results of those studies have been reviewed extensively before, here we will restrict the review to data obtained using fluorescent reporter groups that have been covalently bound to the AChR protein, and describe the usefulness of this approach in characterizing both high and low affinity sites for cholinergic ligands.

The initial logic in these experiments was to introduce a fluorophore close to known binding sites by covalently attaching a small sulfhydryl-selective label, 5-iodoacetamidosalicylic acid (IAS) into the reduced disulfide at positions 192 and 193 on the AChR α-subunits (Dunn *et al.*, 1980, and unpublished observations). This logic was shown to be justified since the binding of both agonists (Dunn *et al.*, 1980) and competitive antagonists (S. M. J. Dunn and M. A. Raftery, unpublished observations), known to bind close to the site of labeling, produced a saturable enhancement of the IAS fluorescence, and K_ds obtained from equilibrium fluorescence titrations were similar to those obtained in direct radiolabeled binding studies using desensitized AChR. Despite the proximity of the IAS label to these binding sites, the label did not perturb the binding of radiolabeled ligands and the flux responses of the AChR were unaffected (Dunn *et al.*, 1980). Local anesthetics, which are noncompetitive inhibitors (NCIs) of the AChR, did not cause any change in IAS fluorescence (with one exception, lidocaine; see Dunn *et al.*, 1981), suggesting that the IAS fluorescence changes specifically monitor the binding of agonists and competitive agonists to the sites close to the cysteines 192 and 193 of the α subunit.

Detailed studies of the kinetics of agonist binding to IAS-labeled AChR have shown that the binding mechanism is complex, involving multiple conformational transitions of the receptor–agonist complex. These transitions were all found to be slow and consistent, not with channel activation, but with desensitization processes previously described in electrophysiological experiments (Hille, 1984). Noncompetitive inhibitors increased the rate of the agonist-induced conformational changes, consistent with the possibility that these ligands accelerate desensitization processes (Dunn *et al.*, 1981), as suggested previously by electrophysiological studies (Magazanik and Vyskocil, 1976). A further characteristic of the mechanism of agonist binding to IAS-labeled AChR is that the affinity of the resting state for ACh, as extrapolated from the kinetic constants for binding, was approximately 1 mM (Dunn *et al.*, 1980), a much higher affinity that expected if binding of ACh to these sites is involved in channel activation.

The results of the above experiments of agonist binding to IAS-labeled AChR presented a dilemma. Despite the proximity of the probe to known cholinergic binding sites, all of the conformational changes detected were too slow to account for channel activation, and agonist affinities did not agree with the dose dependencies of flux responses. This led to additional studies to identify other fluorescent probes that might shed further light on agonist binding mechanisms. One such probe, 4-[N-(iodoacetoxy)ethyl-N-methyl]-amino-7-nitrobenz-2-oxa-1,3-diazole (IANBD), has proved very useful in the characterization of distinct low affinity agonist binding sites on AChR (Dunn and Raftery, 1982a,b; Conti-Tronconi et al., 1982c; Dunn et al., 1983; Raftery et al., 1983). Reaction of AChR with the sulfydryl-selective probe IANBD results in significant incorporation of the label into the α, β, and γ subunits (Conti-Tronconi et al., 1982c), although the NBD group does not label the same disulfide at positions α_{192} and α_{193} that is labeled by IAS (S. M. J. Dunn and M. A. Raftery, unpublished observations). In initial experiments it was intriguing that, under equilibrium conditions, no changes in NBD (nitrobenz-2-oxa-1,3-diazole) fluorescence accompanied the saturation of the high affinity sites by agonists. The NBD fluorescence was, however, specifically and saturably increased upon the binding of relatively high (10^{-5} to 10^{-2} M) concentrations of agonists. Furthermore, the K_d values obtained in fluorescence titrations of NBD-labeled AChR were almost identical to those obtained in flux experiments for a wide range of cholinergic agonists (Dunn and Raftery, 1982b). The fluorescence enhancement appeared to be specific for agonists (but see also below), and prior affinity labeling of Cys_{192} and Cys_{193} on the α subunits by reaction with BAC did not affect the ability of agonists to induce the NBD fluorescence enhancement (Conti-Tronconi et al., 1982c; Dunn et al., 1983). The fluorescence changes induced by agonists were unaffected by physiologically relevant concentrations of d-tubocurarine or NCIs but were completely inhibited by pretreatment of AChR with saturating concentrations of α-BGT.

Kinetic experiments demonstrated that the NBD fluorescence enhancement occurring upon agonist binding was rapid, and indicated a conformational transition of the receptor–agonist complex sufficiently rapid to be implicated in channel opening (Dunn and Raftery, 1982a,b; Raftery et al., 1983). These results suggest that covalently bound NBD monitors binding to sites that may be involved in channel activation, but which are distinct from the sites monitored by IAS attached to Cys_{192} and Cys_{193} on the α subunits and by direct binding of ^3H-AcCh to desensitized AChR. With regard to the stoichiometry of these low affinity sites, recent results (S. M. J. Dunn and M. A. Raftery, in preparation) indicate that there are two low affinity sites for ACh and Carb. Since it has previously been established that there are also two high affinity sites for these ligands (see above), then each AChR molecule must carry at least four binding sites for cholinergic agonists. All four sites have been shown to be inhibited by α-BGT (see above) whereas d-tubocurarine inhibits binding to only the two sites that have high affinity in the desensitized state. As described below, a fifth site, which is not inhibited by a α-BGT, has also been identified.

C. Electrophysiological and Spectroscopic Approaches Reveal an Additional Binding Site Involved in AChR Activation

Early electrophysiological studies demonstrated that anticholinesterase carbamate drugs, such as eserine and neostigmine, have direct effects on

AChR at the neuromuscular junction (Harvey and Dreyden, 1974). It was realized that one of the effects of these drugs was to elicit AChR-mediated depolarization in the absence of receptor agonists such as ACh (Albuquerque *et al.*, 1984; Pascuzzo *et al.*, 1984; Akaike *et al.*, 1984). Similar effects have been observed for the actions of these compounds in the central nervous system (Pereira *et al.*, 1991). It has also been shown that eserine is an agonist for *Torpedo* AChR (Okonjo *et al.*, 1991; Kuhlmann *et al.*, 1991) since it stimulated ion flux in membrane vesicles as monitored in spectrofluorimetric assays (Moore and Raftery, 1980).

From the structural and mechanistic viewpoint, the most interesting aspect of eserine-mediated ion flux in *Torpedo* AChR is that the flux response is not inhibited by *d*-tubocurarine, α-BGT, or ACh-induced receptor desensitization (Okonjo *et al.*, 1991; Kuhlmann *et al.*, 1991). These results clearly indicate that eserine acts at a site distinct from those for natural and synthetic cholinergic ligands. The question arises whether eserine and the natural transmitter ACh operate by means of a common ion permeation pathway. Recent results have shown that eserine, like conventional AChR agonists, induces a saturable enhancement of the fluorescence of NBD-labeled AChR which, as described above, appears to be a characteristic of channel activation (M. A. Raftery and S. M. J. Dunn, unpublished observations). However, unlike other agonists, the NBD fluorescence enhancement induced by eserine was not blocked by α-BGT (M. A. Raftery and S. M. J. Dunn, unpublished observations) which, as noted above, also did not block the flux response. Thus it seems that eserine and the natural transmitter, while binding to separate sites, induce ion translocation by a common pathway.

D. Binding of Neurotoxins: Affinity and Stoichiometry

A vast body of information on the binding of neurotoxins to AChR has previously been reviewed (Katz and Miledi, 1971; Land *et al.*, 1981; Adams, 1981; Conti-Tronconi and Raftery, 1982; Karlin *et al.*, 1986; Lentz and Wilson, 1988; Claudio, 1989; Stroud *et al.*, 1990; Galzi *et al.*, 1991). α-Bungarotoxin has been of major importance in the isolation and characterization of peripheral AChRs, and of certain neuronal AChRs. In *Torpedo* AChR, α-BGT binds to two sites per AChR molecule. For membrane-bound AChR, these sites have different affinities (Conti-Tronconi *et al.*, 1990b), one being of very high affinity and dissociating with a half-life of ~250 h (Conti-Tronconi and Raftery, 1986) and the other being readily reversible (Conti-Tronconi *et al.*, 1990b). Other neurotoxins, including α-cobrotoxin (α-NTX) from *Naja naja siamensis* (Conti-Tronconi and Raftery, 1986) and the coral neurotoxin lophotoxin (Culver *et al.*, 1984), also bind to two sites on the AChR. By contrast, α-dendrotoxin (α-DTX) from *Dendroaspis viridis* binds to the membrane-bound *Torpedo* AChR with a stoichiometry of four per receptor molecule (Conti-Tronconi and Raftery, 1986). Further-more, when α-DTX binds to preformed AChR/α-BGT complexes the rate of dissociation of α-BGT is increased from a $t_{1/2}$ of 250 h to 4.5 h, suggesting that there are cooperative interactions between the additional sites for α-DTX and the sites for α-BGT (Conti-Tronconi and Raftery, 1986).

E. The Use of Peptide Neurotoxins as Affinity Labels Suggests That Subunits Other than the α Are Involved in Formation of Cholinergic Sites

Snake α-neurotoxins have been used to affinity label AChR (Lentz and Wilson, 1988). In all studies using native or derivatized α-BGT, crosslinking to the α subunits was observed. This has been corroborated in several studies in which α-BGT was shown to bind to isolated α-subunits, albeit with lower affinity (Lentz and Wilson, 1988; see also Section II). Labeling of other *Torpedo* AChR subunits was also observed when α-BGT was used. In some studies α-BGT crosslinked the α and the δ subunits (Witzemann and Raftery, 1979; Witzemann *et al.*, 1979). In another study, photolabeling with an α-BGT derivative labeled the α, γ, and δ subunits (Oswald and Changeux, 1982). These results have been widely interpreted to indicate that the α-neurotoxins bind mainly to the α subunits with perhaps some overlap with respect to adjacent subunits. However, as it will became evident from the results of the studies summarized below and in Section IV,F, labeling by α-BGT of subunits other than the α, and of the γ and δ subunits in particular, may indicate that the two sites for α-BGT are at interfaces between subunits, and in particular at the α/γ and α/δ interfaces.

Recently a departure from toxin labeling of the α subunits has been demonstrated in affinity labeling studies using α-conotoxins, which block AChR function (Olivera *et al.*, 1990). Crosslinking of AChR by derivatized α-conotoxins or photoaffinity labeling resulted in the covalent labeling of mainly β and γ subunits, with little, if any, labeling on the α or δ subunits (Myers *et al.*, 1991). These results clearly demonstrate the importance of subunits other than the α in ligand binding site formation. Other subunits, notably γ and δ, have also been implicated in forming at least part of the binding sites for [3H]tubocurare (Blount and Merlie, 1989; Pederson and Cohen, 1990) and [3H]nicotine (Middleton and Cohen, 1991), and it has been suggested that binding sites for these ligands lie at the α/γ and the α/δ subunit interfaces (see Section IV,F).

F. Cholinergic Binding Sites May Be at the Interface of Subunit Domains

As summarized above, there is evidence that as many as four, and possibly five, binding sites for cholinergic ligands exist on the AChR molecule.

The pseudosymmetric pentameric arrangement of the AChR subunits (see Section I,D) would allow for the formation of five homologous binding sites. This is particularly suggestive since some neuronal AChRs are symmetric pentamers of one type of subunit (Couturier *et al.*, 1990b; Sawruk *et al.*, 1990b; Marshall *et al.*, 1990), and therefore have five identical domains, each of which could be associated with a cholinergic site.

There is indirect support for a model of the AChR with multiple binding sites from the results of studies which indicated that the cholinergic site are likely to reside at the interface of subunit pairs. Since the results from studies using peptide neurotoxins as affinity labels are consistent with the possibility that each AChR subunit resides in close proximity to, or contains part of, a cholinergic site (see Section IV,E), a possible model of the ligand binding site arrangement on the AChR molecule may have five such sites, at each subunit/subunit interface. In the neuronal AChRs which are homooligomers of identical α subunits, the five sites must be identical in structure, and presumably in pharmacological properties. On the other hand, AChR composed of two (α and β) of four (α, β, γ, and δ) types of subunits would have a spectrum of different cholinergic sites, perhaps involved with different functions (e.g., activation, desensitization).

For most AChR subtypes, the α subunit alone is not sufficient for formation of AChRs with pharmacologically relevant cholinergic sites. The evidence for the involvement of other subunits in addition to the α subunit, and the location of the binding sites for different cholinergic ligands at subunit/subunit interfaces, comes from three types of studies: (i) expression studies using different combinations of AChR subunits, (ii) affinity labeling studies, and (iii) differential protection by cholinergic ligands of the labeling of particular subunit pairs by DAPA, a cholinergic ligand able to bind all subunits of the *Torpedo* AChR (Witzemann and Raftery, 1977).

1. Expression Studies Indicate That Subunits Other than α Contribute to Formation of Cholinergic Sites

For all AChR subunits studied so far, expression of α subunit is necessary to reproduce ligand binding activity (Claudio *et al.*, 1989). For reconstitution of a functional AChR, however, other subunits are needed (Blount and Merlie, 1989; Kurosaki *et al.*, 1987), with the exception of the vertebrate neuronal α7 subunit (Couturier *et al.*, 1990b) and of α subunits from insect neurons (see Section I,A,3). Experiments that tested expression of different types of neuronal β subunits in association with the same α subunit indicated that the β subunit influences the pharmacologic properties of the resulting AChRs (Goldman *et al.*, 1987; Wada *et al.*, 1988; Duvoisin *et al.*, 1989; Luetje *et al.*, 1990a,b)—e.g., κ-BGT blocks AChR formed by coexpression of α3 and β2 subunits, but not β4 subunit (Duvoisin *et al.*, 1989; Luetje *et al.*, 1990a,b). In muscle AChRs, coexpression of α plus γ and/or δ subunits is necessary for *d*-tubocurare binding, and expression of α/γ or α/δ subunit pairs determines the binding affinity for d-tubocurare (Blount and Merlie, 1989; Blount *et al.*, 1990; Gu *et al.*, 1991; Sine and Claudio, 1991). Thus, these two subunit pairs well account for the two nonequivalent d-tubocurare binding sites of native AChR, which must be at the subunit/subunit interface.

Cooperativity of agonist binding is also a function of intersubunit interactions. Expression of triplet pairs of muscle AChR subunits on the surface mammalian cells demonstrated that cooperative agonist binding is lost if the γ or δ subunits are not expressed (Sine and Claudio, 1991).

2. Labeling of α/γ and α/δ Subunit Pairs with Affinity Labels

Experiments in which *Torpedo* AChR was affinity photolabeled with d-[^{3}H]tubocurare further proved that the two nonequivalent d-tubocurare binding sites are at the α/γ and α/δ subunit interfaces; the α, γ, and δ subunits were specifically labeled, with dose dependence consistent with a high affinity site at the α/γ and a low affinity site at the α/δ interface (Pedersen and Cohen, 1990). Similar studies with [^{3}H]nicotine photoaffinity labeling agree with this interpretation (Middleton and Cohen, 1991). Formation of binding sites at the interfaces of two subunit domains follows the same structural motif as several enzyme active sites.

Using particular conditions for disulfide bond reduction, ^{3}H-BrACh labels both α and γ subunits (Dunn *et al.*, 1993), further implicating the γ subunit as contributing to an agonist binding site. An earlier study also suggested that the γ subunit was in close proximity to the binding site for cholinergic affinity label DDF, which is incorporated into the γ subunit, as well as the α subunit (Dennis *et al.*, 1988).

The involvement of the δ subunit in forming a cholinergic binding site is supported by the cross-linking of Cys_{192}/Cys_{193} with a carboxylate group within the sequence segment $\delta 164-224$ (Czajkowski and Karlin, 1991). Another conclusion from that study was that a carboxylate group from Asp and/or Glu residue(s) of the δ subunit could act to stabilize the binding of the quaternary ammonium group of acetylcholine bound to the α subunit in the region containing Cys_{192}/Cys_{193}. The requirement for a negatively charged group for stabilizing the quaternary ammonium ion of acetylcholine, however, is currently being questioned, as discussed in Section II,E.

3. Studies on Differential Protection by Cholinergic Ligands of the Labeling of Particular Subunit Pairs by DAPA, a Cholinergic Affinity Ligand Able to Label All AChR Subunits

The findings summarized above led us to further examine the photoaffinity labeling of AChR by ^{3}H-DAPA, which in early experiments was shown to label multiple subunits, in an α-BGT-sensitive manner (Witzemann and Raftery, 1977). Recent results have shown that all four *Torpedo* AChR subunits are labeled by ^{3}H-DAPA (S-J. Tine and M. A. Raftery, unpublished). Labeling of the α subunits was the least sensitive to inhibition by cholinergic agonists such as carbamylcholine. Furthermore, if AChR was previously saturated by covalent labeling of the α subunits by BAC (2 mol/mol AChR), labeling of all subunits, except α, was inhibited. These findings suggest that the α subunits contain binding sites for cholinergic ligands that are unaffected by covalently bound BAC. These results are consistent with the fluorescence studies of NBD-labeled AChR described above, in which low affinity binding sites for agonists, which may be important for channel activating, have been shown to be present in AChR prelabeled by BAC.

G. Multiple Ligand Binding Sites of Other Members of the AChR Superfamily

Multiple ligand binding sites may exist on the other members of the AChR superfamily. Individual subunits of GABA$_A$ (Blair *et al.*, 1988), glycine (Schmieden *et al.*, 1989) and 5-HT$_3$ (Maricq *et al.*, 1991) receptors form functional homooligomeric ligand-gated ion channels (see Section I,D), implying that the different subunits can substitute for each other in forming a receptor complex, and that all must be capable of forming binding sites for the neurotransmitter (Cockcroft *et al.*, 1990).

One homologous sequence region of the GABA$_A$, glycine, and acetylcholine receptors contributes to an agonist binding site (region α181–200, see Section III). That structural features required for ligand binding may be conserved between different members of the ionotropic receptor family is suggested by the occurrence of "mixed" binding characteristics in invertebrate AChRs. For instance, bicuculline and picrotoxin (GABA$_A$ receptor antagonists) and strychnine (a glycine receptor blocker) inhibit AChR function in insect central nervous system (Benson, 1988; Pinnock *et al.*, 1988; Marshall *et al.*, 1990; Walker and Holden-Dye, 1991; Gundelfinger, 1992) and nematode muscle (Holden-Dye and Walker, 1990). Some AChR subtypes of the mammalian brain are blocked by bicuculline (Zhang and Feltz, 1991), and *d*-tubocurare blocks several members of the AChR family, including the 5-HT-3 receptor (Neijt *et al.*, 1988).

A number of different compounds, from local anesthetics to quaternary cations, act as NCIs of AChR function with regard to acetylcholine activation, but compete with each other for a common binding site believed to reside within the ion channel (Galzi *et al.*, 1990). Noncompetitive antagonists, such as PCP and MK-801, block the ion channels of several members of the AChR family, and of distant relatives, such as the NMDA receptor (Schwartz and Mindlin, 1988; Ramoa and Albuquerque, 1988; Ramoa *et al.*, 1990). Given the diverse nature of the NCIs, these compounds may cross-react with different receptors by virtue of common quaternary structure of these ion channel complexes, rather than conservation of primary sequence *per se*. The promiscuous binding of NCIs is indicative of conserved structural features of the ion channels of all ionotropic receptors (see Section V).

V. MOLECULAR APPROACHES FOR THE IDENTIFICATION OF THE ION CHANNEL

A. Identification of Sequence Regions Forming the Ion Channel of the AChR

Secondary structural and hydropathicity analysis of the deduced sequences of the AChR subunits identified putative transmembrane segments (see Section I,C,2) and led to suggestions regarding the possible participation of the putative transmembrane segments M1 and M2 (Noda *et al.*, 1983c) or M3 (Devillers-Thiery *et al.*, 1983) in formation of the ion channel. Two approaches were used to test these models: (i) construction and expression of mutant subunits using molecular genetic approaches (Claudio, 1989); (ii) affinity labeling of *Torpedo* AChR with channel blockers (Karlin *et al.*, 1986; Claudio, 1989; Stroud *et al.*, 1990; Galzi *et al.*, 1991).

The role of M1 to M4 in channel formation was first studied by making deletion mutations in the α subunit. Mutation of any of these segments yielded nonfunctional AChRs (Mishina *et al.*, 1985). Using chimeric *Torpedo*–calf AChRs and/or mutated δ subunits, it was shown that mutations in the M2 region and the adjacent segment linking M2 and M3 profoundly affect channel conductance, and confer species-specific channel properties (Imoto *et al.*, 1986; Hucho *et al.*, 1986), suggesting M2 as a channel-forming domain. Removal of a negative charge at the N-terminal (presumably intracellular) end of M2 caused greater reduction of the outward current through the channel, than the inward current. Conversely, removal of one or more negative charges at the C-terminal (presumably extracellular) end of M2 caused a greater reduction of the inward than the outward current (Imoto *et al.*, 1988). Thus rings of negative charge at either end of M2 are important determinants of channel conductance.

Leonard *et al.* (1988) investigated the effect of mutations of serine residues in the M2 of mouse AChR on the inhibition by the open channel blocker QX-222. The apparent affinity of QX-222 decreased proportionally to the number of mutated serines, implicating M2 in channel formation, and suggesting a model in which QX-222 binds within an ion channel formed by the M2 helices contributed by each of the AChR subunits (Leonard *et al.*, 1988; Charnet *et al.*, 1990). Imoto *et al.* (1991) altered the size and polarity of uncharged polar amino acids between the presumed cytoplasmic and extracellular negatively charged rings of M2, and concluded that the threonines and serines contained in the M2 segment of α, β, γ, and δ subunits

A   Monovalent Cation Channel Subunits:

Torpedo AChR	α	TLSISULLSLTUFLLUIUELIP
Mouse AChR	α1	TLSISULLSLTUFLLUIUELIP
Rat AChR	α2	TLCISULLSLTUFLLLITEIIP
Rat AChR	α3	TLCISULLSLTUFLLUITETIP
Rat AChR	α4	TLCISULLSLTUFLLLITEIIP
Rat AChR	α5	SLCTSULUSLTUFLLUIEEIIP
Chick AChR	α7	SLGITULLSLTUFMLLUREIMP
Rat AChR	β1	GLSIFALLTLTUFLLLLAIKUP
Rat AChR	β2	TLCISULLALTUFLLLISKIUP
Rat AChR	β3	SLSTSULUSLTUFLLUIEEIIP
Rat AChR	β4	TLCISULLALTFFLLLISKIUP
Bovine AChR	γ	TUAINULLAQTUFLFLUAKKUP
Bovine AChR	δ	SMAISULLAQSUFLLLISKRLP
Bovine AChR	ε	TUSINULLAQTUFLFLIAQKTP
5HT-3	α1	SFKITLLLGYSUFLIIUSDTLP

Monovalent Anion Channel Subunits:

Glycine	α1/α2/α2*	GLGITTULTMTTQSSGSRASLP
GABA-a	α1/α2/α3/α5	UFGUTTULTMTTLSISARPILP
GABA-a	α4/α6	UFGITTULTMTTLSISARPILP
GABA-a	β1	UALGITTULTMTTISTHLRE
GABA-a	β2/β3	UALGITTULTMTTINTHLRE
GABA-a	γ	TSLGITTULTMTTLSTIA
GABA-a	δ	TSLGITTULTMTTLMUSA

FIGURE 8   The M2 segment of different members of the AChR superfamily. (A) Deduced amino acid sequences of the segment M2 for different members of the AChR superfamily: the *Torpedo* AChR α subunit; the mammalian AChR α1, α2, α3, α4, α5, α7, β1, β2, β3, β4, γ, δ, and ε subunits; the 5HT-3 α1 subunit; the glycine α1, α2, α2* subunits; and the $GABA_A$ α1, α2, α3, α4, α5, α6, β1, β2, β3, γ and δ subunits. (B) Deduced amino acid sequence of the homologue M2 segment of the glutamate receptor subunits: the AMPA/kainate subfamily subunits GluR1, R2, R3, R4, R5, R6, and R7; KA-1, KA-2, and γ2; and the NMDA receptor subunits NMDAR1 (ζ1), ε1, ε2, and ε3. (References for sequences are given in the text, Section I.B.)

form a short, narrow channel close to the cytoplasmic side of the membrane. Results from mutations of charged and polar amino acids in three anionic rings of M2 (extracellular, intermediate, and cytoplasmic) indicate that amino acids in the intermediate ring may be part of a cation selectivity filter (Konno *et al.*, 1991).

M2 has been implicated in channel formation in the neuronal α7 receptor. Mutation of Leu_{247} to threonine altered activation and desensitization

B   **Monovalent/Divalent Cation Channel Subunits:**

AMPA/Kainate Receptors:

Subfamily

	GluR1	FGIFNSLWFSLGAFM.QGCDISPRS
α	GluR2	FGIFNSLWFSLGAFMRQGCDISPRS
	GluR3, R4	FGIFNSLWFSLGAFMQQGCDISPRS
	GluR5	FTLLNSFWFGUGARLMQQGSELMPKA
β	GluR6	FTLLNSFWFGUGARLMRQGSELMPKA
	GluR7	FTLLNSFWFGMGSLMQQGSELMPKA
	KA-1	VSLGNSLWFPUGGFMQQGSEIMPRA
γ	KA-2	VTLGNSLWFPUGGFMQQGSEIMPRA
	γ2	VTLGNSLWFPUGGFMQQGSEUMPRA

NMDA Receptor Subunits:

NMDAR1 (ζ1)	EEEEEDARLTLSSAMWFSWGULLNSGIGEGA
ε1	KAPHGPSFTIGKAIWLLWGLUFNNSUPUQH
ε2	REPGGPSFTIGKAIWLLWGLUFNNSUPUQH
ε3	KKSGGPSFTIGKSUWLLWALUFNNSUPIEH

FIGURE 8   (*Cont.*)

properties and changed the pharmacological profile of the AChR (Revah *et al.*, 1991; Bertrand *et al.*, 1992). It was suggested that in the desensitized "wild type" α7 receptor Leu_{247} blocks the ion channel, and that mutation to a threonine renders this state conductive.

Noncompetitive inhibitor photoaffinity labeling studies have also implicated M2 in an ion channel formation. The local anesthetic ^3H-TPMP+ labels Ser_{262} of the δ subunit, Ser_{254} of the β subunit, and Ser_{248} of the α subunit, all of which are within M2 (Oberthür and Hucho, 1988; Hucho *et al.*, 1986). [^3H]Chlorpromazine labels serine residues within the M2 domains of all AChR subunits (serines δ-262, β-254, α-248, and γ-257, in addition to leucines β-257 and γ-260 and threonine at γ-253) (Giraudat *et al.*, 1986, 1987, 1989; Revah *et al.*, 1990). These studies were carried out using AChR in the resting or desensitized states rather than the open channel state. Time-resolved photoaffinity labeling by [^3H]quinacrine azide was used in an attempt to label AChR in the open channel conformation (DiPaola *et al.*, 1990). Under these conditions, the sequence segment α208–243 was labeled, which encompasses the putative membrane-spanning domain M1, not M2. The time frame of these experiments was sufficiently fast to evade the faster phases of desensitization; however, the saturating concentrations of ACh used make it likely that the flux response would be extremely fast, rendering it debatable whether the channels would still be open at the time of irradiation.

In summary, although the results of the NCI photoaffinity labeling studies are somewhat contradictory, most recent mutagenesis results and also predictions based on the homology of AChR with other ligand-gated ion channels (see below) point to the involvement of M2 in channel formation.

B. Structure of the Ion Channels of Other Members of the AChR Superfamily

The AChR and $5HT_3$ receptors are cation-selective channels (Yang, 1990), whereas the $GABA_A$ and glycine receptors are chloride channels. The M2 domain is likely implicated in forming the lining of the ion channel of the AChR, and is a highly conserved region in all members of the AChR superfamily. Figure 8 aligns the M2 segment of different members of the AChR superfamily, including the more distantly related glutamate receptor subunits: several Ser/Thr residues are highly conserved within receptor subtypes and between receptor types.

The cationic AChR, 5HT-3, and glutamate receptors can be distinguished from anionic channels of the $GABA_A$ and glycine receptors by the complementary charges of the amino acid residues surrounding the transmembrane segments believed to line the ion pore. The M2 segments of the different subunits of these receptor complexes form a ring of charged residues, postulated to act as an ion filter. Like AChR, the cation-selective glutamate and 5HT-3 receptors have both positively and negatively charged amino acids at each end of M2 (Maricq *et al.*, 1991; Moriyoshi *et al.*, 1991). In addition to the cationic ion filter of other glutamate receptor subunits, the NMDAR1 ($\zeta 1$) subunit contains a stretch of glutamic acid residues preceding M2 that has been proposed to be involved in voltage-dependent Mg^{2+} block of this receptor (Moriyoshi *et al.*, 1991).

The M2 segments of the $GABA_A$ and glycine receptors, which conduct the Cl^- anion, have been compared in several recent reviews (Stroud *et al.*, 1990; Betz, 1990a,b). These M2 segments share an abundance of Ser/Thr or small nonpolar amino acid residues. A bulky hydrophobic residue occurs at every fourth position, followed by an amino acid with a small or polar side chain predicted to line the channel lumen (Betz, 1990a,b). At each end of the M2 domains of the $GABA_A$ and glycine receptors, positively charged residues are found, which may be important in conferring anion selectivity.

VI. SUMMARY

The members of the AChR superfamily have likely evolved from an ancestor ligand-gated ion channel, which probably existed as a homomeric complex. Duplication of its gene, and its subsequent divergence, yielded different subunits constituting a single receptor complex, then various receptor subtypes, and ultimately receptors that bind different ligands. Acetylcholine functions as an excitatory neurotransmitter in all phyla studied, whereas the $GABA_A$ and glycine receptors are of more limited distribution phylogenetically (Walker and Holden-Dye, 1991). It should be possible at some time in the future to trace the subsequent divergence and emergence of these other receptors and their subtypes by comparison of their deduced amino acid sequences.

The AChR subunits were highly conserved through the course of evolution. For instance, corresponding subunits of *Torpedo* and mammalian muscle AChRs share ~60–70% amino acid sequence identity, the different subunits forming muscle AChRs ($\alpha 1$, $\beta 1$, γ, ϵ, δ) share ~31–49% identity, and the neuronal AChR subunits so far sequenced, $\alpha 1$ to $\alpha 8$ and $\beta 2$ to $\beta 4$, share ~37–68% sequence identity. On the other hand, the subunits of the glycine and $GABA_A$ receptors share only 22–34% amino acid sequence identity with the muscle AChR subunits. More distance relatives of the AChR superfamily—the NMDA and non-NMDA receptors—share <20% sequence identity with the AChR subunits.

Despite the different degrees of sequence divergence among the different members of this superfamily of proteins, several structural features, such as the transmembrane folding of their constituent subunits and their pentameric structure, formed by homologous or identical subunits similarly arranged around the central ion channel, have been preserved in all members of the AChR superfamily. This therefore offers an excellent system to study the structural features required for ion channel gating, and the underlying process of the evolution of these important components of the nervous system.

ACKNOWLEDGMENTS

The work from the laboratories of the Authors was supported by NIH Grant 5R01-NS10294 (to M. A. R.), the ARO contract DAMD 17-88-C-8120 to M. A. R. and B. M. C. T.), and the NIDA program Project Grant 5PO1-DA05698 (to M. A. R. and B. M. C. T.). S. M. J. D. is a Scholar of the Alberta Heritage Foundation for Medical Research.

REFERENCES

Abramson, S. N., Li, Y., Culver, P., and Taylor, P. (1989) *J. Biol. Chem.* **264**, 12666–12672.
Adams, P. R. (1981) *J. Membr. Biol.* **58**, 161–174.
Agaki, H., Hirai, K., and Hishinuma, F. (1991) *Neurosci. Res.* **11**, 28–40.
Aharonov, A., Tarrab-Hazdai, R., Abramsky, O, and Fuchs, S. (1975) *Proc. Natl. Acad. Sci. U.S.A.* **72**, 1456–1459.

Akaike, A., Ikeda, S. R., Brooks, N., Pascuzzo, G. J., Richett, D. L., and Albuquerque, E. X. (1984) *Mol. Pharmacol.* **25**, 102–112.

Albuquerque, E. X. Akaike, A., Shaw, K. P., and Richett, D. L. (1984) *Fundam. Appl. Toxicol.* **4**, S27–S33.

Anand, R., Conroy, W. G., Schoepfer, R., Whiting, P., and Lindstrom, J. (1991) *J. Biol. Chem.* **266**, 11192–11198.

Anderson, D. J., Blobel, G., Tzartos, S. J., Gullick, W., and Lindstrom, J. (1983) *J. Neurosci.* **3**, 1773–1784.

Anderson, D. J. Walter, P., and Blobel, G. (1982) *J. Cell Biol.* **93**, 501–506.

Aronheim, A., Eschel, Y., Mosckovitz, R., and Gershoni, J. M. (1988) *J. Biol. Chem.* **263**, 9933–9937.

Atassi, M. Z., Manshouri, T., and Yokoi, T. (1988) *FEBS Lett.* **228**, 295–300.

Baldwin, T. J., Yoshihara, C. M., Blackmer, K., Kinter, C. R., and Burden, S. J. (1988) *J. Cell Biol.* **106**, 469–478.

Ballivet, M., Nef, P., Couturier, S., Rungger, D., Bader, C. R., Bertrand, D., and Cooper, E. (1988) *Neuron* **1**, 847–852.

Barkas, T., Mauron, A., Roth, B., Alliod, C., Tzartos, S. J., and Ballivet, M. (1987). *Science* **235**, 77–80.

Barnard, E., Beeson, D., Cockcroft, V., Darlison, M., Hicks, A., Lai, F., Moss, S., and Squire, M. (1986) In *Nicotinic Acetylcholine Receptor: NATO-ASI, Series H* (A. Maelicke, ed.), Vol. 3, pp. 389–415. Springer-Verlag, Heidelberg.

Bellone, M., Tang, F., Milius, R., and Conti-Tronconi, B. M. (1989) *J. Immunol.* **143**, 3568–3579.

Benson, J. A. (1988) In *Molecular Basis of Drug and Pesticide Action* (G. G. Lunt, ed.), pp. 193–206. Elsevier Science, Amsterdam.

Bertazzon, A., Conti-Tronconi, B. M., and Raftery, M. A. (1992) *Proc. Natl. Acad. Sci. U.S.A.* **89**, 9632–9636.

Bertrand, D., Devillers-Thiery, A., Revah, F., Galzi, J.-L., Hussy, N., Mulle, C., Bertrand, S., Ballivet, M., and Changeux, J.-P. (1992) *Proc. Natl. Acad. Sci. U.S.A.* **89**, 1261–1265.

Bettler, B., Boulter, J., Hermans-Borgmeyer, I., O'Shea-Greenfield, A., Deneris, E. S., Moll, C., Borgmeyer, U., Hollmann, M., and Heinemann, S. (1990) *Neuron* **5**, 583–595.

Betz, H. (1990a) *Neuron* **5**, 383–392.

Betz, H. (1990b) *Biochemistry* **29**, 3591–3599.

Betz, H. (1991) *Trends Neurosci.* **14**, 458–461.

Blair, L. A. C., Levitan, E. S., Marshall, J., Dionne, V., and Barnard, E. A. (1988) *Science* **242**, 577–579.

Blount, P., and Merlie, J. P. (1989) *Neuron* **3**, 349–357.

Blount, P., Smith, M. M., and Merlie, J. P. (1990) *J. Cell Biol.* **111**, 2601–2611.

Bossy, B., Ballivet, M., and Spierer, P. (1988) *EMBO J.* **7**, 611–618.

Boulter, J., Luyten, W., Evans, K., Mason, P., Ballivet, M., Boldman, D., Stengelin, S., Martin, G., Heinemann, S., and Patrick, J. (1985) *J. Neurosci.* **5**, 2545–2552.

Boulter, J., Evans, K., Goldman, D., Martin, G., Treco, D., Heinemann, S., and Patrick, J. (1986) *Nature (London)* **319**, 368–374.

Boulter, J., Connolly, J., Deneris, E., Goldman, D., Heinemann, S., and Patrick, J. (1987) *Proc. Natl. Acad. Sci. U.S.A.* **84**, 7763–7767.

Boulter, J., O'Shea-Greenfield, A., Duvoisin, R. M., Connolly, J. G., Wada, E., Jensen, A., Gardner, P. D., Ballivet, M., Deneris, E. S., McKinnon, D., Heinemann, S., and Patrick, J. (1990a) *J. Biol. Chem.* **265**, 4472–4482.

Boulter, J., Hollmann, M., O'Shea-Greenfield, A., Hartley, M., Deneris, E., Maron, C., and Heinemann, S. (1990b) *Science* **249**, 1033–1037.

Brisson, A., and Unwin, P. N. T. (1985) *Nature (London)* **315**, 474–477.

Browning, M. D., Bureau, M., Dudek, E. M., and Losen, R. W. (1990) *Proc. Natl. Acad. Sci. U.S.A.* **87**, 1315–1318.

Burnashev, N., Monyer, H., Seeburg, P. H., and Sakmann, B. (1992) *Neuron* 8, 189–198.

Cachelin, A. B., and Jaggi, R. (1991) *Pflugers Arch.* 419, 579–582.

Carlin, B. E., Lawrence, J. C., Jr., Lindstrom, J. M., and Merlie, J. P. (1986) *Proc. Natl. Acad. Sci. U.S.A.* 83, 498–502.

Casalotti, S. O., Stephenson, F. A., and Barnard, E. A. (1986) *J. Biol. Chem.* 261, 15013–15016.

Cauley, K., Agranoff, B. W., and Goldman, D. (1989) *J. Cell Biol.* 108, 637–645.

Cauley, K., Agranhoff, B. W., and Goldman, D. (1990) *J. Neurosci.* 10, 670–683.

Charnet, P., Labarca, C., Leonard, R. J., Vogelaar, N. J., Czyzk, L., Gouin, A., Davidson, N., and Lester, H. A. (1990) *Neuron* 4, 87–95.

Chaturvedi, V., Donnelly-Roberts, D. L., and Lentz, T. L. (1992) *Biochemistry* 31, 1370–1375.

Chavez, R. A., and Hall, Z. W. (1991) *J. Biol. Chem.* 266, 15532–15538.

Chavez, R. A., and Hall, Z. W. (1992) *J. Cell Biol.* 116, 385–393.

Chiappinelli, V. A. (1985) *Pharmac. Ther.* 31, 1–32.

Chini, B., Clementi, F., Hukovic, N., and Sher, E. (1992) *Proc. Natl. Acad. Sci. U.S.A.* 89, 1572–1576.

Clarke, J. H., and Martinez-Carrion, M. (1986) *J. Biol. Chem.* 261, 10063–10072.

Claudio, T. (1989) In *Frontiers in Molecular Biology* (D. M. Glover and B. D. Hammes, Eds.), pp. 63–142, IRL Press, Oxford.

Cockcroft, V. B. Osguthorpe, D. J. Barnard, E. A., and Lunt, G. G. (1990) *Proteins Struct. Funct. Genet.* 8, 386–397.

Cohen, J. B., Sharp, S. D., and Liu, W. S. (1991) *J. Biol. Chem.* 266, 23354–23364.

Conti-Tronconi, B. M., and Raftery, M. A. (1982) *Annu. Rev. Biochem.* 51, 491–530.

Conti-Tronconi, B. M., Hunkapillar, M. W., Lindstrom, J. M., and Raftery, M. A. (1982a) *Proc. Natl. Acad. Sci. U.S.A.* 79, 6489–6493.

Conti-Tronconi, B. M., Hunkapillar, M. W., Gotti, C., and Raftery, M. A. (1982b) *Science* 218, 1227–1229.

Conti-Tronconi, B. M. Dunn, S. M. J., and Raftery, M. A. (1982c) *Biochem. Biophys. Res. Commun.* 107, 123–129.

Conti-Tronconi, B. M., Hunkapillar, M. W., Lindstrom, J. M., and Raftery, M. A. (1984) *J. Receptor Res.* 4, 801–816.

Conti-Tronconi, B. M., Dunn, S. M. J., Barnard, E. A., Dolly, J. O., Lai, F. A., Ray, N., and Raftery, M. A. (1985) *Proc. Natl. Acad. Sci. U.S.A.* 82, 5208–5212.

Conti-Tronconi, B. M., and Raftery, M. A. (1986) *Proc. Natl. Acad. Sci. U.S.A.* 83, 6646–6650.

Conti-Tronconi, B. M., Tzartos, S., Spencer, S. R., Kokla, A., Tang, F., and Maelicke, A. (1988) In *Nicotinic Acetylcholine Receptors in the Nervous System NATO-ASI, Series H* (F. Clementi, C. Gotti, and E. Sher, Eds.), Vol. 25, pp. 119–136. Springer-Verlag, Berlin.

Conti-Tronconi, B. M., Fels, G., McLane, K., Tang, F., Bellone, M., Kokla, A., Tzartos, S., Milius, R., and Maelicke, A. (1989) In *Molecular Neurobiology of Neuroreceptors and Ion Channels: NATO-ASI, Series H* (A. Maelicke, Ed.), Vol. 32, pp. 291–310. Springer-Verlag, Berlin.

Conti-Tronconi, B. M., Tang, F., Diethelm, B. M., Spencer, S. R., Reinhardt-Maelicke, S., and Maelicke, A. (1990a) *Biochemistry* 29, 6221–6230.

Conti-Tronconi, B. M., Tang, F., Walgrave, S., and Gallagher, W. (1990b) *Biochemistry* 29, 1046–1054.

Conti-Tronconi, B. M., Diethelm, B. M., Wu, X., Tang, F., and Bertazzon, T. (1991) *Biochemistry* 30, 2575–2584.

Cooper, E., Couturier, S., and Ballivet, M. (1991) *Nature (London)* 350, 235–238.

Couturier, S., Erkman, L., Vanera, S., Rungger, D., Bertrand, S., Boulter, J., Ballivet, M., and Bertrand, D. (1990a) *J. Biol. Chem.* 265, 17560–17567.

Couturier, S., Bertrand, D., Matter, J.-M. Herandez, M.-C., Bertrand, S., Millar, N., Valera, S., Barkas, T., and Ballivet, M. (1990b) *Neuron* 5, 847–856.

Criado, M., Hochschwender, S., Sarin, V., Fox, V. L., and Lindstrom, J. (1985) *Proc. Natl. Acad. Sci. U.S.A.* **82**, 2004–2008.

Culver, P., Burch., Potenza, C., Wasserman, L., Fenical, W., and Taylor, P. (1985) *Mol. Pharmacol.* **28**, 436–444.

Culver, P., Fenical, W., and Taylor, P. (1984) *J. Biol. Chem.* **259**, 3763–3770.

Czajkowski, C., and Karlin, A. (1991) *J. Biol. Chem.* **266**, 22603–22612.

Dani, J. A. (1989) *Trends Neurosci.* **12**, 125–128.

Davies, D. R., Sheriff, S., and Padlan, E. A. (1988) *J. Biol. Chem.* **263**, 10541–10544.

Dawson, T. L., Nicholas, R. A., and Dingledine, R. (1990) *Mol. Pharmacol.* **38**, 779–784.

Dayhoff, M. D. (1976) In *Atlas of Protein Structure,* Vol. 5, pp. 77–268.

Deneris, E. S., Boulter, J., Swanson, L. W., Patrick, J., and Heinemann, S. (1989) *Biol. Chem.* **264**, 6268–6272.

Deneris, E. S., Connolly, J., Boulter, J., Wada, E., Wada, K., Swanson, L. W., Patrick, J., and Heinemann, S. (1988) *Neuron* **1**, 45–54.

Deneris, E. S., Connolly, J., Rogers, S. W., and Duvoisin, R. (1991) *Trends Pharmacol. Sci.* **12**, 34–40.

Dennis, M., Giraudat, J., Kotzba-Hibert, F., Goeldner, M., Hirth, C., Chang, J.-Y., Lazure, C., Chretien, M., and Changeux, J.-P. (1988) *Biochemistry* **27**, 2346–2357.

Devillers-Thiery, A., Changeux, J.-P., Paroutaud, P., and Strosberg, A. D. (1979) *FEBS Lett.* **104**, 99–105.

Devillers-Thiery, A., Giraudat, J., Bentaboulet, M., and Changeux, J.-P. (1983) *Proc. Natl. Acad. Sci. U.S.A.* **80**, 2067–2071.

Dice, J. F. (1990) *Trends Biochem. Sci.* **15**, 305–309.

Dingledine, R. (1991) *Trends Pharmacol. Sci.* **12**, 360–362.

DiPaola, M., Czajkowski, C., Bodkin, M., and Karlin, A. (1988) *Soc. Neurosci. Abstr.* **14**, 640 (260.5).

DiPaola, M., Czajkowski, C., and Karlin, A. (1989) *J. Biol. Chem.* **264**, 15457–15463.

DiPaola, M., Kao, P. N., and Karlin, A. (1990) *J. Biol. Chem.* **265**, 11017–11029.

Donnelly-Roberts, D. L., and Lentz, T. L. (1991) *Biochemistry* **30**, 7484–7491.

Dougherty, D. A., and Stauffer, D. A. (1990) *Science* **250**, 1558–1560.

Dunn, S. M. J., Blanchard, S. G., and Raftery, M. A. (1980) *Biochemistry* **19**, 5645–5652.

Dunn, S. M. J., Blanchard, S. G., and Raftery, M. A. (1981) *Biochemistry* **20**, 5617–5624.

Dunn, S. M. J., and Raftery, M. A. (1982a) *Proc. Natl. Acad. Sci. U.S.A.* **79**, 6757–6761.

Dunn, S. M. J., and Raftery, M. A. (1982b) *Biochemistry* **21**, 6264–6271.

Dunn, S. M. J., Conti-Tronconi, B. M., and Raftery, M. A. (1983) *Biochemistry* **22**, 2512–2518.

Dunn, S. M. J., Conti-Tronconi, B. M., and Raftery, M. A. (1986) *Biochem. Biophys. Res. Commun.* **139**, 830–837.

Dunn, S. M. J., and Raftery, M. A. (1993) *Biochemistry* **32**, 8608–8615.

Duvoisin, R. M., Deneris, E. S., Patrick, J., and Heinemann, S. (1989) *Neuron* **3**, 487–496.

Dwyer, B. P. (1991) *Biochemistry* **30**, 4105–4112.

Egebjerg, J., Bettler, B., Hermans-Borgmeyer, I., and Heinemann, S. (1991) *Nature (London)* **351**, 745–748.

Elliott, J., Dunn, S. M. J., Blanchard, S. G., and Raftery, M. A. (1979) *Proc. Natl. Acad. Sci. U.S.A.* **76**, 2576–2579.

Engel, W. K., Trotter, J. L., McFarlin, D. E., and McIntosh, C. L. (1977) *Lancet* **1**, 1310–1311.

Fairclough, R. H., Finer-Moore, J., Love, R. A., Kristofferson, D., Desmeules, P. J., and Stroud, R. M. (1983) *Cold Spring Harbor Symp. Quant. Biol.* **48**, 9–20.

Finer-Moore, J., and Stroud, R. (1984) *Proc. Natl. Acad. Sci. U.S.A.* **81**, 155–159.

Fraenkel, Y., Gershoni, J. M., and Navon, G. (1991) *FEBS Lett.* **291**, 225–228.

Galzi, J.-L., Revah, F., Bessis, A., and Changeux, J.-P. (1991) *Annu. Rev. Pharmacol.* **31**, 37–72.

Galzi, J.-L., Revah, F., Black, D., Goeldner, M., Hirth, C., and Changeux, J.-P. (1990) *J. Biol. Chem.* **265**, 10430–10437.

Gasic, G. P., and Heinemann, S. (1991) *Curr. Opinion Neurobiol.* **1**, 20–26.

Gehle, V. M., and Sumikawa, K. (1991) *Mol. Brain Res.* **11**, 17–25.

Gershoni, J. M. (1987) *Proc. Natl. Acad. Sci. U.S.A.* **84**, 4318–4321.

Gershoni, J. M., and Hawrot, E., and Lentz, T. L. (1983) *Proc. Natl. Acad. Sci. U.S.A.* **80**, 4973–4977.

Ghosh, P., and Stroud, R. M. (1991) *Biochemistry* **30**, 3551–3557.

Giraudat, J., Dennis, M., Heidmann, T., Chang, J.-Y., and Changeux, J.-P. (1986) *Proc Natl. Acad. Sci. U.S.A.* **83**, 2719–2723.

Giraudat, J., Dennis, M., Heidmann, T., Haumont, P.-Y., Lederer, F., and Changeux, J.-P. (1987) *Biochemistry* **26**, 2410–2418.

Giraudat, J., Galzi, J.-L., Revah, F., Changeux, H.-P., Haumont, P.-V., and Lederer, F. (1989) *FEBS Lett.* **253**, 190–198.

Goldman, D., Deneris, E., Luyten, W., Kochlar, A., Patrick, J., and Heinemann, S. (1987) *Cell* **48**, 965–973.

Goldman, D., Simmons, D., Swanson, L. W., Patrick, J., and Heinemann, S. (1986) *Proc. Natl. Acad. Sci U.S.A.* **86**, 4076–4080.

Gotti, C., Frigerio, F., Bolognesi, R., Longhi, R., Racchetti, G., and Clementi, F. (1988) *FEBS Lett.* **228**, 118–122.

Gotti, C., Mazzola, G., Longhi, R., Fornasari, D., and Clementi, F. (1987) *Neurosci. Lett.* **82**, 113–119.

Graham, D., Pfeiffer, F., and Betz, H. (1983) *Eur. J. Biochem.* **131**, 519–525.

Graham, D., Pfeiffer, F., Simler, R., and Betz, H. (1985) *Biochemistry* **24**, 990–994.

Grant, G. A., Frazier, M. W., and Chiappineli, V. A. (1988) *Biochemistry* **27**, 3794–3798.

Gray, W. R., Olivera, B. M., and Cruz, L. J. (1988) *Annu. Rev. Biochem.* **57**, 665–700.

Grenningloh, G., Rienitz, A., Scmitt, B., Methfessel, C., Zenson, M., Beyreuther, K., Gundelfinger, E. D., and Betz, H. (1987) *Nature (London)* **328**, 215–220.

Grenningloh, G., Schmieden, V., Schofield, P. R., Seeburg, P. H., Siddique, T., Mohandas, T. K., Becker, C. M., and Betz, H. (1990) *EMBO J.* **9**, 771–776.

Griesman, G. E., McCormick, D. J., De Aizpurua, H. J., and Lennon, V. A. (1990) *J. Neurochem.* **54**, 1541–1547.

Gross, A., Ballivet, M., Rungger, D., and Bertrand, D. (1991) *Pflugers Arch.* **419**, 545–551.

Gu, Y., and Hall, Z. W. (1988) *Neuron* **1**, 117–125.

Gu, Y., Forsayeth, J. R., Verrall, S., Yu, X. M., and Hall, Z. W. (1991) *J. Cell Biol.* **114**, 799–807.

Gundelfinger, E. D. (1992) *Trends Neurosci.* **15**, 206–211.

Guy, R. (1983) *Biophys. J.* **45**, 249–261.

Haggerty, J. G., and Froehner, S. C. (1981) *J. Biol. Chem.* **256**, 8294–8297.

Hall, Z. W., Gorin, P. D., Silberstein, L., and Bennet, C. (1985) *J. Neurosci.* **5** 730–734.

Harvey, A. L., and Dryden, W. F. (1974) *J. Pharmacol.* **26**, 865–870.

Herb, A., Wisden, W., Luddens, H., Puia, G., Vicini, S., and Seeburg, P. H. (1992a) *Proc. Natl. Acad. Sci. U.S.A.* **89**, 1433–1437.

Herb, A., Burnashev, N., Werner, P., Sakmann, B., Wisden, W., and Seeburg, P. H. (1992b) *Neuron* **8**, 775–785.

Hermsen, B., Heiermann, R., and Maelicke, A. (1991) *Biol. Chem. Hoppe-Seyler* **372**, 891–899.

Hille, B. (1984) In *Ionic Channels of Excitable Membranes*. Sinauer, Suderland, MA.

Holden-Dye, L., and Walker, R. J. (1990) *Ascaris. Neurosci. Lett. Suppl* **38**, 5111.

Hollmann, M., O'Shea-Greenfield, A., Rogers, S. W., and Heinemann, S. (1989) *Nature (London)* **342**, 643–648.

Hucho, F. (1986) *Eur. J. Biochem.* **158**, 211–226.

Hucho, F., Oberthür, W., and Lottspeich, F. (1986) *FEBS Lett.* **205**, 137–142.

Huganir, R. L., and Racker, E. (1982) *J. Biol. Chem.* **257**, 9372–9378.

Hulme, E. C., Birdsall, N. J. M., and Buckley, N. J. (1990) *Annu. Rev. Pharmacol. Toxicol.* **30**, 633–673.

Hunkapiller, M. W., Strader, C. D., Hood, L., and Raftery, M. A. (1979) *Biochem. Biophys. Res. Commun.* **91**, 164–169.

Imoto, K., Busch, C., Sakmann, B., Mishina, M., Konno, T., Nakai, J., Bujo, H., Mori, Y., Fukada, K., and Numa, S. (1988) *Nature (London)* **335**, 645–648.

Imoto, K. J., Konno, T., Nakai, J., Wang, F., Misha, M., and Numa, S. (1991) *FEBS Lett.* **289**, 193–200.

Imoto, K., Methfessel, C., Sakmann, B., Mishina, M., Moti, Y., Konno, T., Fukuda, K., Kurasaki, M., Bujo, H., Fujita, Y., and Numa, S. (1986) *Nature (London)* **324**, 670–674.

Julius, D. (1991) *Annu. Rev. Neurosci.* **14**, 335–360.

Kao, I., and Drachman, D. B. (1977) *Science* **195**, 74–75.

Kao, P. N., Dwork, A. J., Kaldany, R.-RJ, Silver, M. L., Wideman, J., Stein, S., and Karlin, A. (1984) *J. Biol. Chem.* **259**, 11662–11665.

Kao, P. N., and Karlin, A. (1986) *J. Biol. Chem.* **261**, 8085–8088.

Karlin, A., Kao, P. N., and Dipaola, M. (1986) *Trends Pharmacol. Sci.* **7**, 304–308.

Katz, B., and Miledi, R. (1971) *Nature (London)* **232**, 124–126.

Kawanammi, S., Conti-Tronconi, B. M., Racs, J., and Raftery, M. A. (1988) *J. Neurol. Sci.* **87**, 195–209.

Keinanen, K., Wisden, W., Sommer, B., Werner, P., Herb, A., Verdoorn, T. A., Sakmann, B., and Seeburg, P. H. (1990) *Science* **249**, 556–560.

Kellaris, K. V., Ware, D. K., Smith, S., and Kyte, J. (1989) *Biochemistry* **28**, 3469–3482.

Kirchner, T., Tzartos, S., Hoppe, F., Schalke, B., Wekerle, H., and Muller-Hermelink, H. K. (1988) *Amer. J. Pathol.* **130**, 268–280.

Kistler, J., Stroud, R. M., Klymkowsky, M. W., Lalancette, R. A., and Fairclough, R. H. (1982) *Biophys. J.* **37**, 371–383.

Knoflach, F., Rhyner, T., Villa, M., Kellenberger, S., Drescher, U., Malherbe, P., Sigel, E., and Mohler, H. (1991) *FEBS Lett.* **293**, 191–194.

Konno, T., Busch, C., Von Kitzing, E., Imoto, K., Wang, F., Nakai, J., Mishina, M., Numa, S., and Sakmann, B. (1991) *Proc. R. Soc. London B: Bio. Sci.* **244**, 69–79.

Kordossi, A. A., and S. J. Tzartos. (1987) *EMBO J.* **6**, 1605–1610.

Kubo, T., Noda, M., Takai, T., Tanabe, T., Kayano, T., Shimizu, S., Tanaka, K., Takahashi, H., Hirose, T., Inayama, S., Kikuno, R., Miyata, T., and Numa, S. (1985) *Eur. J. Biochem.* **149**, 5–13.

Kulhmann, J., Okonjo, K. O., and Maelicke, A. (1991) *FEBS Lett.* **279**, 216–218.

Kuhse, J., Schmieden, V., and Betz, H. (1990a) *Neuron* **5**, 867–873.

Kuhse, J., Schmieden, V., and Betz, H. (1990b) *J. Biol. Chem.* **265**, 22317–22320.

Kuhse, J., Schmieden, V., and Betz, H. (1991) *FEBS Lett.* **283**, 73–77.

Kurosaki, T., Fukuda, K., Konno, T., Mori, Y., Tanaka, K.I., Misha, M., and Numa, S. (1987) *FEBS Lett.* **U14**, 253–258.

Kutsuwada, T., Kashiwabuchhi, N., Mori, H., Sakimura, K., Kushiya, E., Araki, K., Meguro, H., Masaki, H., Kumanishi, T., Arakawa, M., and Mishina, M. (1992) *Nature (London)* **358**, 36–41.

Land, B. R., Saltpeter, E. E., and Saltpeter, M. M. (1981) *Proc Natl Acad Sci U.S.A.* **78**, 7200–7204.

Langosch, D., Thomas, L., and Betz, H. (1988) *Proc. Natl. Acad. Sci. U.S.A.* **85**, 7394–7398.

LaRochelle, William J., Wray, B. E., Sealock, R., and Froehner, S. C. (1985) *J. Cell Biol.* **100**, 684–691.

Laurie, B. J., Seeburg, P. H., and Wisden, W. (1992) *J. Neurosci.* **12**, 1063–1076.

Lei, S. J., Raftery, M. A., and Conti-Tronconi, B. M. (1993) *Biochemistry* **32**, 91–98.

Lentz, T. L., and Wilson, P. T. (1988) *Int. Rev. Neurobiol.* **29**, 117–160.

Leonard, R. J., Labarca, C. G., Charnet, P., Davidson, N., and Lester, H. A. (1988) *Science* **242**, 1578–1581.

Levitan, E. S., Schofield, P. R., Burt, D. R., Rhee, L. M., Wisden, W., Kohler, M., Fujita, N., Rodriquez, H. F., Barnard, E. A., and Seeburg, P. H. (1988a) *Nature (London)* **335**, 76–79.

Levitan, E. S., Blair, L. A. C., Dionne, V. E., and Barnard, E. A. (1988b) *Neuron* **1**, 773–781.

Lindstrom, J., Criado, M., Hochschwener, S., Fox, J. L., and Sarin, V. (1984) *Nature (London)* **311**, 573–575.

Lindstrom, J., Schoepfer, R., and Whiting, P. (1987) *Mol. Neurobiol.* **1**, 281–337.

Lindstrom, J., Shelton, D., and Fujii, Y. (1988) *Adv. Immunol.* **42**, 233–284.

Lindstrom, J., Whiting, P., Schoepfer, R., Luther, M., and Das, M. (1989) In *Computer-Assisted Modeling of Receptor–Ligand Interactions*, pp. 245–266. A. R. Liss, New York.

Listerud, M., Brussaard, A. B. Devay, P., Colman, D. R., and Role, L. W. (1991) *Science* **254**, 1518–1521.

Lolait, S. J., O'Carroll, A., Kusano, K., Muller, J., Brownstein, M. J., and Mahan, L. C. (1989) *FEBS Lett.* **246**, 145–148.

Loring, R. H., Chiappinelli, V. A., Zigmond, R. E., and Cohen, J. B. (1984) *Neuroscience* **11**, 989–999.

Luddens, H., Killisch, I., and Seeburg, P. H. (1991) *J. Receptor Res.* **11**, 535–551.

Luddens, H., Pritchett, D. B., Kohler, M., Killisch, I., Keinanen, K., Monyer, H., Sprengel, R., and Seeburg, P. H. (1990) *Nature (London)* **346**, 648–651.

Luddens, H., and Wisden, W. (1991) *Trends Pharmacol. Sci.* **12**, 49–51.

Luetje, C. W., and Patrick, J. (1991) *J. Neurosci.* **11**, 837–845.

Luetje, C. W., Wada, K., Rogers, S., Abramson, S. N., Tsuji, K., Heinemann, S., and Patrick, J. (1990a) *J. Neurochem.* **55**, 632–640.

Luetje, C. W., Patrick, J., and Seguela, P. (1990b) *FASEB J.* **4**, 2753–2760.

Luetje, C. W., Piattoni, M., and Patrick, J. (1992) *Mol. Pharmacol.* **44**, 657–666.

Maelicke, M. A. (1988) In *Handbook of Experimental Pharmacology: The Cholinergic Synapse* (V. P. Whittaker, Ed.), Vol. 86, pp. 267–300. Springer-Verlag, Berlin.

Maelicke, M. A., Plumer-Wilk, R., Fels, G., Spencer, S. R., Engelhard, M., Veltel, D., and Conti-Tronconi, B. M. (1989) *Biochemistry* **29**, 1396–1405.

Magazanik, L. G., and Vyskocil, F. (1976) In *Motor Innervation of Muscle* (S. Thesless, Ed.), pp. 151–176.

Malosio, M.-L., Marqueze-Pouey, B., Kuhse, J., and Betz, H. (1991) *EMBO J.* **10**, 2401–1409.

Mamalaki, C., Stephenson, F. A., and Barnard, E. A. (1987) *EMBO J.* **6**, 561–565.

Maricq, A. V., Peterson, A. S., Brake, A. J., Myers, R. M., and Julius, D. (1991) *Science* **254**, 432–437.

Marshall, J., Buckingham, S. D., Shingai, R., Lunt, G. G., Goosey, M. W., Darlison, M. G., Sattelle, D. B., and Barnard, E. A. (1990) *EMBO J.* **9**, 4391–4398.

McLane, K. E., Tang, F., and Conti-Tronconi, B. M. (1990a) *J. Biol. Chem.* **265**, 1537–1544.

McLane, K. E., Wu, X., and Conti-Tronconi, B. M. (1990b) *J. Biol. Chem.* **265**, 9816–9824.

McLane, K. E., Wu, X., Diethelm, B. M., and Conti-Tronconi, B. M. (1991a) *Biochemistry* **30**, 4925–4934.

McLane, K. E., Wu, X., and Conti-Tronconi, B. M. (1991b) *Biochem. Biophys. Res. Commun.* **176**, 11–18.

McLane, K. E., Schoepfer, R., Wu, X., Lindstrom, J. M., and Conti-Tronconi, B. M. (1991c) *J. Biol. Chem.* **266**, 15230–15239.

McLane, K. E., Wu, X., and Conti-Tronconi, B. M. (1991d) *Biochemistry* **30**, 10730–10738.

McLane, K. E., Weaver, W., Lei, S., Chiappinelli, V. A., and Conti-Tronconi, B. M. (1993) *Biochemistry* **32**, 6988–6994.

McCormick, D. J., and Atassi, M. Z. (1984) *Biochem. J.* **224**, 995–1000.

McCrea, P. D., Popot, J.-L., and Engelman, D. M. (1987) *EMBO J.* **6**, 3619–3626.

Meguro, H., Mori, H., Araki, K., Kushiya, E., Kutsuwada, T., Yamazaki, M., Kumanishi, T., Arakawa, M., Sakimura, K., and Mishina, M. (1992) *Nature (London)* 357, 70–74.

Middleton, R. E., and Cohen, J. B. (1991) *Biochemistry* 30, 6987–6997.

Mihovilovic, M., and Richman, D. P. (1987) *J. Biol. Chem.* 262, 4978–4986.

Miller, R. J. (1989) In *Second Messengers in the Central Nervous System*, pp. 23–35. Soc. Neurosci., Washington, DC.

Mishina, M., Kurosaki, T., Tobimatsu, T., Morimoto, Y., Noda, M., Yamamoto, T., Terao, M., Lindstrom, J., Takahashi, T., Kuno, M., and Numa, S. (1984) *Nature (London)* 307, 604–608.

Mishina, M., Tobimatsu, T., Imoto, K., Tanaka, K., Fujita, Y., Fukuda, K., Kurasaki, M., Takahashi, H., Morimoto, Y., Hirose, T., Inayama, S., Takahashi, T., Kuno, M., and Numa, S. (1985) *Nature (London)* 313, 364–369.

Mishina, M., Toshiyuki, T., Imoto, K., Noda, M., Takahashi, T., Numa, S., Methfessel, C., and Sakmann, B. (1986) *Nature (London)* 321, 406–411.

Mitra, A. K., McCarthy, M. P., and Stroud, R. M. (1989) *J. Cell Biol.* 109, 755–774.

Monaghan, D. T., Bridges, R. J., and Cotman, C. W. (1989) *Annu. Rev. Pharmacol. Toxicol.* 29, 365–402.

Monyer, H., Sprengel, R., Schoepfer, R., Herb, A., Higuchi, M., Lomeli, H., Burnashev, N., Sakmann, B., and Seeburg, P. H. (1992) *Science* 256, 1217–1221.

Moore, C. R., Yates, J. R., III, Griffin, P. R., Shabanowitz, J., Martino, P. A., Hunt, D. F., and Cafiso, D. S. (1989) *Biochemistry* 28, 9184–9191.

Moore, H.-P. H., and Raftery, M. A. (1979) Ligand induced interconversion of affinity states in membrane-bound acetylcholine receptor from *Torpedo californica*. Effects of sulfydryl and disulfide reagents. *Biochemistry* 10, 1862–1867.

Moore, H.-P. H., and Raftery, M. A. (1980) *Proc. Natl. Acad. Sci. U.S.A.* 77, 4509–4513.

Moriyoshi, K., Masu, M., Ishii, T., Shigemoto, R., Mizuno, N., and Nakanishi, S. (1991) *Nature (London)* 354, 31–37.

Moschovitz, R., and Gershoni, J. M. (1988) *J. Biol. Chem.* 263, 1017–1022.

Mulac-Jericevic, B., and Atassi, M. Z. (1987) *Biochem. J.* 248, 847–852.

Myers, R. A., Zafaralla, G. C., Gray, W. R., Abbott, J., Cruz, L. J., and Olivera, B. M. (1991) *Biochemistry* 30, 9370–9377.

Nakanishi, N., Schneider, N. A., and Axel, R. A. (1990) *Neuron* 5, 569–581.

Nef, P., Oneyser, C., Alliod, C., Couturier, S., and Ballivet, M. (1988) *EMBO J.* 7, 595–601.

Nef, P., Oneyser, C., Barkas, T., and Ballivet, M. (1986) In *Nicotinic Acetylcholine Receptor: NATO-ASI, Series H*, (A. Maelicke, Ed.), Vol. 3, pp. 389–415, Springer-Verlag, Heidelberg.

Neijt, H. C., Duits, I. J., and Vijverberg, H. P. M. (1988) *Neuropharmacology* 27, 301–307.

Nelson, S., and Conti-Tronconi, B. M. (1990) *J. Neuroimmunol.* 29, 81–92.

Neubig, R. R., and Cohen, J. B. (1979) *Biochemistry* 18, 5464–5475.

Neumann, D., Fridkin, M., and Fuchs, S. (1984) *Biochem. Biophys. Res. Commun.* 121, 673–679.

Neumann, D., Gershoni, J. M., Fridkin, M., and Fuchs, S. (1985) *Proc. Natl. Acad. Sci. U.S.A.* 82, 3490–3493.

Neumann, D., Barchan, D., Safran, A., Gershoni, J. M., and Fuchs, S. (1986a) *Proc. Natl. Acad. Sci. U.S.A.* 83, 3008–3011.

Neumann, D., Barchan, D., Fridkin, M., and Fuchs, S. (1986b) *Proc. Natl. Acad. Sci. U.S.A.* 83, 9250–9253.

Neumann, D., Barchan, D., Horowitz, M., Kochva, E., and Fuchs, S. (1989) *Proc. Natl. Acad. Sci. U.S.A.* 86, 7255–7259.

Noda, M., Takahashi, H., Tanabe, T., Toyosato, M., Furutani, Y., Hirose, T., Asai, M., Inayama, S., Miyata, T., and Numa, S. (1982) *Nature (London)* 299, 793–797.

Noda, M., Takahashi, H., Tanabe, T., Toyoto, M., Kikyotani, S., Hirose, T., Asai, M., Takashima, H., Inayama, S., Miyata, T., and Numa, S. (1983a) *Nature (London)* 301, 251–255.

Noda, M., Takahashi, H., Tanabe, T., Toyosato, M., Kikyotani, S., Furutani, Y., Hirose, T., Takashima, H., Inayama, S., Miyata, T., and Numa, S. (1983b) *Nature (London)* 302, 528–532.

Noda, M., Furutani, Y., Takahashi, H., Toyosato, M., Tanabe, T., Shimizu, S., Kikyotani, S., Kayano, T., Hirose, T., Inayama, S., and Numa, S. (1983c) *Nature (London)* 302, 818–823.

Oberthür, W., and Hucho, F. (1988) *J. Protein Chem.* 7, 141–150.

Oblas, B., Singer, R. H., and Boyd, N. D. (1986) *Mol. Pharmacol.* 29, 649–656.

Ochoa, E. L. M., Chattopadhyay, A., and McNamee, M. G. (1989) *Cell. Mol. Neurobiol.* 9, 141–178.

Ohana, B., Fraenkel, Y., Navon, G., and Gershoni, J. M. (1991) *Biochem. Biophys. Res. Commun.* 179, 648–654.

Ohana, B., and Gershoni, J. M. (1990) *Biochemistry* 29, 6409–6415.

Oiki, S., Danho, W., Madison, V., and Montal, M. (1988) *Proc. Natl. Acad. Sci. U.S.A.* 85, 8703–8707.

Okonjo, K. O., Kuhlmann, J., and Maelicke, A. (1991) *Eur. J. Biochem.* 200, 671–677.

Olivera, B. M., Rivier, J., Clark, C., Ramilo, C. A., Corpuz, G. P., Abogadie, F. C., Mena, E. E., Woodward, S. R., Hillyard, D. R., and Cruz, L. J. (1990) *Science* 249, 257–263.

Oswald, R. E., and Changeux, J.-P. (1982) *FEBS Lett.* 139, 225–229.

Ovchinnikov, Y. A., Lipkin, V. M., Shuvaeva, T. M., Bogachuk, A. P., and Shemyakin, V. V. (1985) *FEBS Lett.* 179, 107–110.

Papke, R. L., Boulter, J., Patrick, J., and Heinemann, S. (1989) *Neuron* 3, 589–596.

Papke, R. L., and Heinemann, S. F. (1991) *J. Physiol.* 440, 95–112.

Pascuzzo, G. J., Akaide, A., Maleque, M. A., Shaw, K.-P., Aronstam, R. S., Rickett, D. L., and Albuquerque, E. X. (1984) *Mol Pharmacol.* 25, 92–101.

Pederson, S. E. Bridgman, P. C., Sharp, S. D., and Cohen, J. B. (1990) *J. Biol. Chem.* 265, 569–581.

Pedersen, S. E., and Cohen, J. B. (1990) *Proc. Natl. Acad. Sci. U.S.A.* 87, 2785–2789.

Pedersen, S. E., Dreyer, E. B., and Cohen, J. B. (1986) *J. Biol. Chem.* 261, 13735–13742.

Pereira, E. F. R., Alkondon, M., and Albuquerque, E. X. (1991) In *Proceedings of the 1991 Medical Defense Bioscience Review,* pp. 229–233.

Peroutka, S. J. (1988) *Annu. Rev. Neurosci.* 11, 45–60.

Pinnock, R. D., Lummis, S. C. R., Chiappinelli, V. A., and Sattelle, D. B. (1988) *Brain Res.* 458, 45–52.

Poulter, L., Earnest, J. P., Stroud, R. M., and Burlingame, A. L. (1989) *Proc. Natl. Acad. Sci. U.S.A.* 86, 6645–6649.

Pritchett, D. B., Sontheimer, H., Shivers, B. D., Ymer, S., Kettenmann, H., Schofield, P. R., and Seeburg, P. H. (1989a) *Nature (London)* 338, 582–585.

Pritchett, D. B., Luddens, H., and Seeburg, P. H. (1989b) *Science* 245, 1389–1392.

Pritchett, D. B., and Seeburg, P. H. (1990) *J. Neurochem.* 54, 1802–1804.

Pritchett, D. B., and Seeburg, P. H. (1991) *Proc. Natl. Acad. Sci. U.S.A.* 88, 1421–1425.

Radding, W., Corfield, P. W. R., Levinson, L. S., Hashim, G. A., and Low, B. W. (1988) *FEBS Lett.* 231, 212–216.

Raftery, M. A., Conti-Tronconi, B. M., and Dunn, S. M. J. (1985) *Fundam. Appl. Toxicol.* 5, 539–546.

Raftery, M. A., Dunn, S. M. J., Conti-Tronconi, B. M., Middlemas, D. S., and Crawford, R. D. (1983) *Cold Spring Harbor Symp. Quant. Biol.* 48, 21–33.

Raftery, M. A. Hunkapillar, M. W., Strader, C. D., and Hood, L. E. (1980) *Science* 208, 1454–1457.

Ralston, S., Sarin, V., Thanh, H. L., Rivier, J., Fox, L., and Lindstrom, J. (1987) *Biochemistry* **26**, 3261–3266.

Ramoa, A. S., and Albuquerque, E. X. (1988) *FEBS Lett.* **235**, 156–162.

Ramoa, A. S., Alkondon, M., Arcava, Y., Irons, J., Lunt, G. G., Deshpande, S., S., Wonnacott, S., Aronstam, R. S., and Albuquerque, E. X. (1990) *J. Pharmacol. Exp. Ther.* **254**, 71–82.

Ratnam, M., and Lindstrom, J. (1984) *Biochem. Biophys. Res. Commun.* **12**, 1225–1233.

Ratnam, M., Le Nguyen, D., Rivier, J., Sargent, P. B., and Lindstrom, J. (1986a) *Biochemistry* **25**, 2633–2643.

Ratnam, N., Sargent, P. B., Sarin, V., Fox, J. L., Nguyen, D. L., Rivier, J., Criado, M., and Lindstrom, J. M. (1986b) *Biochemistry* **25**, 2621–2632.

Ravdin, P. M., and Berg, D. K. (1979) *Proc. Natl. Acad. Sci. U.S.A.* **76**, 2072–2076.

Revah, F., Bertrand, D., Galzi, J.-L., Devillers-Thiery, A., Mulle, C., Hussy, N., Bertrand, S., Ballivet, M., and Changeux, J.-P. (1991) *Nature (London)* **353**, 846–849.

Revah, F., Galzi, J.-L., Giraudat, J., Haumon, P.-Y., Lederer, F., and Changeux, J.-P. (1990) *Proc. Natl. Acad. Sci. U.S.A.* **87**, 4675–4679.

Roth, B., Schwendimann, B., Hughes, C. J., Tzartos, S. J., and Barkas, T. (1987) *FEBS Lett.* **221**, 172–178.

Ruiz-Gomez, A., Morato, E., Garcia-Calvo, M., Valdivieso, F., and Mayor, F., Jr. (1990) *Biochemistry* **29**, 7033–7040.

Sakimura, K., Bujo, J., Kushiya, E., Araki, K., Yamazaki, M., Megureo, H., Warashina, A., Numa, S., and Mishina, M. (1990) *FEBS Lett.* **272**, 73–80.

Sakimura, K., Morita, T., Kushiya, E., and Mishina, M. (1992) *Neuron* **8**, 267–274.

Sakmann, B. (1992) *Neuron* **8**, 613–629.

Sakmann, B., Methfessel, C., Mishina, M., Takahashi, T., Takai, T., Kurasaki, M., Fukuda, K., and Numa, S. (1985) *Nature (London)* **318**, 538–543.

Sawruck, E., Udri, C., Betz, H., and Schmitt, B. (1990a) *FEBS Lett.* **273**, 177–181.

Sawruck, E., Scloss, P., Betz, H., and Schmidt, B. (1990b) *EMBO J.* **9**, 2671–2677.

Schleup, M., Willcox, N., Vincent, A., Dhoot, G. K., and Newsom-Davis, J. (1987) *Ann. Neurol.* **22**, 212–222.

Schlosse, P., Hermans-Borgmeyer, I., Betz, H., Gundelfinger, E. D. (1988) *EMBO J.* **7**, 2889–2894.

Schmieden, V., Grenningloh, G., Schofield, P. R., and Betz, H. (1989) *EMBO J.* **8**, 695–700.

Schoepfer, R., Conroy, W. G., Whiting, P., Gore, M., and Lindstrom, J. (1990) *Neuron* **4**, 35–48.

Schoepfer, R., Whiting, P., Esch, F., Blacher, R., Shimasaki, S., and Lindstrom, J. (1988) *Neuron* **1**, 241–248.

Schofield, P. R., Darlison, M. G., Fujita, N., Burt, D. R., Stephenson, F. A., Rodriquez, H., Rhee, L. M., Ramachandran, J., Reale, V., Glencorse, T. A., Seeburg, P. H., and Barnard, E. A. (1987) *Nature (London)* **328**, 221–227.

Schuetze, S. M. (1986) *Trends Neurosci.* **9**, 386–388.

Schumacher, M., Camp, S., Maulet, Y., Newton, M., Macphee-Quigley, K., Taylor, S. S., Friedman, T., and Taylor, P. (1986) *Nature (London)* **319**, 407–409.

Schuster, C. M., Ultsch, A., Schloss, P., Cox, J. A., Schmitt, B., and Betz, H. (1991) *Science* **254**, 112–114.

Schwartz, R. D., and Mindlin, M. C. (1988) Inhibition of the GABA receptor-gated *J. Pharmacol. Exp. Ther.* **244**, 963–970.

Shivers, B. D., Killisch, I., Sprengel, R., Sontheimer, H., Kohler, M., Schofield, P. R., and Seeburg, P. H. (1989) *Neuron* **3**, 327–337.

Sigel, E., Baur, R., Kellenberger, S., and Malherbe, P. (1992) *EMBO J.* **11**, 2017–23.

Sigel, E., Baur, R., Trube, G., Mohler, H., and Malherbe, P. (1990) *Neuron* **5**, 703–711.

Sine, S. M., and Claudio, T. (1991) *J. Biol. Chem.* **266**, 19369–19377.

Smart, L., Meyers, H.-W., Hilginfeld, R., Saenger, W., and Maelicke, A. (1984) *FEBS Lett.* **178**, 64–68.

Smith, M. A., Stllberg, J., Lindstrom, J. M., and Berg, D. K. (1985) *J. Neurosci.* **5**, 2726–2731.

Sommer, B., Burnashev, N., Verdoorn, T. A., Keinanen, K., Sakmann, B., and Seeburg, P. H. (1992) *EMBO J.,* **11**, 1651–1656.

Sommer, B., Keinanen, K., Verdoorn, T. A., Wisden, W., Burnashev, N., Herb, A., Kohler, M., Takagi, T., Sakmann, B., and Seeburg, P. H. (1990) *Science* **249**, 1580–1585.

Sommer, B., Kohler, M., Sprengel, R., and Seeburg, P. H. (1991) *Cell* **67**, 11–19.

Sommer, B., and Seeburg, P. H. (1992) *Trends Pharmacol. Sci.* **13**, 291–296.

Sontheimer, H., Becker, C.-M., Pritchett, D. B., Schofield, P. R., Grenningloh, B., Kettemann, H., Betz, H., and Seeburg, P. H. (1989) *Neuron* **2**, 1491–1497.

Stevens, C. F. (1987) *Nature (London)* **328**, 198–199.

Stollberg, J., Whiting, P. J., Lindstrom, J. M., and Berg, D. K. (1986) *Brain Res.* **378**, 179–182.

Strader, C. B. D., Revel, J.-P., and Raftery, M. A. (1979) *J. Cell Biol.* **83**, 499–510.

Stroud, R. M., McCarthy, M. P., and Shuster, M. (1990) *Biochemistry* **29**, 11009–11023.

Sumikawa, K., and Miledi, R. (1989) *Proc. Natl. Acad. Sci. U.S.A.* **86**, 367–371.

Sumikawa, K., and Gehle, V. M. (1992) *J. Biol. Chem.* **267**, 6286–6290.

Sussman, J. L., Harel, M., Frolow, F., Oefner, C., Goldman, A., Toker, L., and Silman, I. (1991) *Science* **263**, 872–879.

Takai, T., Noda, M., Furutani, Y., Takahashi, H., Notake, M., Shimizu, S., Kayano, T., Tanabe, T., Tanaka, K., Hirose, T., Inayama, S., and Numa, S. (1984) *Eur. J. Biochem.* **143**, 109–115.

Tanabe, T., Noda, M., Furutani, Y., Takai, T., Takahashi, H., Tanaka, K., Hirose, T., Inayama, S., and Numa, S. (1984) *Eur. J. Biochem.* **144**, 11–17.

Takai, T., Noda, M., Mishina, M., Shimizu, S., Furutani, Y., Kayano, T., Ikeda, T., Kubo, T., Takahashi, H., Takahashi, T., Kuno, M., and Numa, S. (1985) *Nature (London)* **315**, 761–764.

Tomaselli, G. F., McLaughlin, J. T., Jurman, M. E., Hawrot, E., and Yellen, G. (1991) *Biophys. J.* **60**, 721–724.

Trautman, A. (1982) *Nature (London)* **298**, 272–275.

Tzartos, S. J., and Changeux, J.-P. (1983) *EMBO J.* **2**, 381–387.

Tzartos, S. J., and Remoundos, M. S. (1990) *J. Biol. Chem.* **265**, 21462–21467.

Ueno, S., Wada, K., Takahashi, M., and Tarui, S. (1980) *Clin. Exp. Immunol.* **42**, 463–469.

Unwin, N., Toyoshima, C., and Kubalek, E. (1988) *J. Cell. Biol.* **107**, 1123–1138.

Wada, K., Ballivet, M., Boulter, J., Connolly, J., Wada, E., Deneris, E. S., Swanson, L. W., Heinemann, S., and Patrick, J. (1988) *Science* **240**, 330–334.

Wada, E., Wada, K., Boulter, J., Deneris, E., Heinemann, S., Patrick, J., and Swanson, L. W. (1989) *J. Comp. Neurol.* **284**, 314–335.

Walker, R. J., and Holden-Dye, L. (1991) *Parasitology* **102**, S7–29.

Walker, J. W., Lukas, R. J., and McNamee, M. G. (1981) *Biochemistry* **20**, 2191–2199.

Watters, D., and Maelicke, A. (1983) *Biochemistry* **22**, 1811–1819.

Werner, P., Voigt, M., Keinanen, K., Wisden, W., and Seeburg, P. H. (1991) *Nature (London)* **351**, 742–744.

Whiting, P., Liu, R., Morley, B. J., and Lindstrom, J. M. (1987) *J. Neurosci.* **7**, 4005–4016.

Whiting, P., Schoepfer, R., Lindstrom, J., and Priestley, T. (1991) *Mol. Pharmacol.* **40**, 463–472.

Whiting, P. J., and Lindstrom, J. M. (1986) *Biochemistry* **25**, 2082–2093.

Whiting, P. J., and Lindstrom, J. M. (1987) *Proc. Natl. Acad. Sci. U.S.A.* **84**, 595–599.

Whiting, P. J., and Lindstrom, J. M. (1988) *J. Neurosci.* **8**, 3395–3404.

Wilson, P. T., Gershoni, J. M., Hawrot, E., and Lentz, T. L. (1984) *Proc. Natl. Acad. Sci. U.S.A.* **81**, 2553–2557.

Wilson, P. T., Hawrot, E., and Lentz, T. L. (1988) *Mol. Pharmacol.* **34**, 643–651.

Wilson, P. T., and Lentz, T. L. (1988) *Biochemistry* **27**, 6667–6674.

Wilson, P. T., Lentz, T. L., and Hawrot, E. (1985) *Proc. Natl. Acad. Sci. U.S.A.* **82**, 8790–8794.

Wisden, W., Laurie, D. J., Monyer, H., and Seeburg, P. H. (1992) *J. Neuroscience* 12, 1039–1062.

Witzemann, V., Muchmore, D., and Raftery, M. A. (1979) *Biochemistry* 18, 5511–5520.

Witzemann, V., and Raftery, M. A. (1977) *Biochemistry* 16, 5862–5868.

Witzemann, V., and Raftery, M. A. (1979) *Biochem. Biophys. Res. Commun.* 85, 623–628.

Yakel, J. L., Shao, X. M., and Jackson, M. B. (1990) *Brain Res.* 533, 46–52.

Yang, J. (1990) *J. Gen. Physiol.* 96, 1177–1198.

Yee, A. S., Corley, D. E., and McNamee, M. G. (1986) *Biochemistry* 25, 2110–2119.

Ymer, S., Schofield, P. R., Draguhn, A., Werner, P., Kohler, M., and Seeburg, P. H. (1989) *EMBO J.* 8, 1665–1670.

Young, E. F., Ralston, E., Blake J., Ramachandran, J., Hall, Z. H., and Stroud, R. M. (1985) *Proc. Natl. Acad. Sci. U.S.A.* 82, 626–630.

Zhang, Z.-W., and Feltz, P. (1991) *Brit. J. Pharmacol.* 102, 19–22.

Index

Printed and bound by CPI Group (UK) Ltd, Croydon, CR0 4YY

08/05/2025

01864882-0001